Genetics of Colorectal Cancer

John D. Potter • Noralane M. Lindor
Editors

Genetics of Colorectal Cancer

Springer

Editors
John D Potter, MD, PhD
Member and Senior Advisor
Fred Hutchinson Cancer Research Center
PO Box 19024
M4-B814
Seattle WA 98109-1024
USA
email: jpotter@fhcrc.org

Noralane M. Lindor, MD
Department of Medical Genetics
Mayo Clinic College of
Medicine
Rochester, MN 55905
USA
email: nlindor@mayo.edu

ISBN: 978-0-387-09567-7 e-ISBN: 978-0-387-09568-4
DOI: 10.1007/978-0-387-09568-4

Library of Congress Control Number: 2008937470

Cover illustration: Design: Clayton Hibbert and John Potter from an original figure by Jeremy Jass; Realisation: Scott Sutherland and Clayton Hibbert; Photography: Philip Meadows

Printed on acid-free paper

springer.com

For Jeremy Jass: friend,
teacher, scientist

Contents

Introduction 1

Section I Epidemiology and Models of Colorectal Cancer

1 **Colorectal Cancer: Epidemiology** 5
John D. Potter and David Hunter

2 **Mouse Models of Intestinal Cancer** 27
Erin M. Perchiniak and Joanna Groden

Section II Pathways to Colorectal Cancer

3 **The Chromosomal-Instability Pathway and *APC* Gene Mutation in Colorectal Cancer** 53
Robert Gryfe

4 **DNA Methylation in Colorectal Cancer: Multiple Facets of Tumorigenesis** 73
Joanne P. Young and Peter W. Laird

5 **Pathways and Pathology** 97
Jeremy R. Jass

Section III Germline Susceptibility – Mendelian and Other Syndromes

6 **Familial Adenomatous Polyposis** 125
Julian A. Sanchez, Graham Casey, and James M. Church

7 **DNA Mismatch Repair and Lynch Syndrome** 141
Brittany C. Thomas, Matthew J. Ferber, and Noralane M. Lindor

8 Additional Syndromes with Hereditary Predisposition to Colorectal Cancer

8.1 *MUTYH*-Associated Polyposis 173
Spring Holter and Steven Gallinger

8.2 Familial Colorectal Cancer Type X 183
Noralane M. Lindor

8.3 Families with Serrated Neoplasia of the Colon 187
Joanne P. Young

8.4 Peutz-Jeghers Syndrome 193
Douglas L. Riegert-Johnson and Lisa A. Boardman

8.5 Juvenile Polyposis 199
Kara A. Mensink, Jeremy R. Jass, and Noralane M. Lindor

8.6 *BLM* mutation and Colorectal Cancer Susceptibility 207
Beatriz Russell and Joanna Groden

8.7 The Role of *p53* in Colorectal Cancer 213
Serena Masciari and Sapna Syngal

8.8 Chromosomes 8q24 and 9p24: Associations with Colorectal Cancer 219
John D. Potter

Section IV Germline Susceptibility – Gene-Environment Interactions

9 Genetic Variability in Folate-Mediated One-Carbon Metabolism and Risk of Colorectal Neoplasia 223
Amy Y. Liu and Cornelia M. Ulrich

10 Genetic Variability in NSAID Targets and NSAID-Metabolizing Enzymes and Colorectal Neoplasia 243
Elizabeth M. Poole, James T. Cross, John D. Potter, and Cornelia M. Ulrich

11 The Role of Chemical Carcinogens and Their Biotransformation in Colorectal Cancer 261
Loïc Le Marchand

12 Calcium and Vitamin D 277
Roberd M. Bostick, Michael Goodman, and Eduard Sidelnikov

Index 299

Contributors

Lisa A. Boardman Division of Gastroenterology and Hepatology, Mayo Clinic College of Medicine

Roberd M. Bostick Department of Epidemiology, Rollins School of Public Health, Emory University
rmbosti@sph.emory.edu

Graham Casey Genetic Epidemiology Program, Department of Preventive Medicine, USC/Norris Comprehensive Cancer Center

James M. Church Department of Colorectal Surgery, Cleveland Clinic

James T. Cross Division of Public Health Sciences, Fred Hutchinson Cancer Research Center

Matthew J. Ferber Department of Laboratory Medicine & Pathology, Mayo Clinic College of Medicine

Steven Gallinger Department of Surgery, Samuel Lunenfeld Research Institute, Mount Sinai Hospital, University of Toronto
Steven.Gallinger@uhn.on.ca

Michael Goodman Department of Epidemiology, Rollins School of Public Health, Emory University

Joanna Groden Department of Molecular Virology, Immunology and Medical Genetics, The Ohio State University College of Medicine.
joanna.groden@osumc.edu

Robert Gryfe Department of Surgery, Samuel Lunenfeld Research Institute, Mount Sinai Hospital, University of Toronto
RGryfe@mtsinai.on.ca

Spring Holter Dr. Zane Cohen Digestive Diseases Clinical Research Centre, Familial Gastrointestinal Cancer Registry, Mount Sinai Hospital, University of Toronto
sholter@mtsinai.on.ca

David Hunter Department of Epidemiology, Harvard University
david.hunter@channing.harvard.edu

Jeremy R. Jass Department of Cellular Pathology, St Mark's Hospital
jeremy.jass@nwlh.nhs.uk

Peter W. Laird USC/Norris Comprehensive Cancer Center, Keck School of Medicine
plaird@usc.edu

Loïc Le Marchand Epidemiology Program, Cancer Research Center of Hawaii, University of Hawaii
loic@crch.hawaii.edu

Noralane M. Lindor Department of Medical Genetics, Mayo Clinic College of Medicine
nlindor@mayo.edu

Amy Y. Liu Division of Public Health Sciences, Fred Hutchinson Cancer Research Center

Serena Masciari Population Sciences Division, Dana-Farber Cancer Institute

Kara A. Mensink Department of Laboratory Medicine & Pathology, Mayo Clinic College of Medicine
Mensink.Kara@Mayo.edu

Erin M. Perchiniak Department of Molecular Virology, Immunology and Medical Genetics, The Ohio State University College of Medicine

Elizabeth M. Poole Division of Public Health Sciences, Fred Hutchinson Cancer Research Center
epoole@fhcrc.org

John D. Potter Cancer Prevention Research Program, Division of Public Health Sciences, Fred Hutchinson Cancer Research Center
jpotter@fhcrc.org

Douglas L. Riegert-Johnson Division of Gastroenterology and Hepatology, Mayo Clinic
riegertjohnson.douglas@mayo.edu

Beatriz Russell Department of Molecular Virology, Immunology and Medical Genetics, The Ohio State University College of Medicine

Julian A. Sanchez Department of Colorectal Surgery, Cleveland Clinic

Eduard Sidelnikov Department of Epidemiology, Rollins School of Public Health, Emory University

Sapna Syngal Familial Gastrointestinal Cancer Program, Dana-Farber Cancer Institute/Brigham and Women's Cancer Center
ssyngal@partners.org

Brittany C. Thomas Department of Laboratory Medicine & Pathology, Mayo Clinic College of Medicine
thomas.brittany@mayo.edu

Cornelia M. Ulrich Division of Public Health Sciences, Fred Hutchinson Cancer Research Center
nulrich@fhcrc.org

Joanne P. Young Cancer Council, Queensland Institute of Medical Research
Joanne.Young@qimr.edu.au

Introduction

Colorectal cancer (CRC) is among the most common cancers in western populations and indeed, increasingly, all over the world. There are several known modifiable risk factors for CRC and, for a number of well-characterized inherited syndromes where the phenotype includes CRC, germline genetic mutations are known. Polyps with varying histology (but mostly adenomas) are established preneoplastic lesions, the accessibility of which, to endoscopy, allows, simultaneously, effective screening, early diagnosis, and treatment. Endoscopy and histopathology have also facilitated the emergence of an increasingly clear picture of the molecular steps to cancer down several paths.

This volume is focused on the current picture of genetics in CRC – both inherited and acquired – and the ways in which the inherited lesions influence risk directly, as well as how they interact with the environment; the ways in which molecular progression occurs; and the possible insights into prevention, early diagnosis, and treatment that the knowledge of genetics provides.

The book is divided into four parts. In the first section, we describe the epidemiology of CRC, paying particular attention to what is known about behavioral, dietary, and host-related risk factors, and we examine the murine models of CRC which provide useful insights into prevention, progression, and potential therapy.

In Sect. 2, we present overviews of the molecular pathways to CRC, describing in detail the major pathways: the chromosomal-instability pathway involving mutations in the *APC* gene, the DNA-methylation pathway involving widespread epigenetic alterations, and the DNA mismatch repair pathway with its signature microsatellite instability. Chapter 3 in this section describes, in detail, the relationships between the pathways to progression and pathology.

The third section provides a detailed discussion of the known major and minor CRC syndromes, not only familial adenomatous polyposis and Lynch syndrome, but also *MUTYH*-associated polyposis (MAP), familial CRC type X, serrated neoplasia of the colon, Peutz–Jeghers syndrome, juvenile polyposis, germline mutations in p53, and *BLM*-related CRC. The recent association studies involving chromosomes 8q24 and 9p24 are also highlighted.

The final section presents what is known about interactions between polymorphisms in several metabolic and nutrition-related pathways and established environmental risk and protection factors for CRC. Specific chapters focus on

J.D. Potter and N.M Lindor (eds.), *Genetics of Colorectal Cancer*,
DOI: 10.1007/978-0-387-09568-4_1,

folate-mediated one-carbon metabolism, on genetic variability in NSAID targets and NSAID-metabolizing enzymes, on biotransformation of chemical carcinogens, and on calcium and vitamin D.

CRC continues to provide enormous challenges because of its high incidence and its important contribution to cancer mortality. However, it also contrasts with a number of the other common cancers, inasmuch as we know that early detection, particularly via colonoscopy, reduces incidence considerably. Finally, we also know a great deal about the roles of genetics and the environment in causing and protecting against CRC.

Section I
Epidemiology and Models of Colorectal Cancer

Chapter 1
Colorectal Cancer: Epidemiology

John D. Potter and David Hunter

Introduction

Cancers of the colon and rectum are among the most common in western populations; there are several known risk factors that are potentially modifiable. Polyps of several histologic subtypes are established preneoplastic lesions; their accessibility by endoscopy has allowed a detailed picture to emerge of many of the molecular steps in several progression pathways. Determining and quantifying the factors – both genetic and environmental – that facilitate and deter progression is now a fundamental challenge for nutritional, environmental, molecular, and genetic epidemiology.

Adenomas are precursors of colorectal cancers. There is increasing evidence that hyperplastic polyps can give rise to serrated polyps and subsequently to cancer. Colorectal cancers have been distinguished mostly by their varying anatomic locations, but molecular classification that is based on the presence of genomic instability versus chromosomal instability has been increasingly used more recently. The pathogenesis and molecular analysis of subtypes is described in Section II.

Descriptive Epidemiology

At the end of the twentieth century, almost one million cases of colorectal cancer occurred worldwide each year, about 9.5% of all new cases of cancer (Stewart et al. 2003). Colorectal cancer is the third most common incident cancer in the United States and the second most common cause of cancer death (ACS 2006). Incidence rates vary more than 20-fold around the world, with the lowest rates in India and the highest in Japan. Rates increase sharply with age. Colon cancer occurs with approximately equal frequency between the sexes, but rectal cancer can be up to twice as common in men as in women.

J.D. Potter (✉)
Cancer Prevention Research Program, Division of Public Health Sciences, Fred Hutchinson Cancer Research Center
e-mail: jpotter@fhcrc.org

J.D. Potter and N.M Lindor (eds.), *Genetics of Colorectal Cancer*,
DOI: 10.1007/978-0-387-09568-4_2,

International variation and migrant data demonstrate clearly that colorectal cancer is highly sensitive to changes in environment. For immigrants and their descendants, incidence rates rapidly reach those of the host country (Haenszel 1961; McMichael and Giles 1988). Diet, exercise, and other lifestyle differences may explain most of the international variation in rates. The highest rates in the world are now seen among Japanese in Japan – long noted as a developed country with low rates; rates among Japanese men are now higher than those in the United States and, among women, they are higher than in many European countries.

Genetic and Molecular Events in Colorectal Cancer

The somatic genetic events in progression are better understood for colorectal cancer than for most other cancers. Probably less than 5% of colorectal cancer occurs in the context of the familial syndromes discussed in Chaps. 6–8. However, these syndromes are crucial to understanding this cancer because, unlike several other common cancers, each of the pathways in those with inherited genetic predisposition has a counterpart in the much more common nonfamilial disease. Specific genetic variation modifies the effects of lifestyle factors, such as diet and smoking, as well as specific preventive agents such as NSAIDs. These interactions provide further insight into pathogenesis and prevention and are discussed in Chaps. 9–12. In this chapter, we will focus on the known environmental causes of colorectal cancer.

The adenoma-carcinoma hypothesis was developed in the 1970s (Hill et al. 1978), stimulated by several observations: (1) early cancers arise in adenomatous polyps; (2) persons with unresected polyps are at higher risk of colorectal cancer; and (3) individuals with syndromes characterized by multiple colorectal polyps are at very high risk of colorectal cancer. The current version of the adenoma-carcinoma sequence proposes that benign adenomas arise from proliferation of abnormal colonic crypt cells that result in aberrant crypt foci and microadenomas. Macroscopic polyps grow in size and undergo neoplastic transformation. See Chap. 5 for details.

Tissues at each stage of this process are accessible to colonoscopy or surgery, which allowed Vogelstein and others to study the molecular events at each stage, from normal epithelium to cancer (Fearon and Vogelstein 1990). They identified several genes that tend to mutate early (especially mutation or loss of *APC*), later (mutations in *KRAS*), and very late in the sequence (loss of p53), providing a clear demonstration in humans of the multihit theory, accurately predicting that humans born with germline mutations in relevant genes – in colorectal cancer, *APC* (see Chap. 3) – are at higher risk at an earlier age. There are other pathways to colorectal cancer, the most important of which has both an inherited and an acquired manifestation: individuals with an inherited defect in DNA mismatch-repair genes are at elevated risk of colorectal cancer (see Chap. 4). Other pathways and phenotypes have also been described (see Chap. 5).

Environmental Risk Factors

Diet

Studies correlating the international variation in per capita consumption of specific foods and nutrients with colorectal cancer incidence and mortality rates resulted in numerous hypotheses about an adverse influence of high red meat and fat intake, and beneficial effects of fruit, vegetable, and fiber intakes. Subsequent evidence from case-control and cohort studies is somewhat mixed (Potter et al. 1993; Steinmetz and Potter 1996; WCRF Panel 1997).

Vegetables, Fruits, Fiber, and Micronutrients

A number of prospective studies of the role of vegetable and fruit consumption in colon cancer have reported modest, and somewhat inconsistent, findings of lower risk with higher consumption (Potter et al. 1993; Steinmetz and Potter 1996; WCRF Panel 1997). More recently, several cohort studies have also been inconsistent: in Seventh-day Adventists, a lower risk of colorectal cancer was observed with higher intake of a variety of plant foods, but this was statistically significant only for legumes (Singh and Fraser 1998); A follow-up study of 61,463 Swedish women showed that low fruit and vegetable consumption (<1.5 servings per day) was associated with a relative risk of 1.65, due largely to low fruit intake (Terry et al. 2001); in contrast, in a prospective study of 88,776 women and 47,325 men, no relation was apparent with intake of vegetables or fruit (Michels et al. 2000).

Almost all case-control studies of vegetables and fruit have shown some degree of lower risk of colon cancer with higher consumption of at least one category of these exposures (Potter et al. 1993; WCRF Panel 1997); the results are particularly consistent for raw, green, and cruciferous vegetables. A meta-analysis of six case-control studies of vegetables (Trock et al. 1990) produced a relative risk of 0.48 for highest versus lowest quintiles. For about half of the studies of rectal cancer, an inverse association was observed for at least one vegetable or fruit category, most consistently for cruciferous vegetables. Five case-control studies of adenomas found an inverse association, not always statistically significant, for vegetables (Potter et al. 1993; WCRF Panel 1997). A recent cohort study noted that any beneficial effects of fruit may be confined to those colon tumors not expressing *MLH1* protein product (Wark et al. 2005). The Nurses' Health Study suggested that fruit and legumes may be associated with a lower risk of polyps (Michels et al. 2006). Neither of these studies found associations with vegetables.

Plant foods contain an extensive variety of compounds: both micronutrients, such as carotenoids, folate, and ascorbate, as well as other bioactive compounds with multiple anticarcinogenic properties, such as phenols, flavonoids, isothiocyanates, and indoles (Wattenberg 1978; Steinmetz and Potter 1991). In 1997, a comprehensive overview concluded: "the evidence that diets rich in vegetables protect against cancers of the colon and rectum is convincing" (WCRF Panel 1997). The much smaller, or even absent, inverse associations seen in cohort studies published subsequent to that overview are troubling. The results suggest that the association with vegetable and fruit intake may be limited to specific foods, may be nonlinear, may be part of a more complex dietary pattern, rather than a simple function of fruit and vegetable intake, or, perhaps, suggest that some aspects of vegetables themselves are changing (Potter 2005).

The relationship of dietary fiber to risk of colorectal cancer is inconsistent, probably both because of the heterogeneous nature of fiber itself, and inconsistencies in the way that its intake is measured (WCRF Panel 1997; Potter 1999b). That dietary fiber – derived from vegetables as well as grains – might reduce the risk of colon cancer was proposed originally by Dennis Burkitt (Burkitt 1969). Cohort studies are only weakly supportive of the fiber hypothesis, with two studies finding no association and two, a weak inverse association; a large cohort study of total dietary fiber reported no association with either incident carcinoma or adenoma in women (Fuchs et al. 1999). Only one prospective study has provided data on rectal cancer, with little evidence of an association. In contrast, a combined analysis of 13 case-control studies found a reduction in colorectal cancer risk with increasing intake of dietary fiber (Howe et al. 1992) as did a meta-analysis of 16 case-control studies (Trock et al. 1990). A high intake of fiber from vegetables and cereals was associated with a halving of the risk for adenomas in a prospective study (Giovannucci et al. 1992). Inverse associations with total fiber, fiber from cereals, and fiber from vegetables and fruits have been seen in case-control studies (WCRF Panel 1997). The most recent cohort results on dietary fiber do not resolve the inconsistencies: the PLCO cohort found that those in the highest quintile of dietary fiber intake had a 27% lower risk of distal colonic (but not rectal) adenoma than those in the lowest quintile of intake (Peters et al. 2003), and the EPIC study of approximately 2 million person years and more than 1,000 cases of colorectal cancer in Europe reported that dietary fiber was associated with a 25% reduction in risk from lowest to highest quintile, with the strongest association in the left colon and the weakest in the rectum (very similar to the PLCO findings) (Bingham et al. 2003). However, in a comparable study (1.8 million person years; approximately 1,000 cases) in the US, a similar analysis to that of the EPIC study showed a hazard ratio of 0.91 (0.87–95) and, after adjusting for additional confounders (including folate, red meat, processed meat, and glycemic load), this association disappeared (Michels et al. 2005). Dietary methods and the differences in food supply in the United States and Europe may both be important to the differences in findings. In a pooled analysis of 13, largely United States, cohort studies with 8,081 cases, findings were similar to the US cohort: an inverse association with dietary fiber intake was no longer statistically significant after adjusting for other risk factors (Park et al. 2005).

Intervention studies are much less consistent with the hypotheses that vegetables and/or dietary fiber reduce risk. Three randomized trials have examined the subsequent development of adenomas in patients with a prior adenoma. In Australia, no reduction in metachronous adenoma incidence was seen after 4 years; however, the risk of large adenomas was reduced among those in the low-fat-plus-wheat-bran arm

(MacLennan et al. 1995). In Arizona, the relative risk of metachronous adenomas in the high-fiber group was not reduced (Alberts et al. 2000). In the largest study, about 2,000 men and women were randomized either to a low-fat, high-fiber, increased-fruit-and-vegetables-diet or to usual diet; after 4 years, the relative risk of having one or more metachronous adenomas was 1.0 (Schatzkin et al. 2000). Thus, short-term interventions with increased fiber or vegetables and fruit do not reduce the occurrence of metachronous adenomas. The multihit theory of colorectal cancer hypothesizes the accumulation of widespread damage in multiple genes over decades; beginning an intervention in individuals who already have many colonic epithelial cells well advanced toward adenoma is not as good a test of that particular hypothesis as is one applied early in life for a longer period.

Interpretability of the randomized studies is somewhat limited; nonetheless, it seems unlikely that studies large enough and of long enough duration to test the influence of fruits, vegetables, and/or fiber on cancer risk will ever be designed and conducted. The Women's Health Initiative clinical trial assessed the effects of a low-fat diet on breast-cancer risk among about 48,000 women; these women were specifically counseled to increase fruit and vegetable intake. The results of this trial showed no benefit for colorectal cancer (Beresford et al. 2006).

Other nutrients have been invoked to explain the possibly reduced risk associated with vegetables (Steinmetz and Potter 1991, 1996). Freudenheim, who first proposed a role for folate in colorectal cancer, reported lower risks of both colon and rectal cancer with high folate intakes in her case-control study (Freudenheim et al. 1991). Folate was not associated with risk of colon cancer in a cohort of men (Giovannucci et al. 1995c); nonetheless, the risk of colon cancer and adenoma was increased in men with, simultaneously, low folate, low methionine, and high alcohol (Giovannucci et al. 1993b). Slattery et al. (1997b) found no association between folate and other methyl-group micronutrients and colon cancer. Sanjoaquin cautiously concluded that folate from foods (but not supplements) may be associated with a reduced risk of colorectal cancer in a meta-analysis of cohort studies (Sanjoaquin et al. 2005). Most recently, a randomized trial of folic acid supplementation showed no overall benefit for colorectal adenomas and statistically significant increases in the occurrence of large polyps and multiple polyps, both phenotypes that incur higher risk of colorectal cancer (Cole et al. 2007). It is increasingly clear that folate may prevent polyps and cancer if it is present at sufficient levels prior to the appearance of the earliest changes, but may accelerate the growth of preexisting abnormal cells (Ulrich and Potter 2007). The role of genetic variability in one-carbon (methyl-group) metabolism is discussed in Chap. 9.

Long-term use of multivitamin supplements has been associated with a halving of the risk of colorectal cancer in two cohort studies (White et al. 1997; Giovannucci et al. 1998), the second of which also showed a lower risk with vitamin E supplement use. In an observational study of metachronous adenoma, multivitamin, vitamin E, and calcium supplements (Whelan et al. 1999) all were associated with about a halving of risk of metachronous adenomas. In the ACS Cancer Prevention Study II (of about 150,000 men and women), regular multivitamin use at baseline was not associated with reduced risk; however, reported use 10 years earlier showed a relative risk of 0.71 (95% CI: 0.57–0.89) (Jacobs et al. 2003). Harvard cohort studies showed a lower risk for men, but not women, who took vitamin E supplements (Wu et al. 2002). Satia-Abouta showed that several micronutrients (including vitamin E) were higher in controls than colon cancer cases; most of the differences were explained by supplements (Satia-Abouta et al. 2003). Higher serum selenium has been associated with a lower risk of polyps (Connelly-Frost et al. 2006). How much these data help explain the plant-food associations with colorectal cancer is not clear.

Meat

The overall evidence suggests that meat eating is associated with an elevated risk of colorectal cancer, though that evidence is not entirely consistent. Women who consumed red meat frequently versus rarely had a 2.5-fold increase in the risk of colon cancer in the Nurses' Health Study (Willett et al. 1990). Male health professionals showed a similar pattern (Giovannucci et al. 1994b). Two cohort studies were focused on meat intake in low-consuming populations: one found no association (Key et al. 1998), and the other an elevated risk with higher consumption (Singh and Fraser 1998). In contrast, the Cancer Prevention Study II of the ACS showed no difference in the risk of colorectal cancer death in men or women in the uppermost versus lowest quintiles of meat consumption (Thun et al. 1992). Two cohorts in Europe and one in Iowa, USA (WCRF Panel 1997), were similarly null. The most recent cohort data, however, from Europe (Norat et al. 2005), Australia (English et al. 2004), and the US (Chao et al. 2005) suggest that fresh and processed meat are each associated with elevated risk. Other studies have shown that higher consumption of processed meat is associated with higher risk of colorectal cancer (WCRF Panel 1997) and adenoma (Robertson et al. 2005). Nonetheless, reduction of meat consumption in a randomized trial did not modify the incidence of metachronous adenomas (Mathew et al. 2004).

As with the cohort studies, almost all risk estimates in the case-control studies are increased or null for higher meat intake; however, the largest case-control study found no association with meat (Kampman et al. 1999). A Swedish case-control study found a relative risk of 2.7 for colon cancer and 6.0 for rectal cancer among heavy consumers of fried meat with a markedly browned surface (Gerhardsson de Verdier et al. 1991). Schiffman and Felton (1990) found similarly elevated risks among those preferring well-done meat, a finding repeated in two case-control studies of adenomas (Probst-Hensch et al. 1997; Sinha et al. 1999) and in a large U.S. case-control study of cancer (Kampman et al. 1999). A later study in Sweden failed to observe an association with cooking or with a heterocyclic-amine consumption (Augustsson et al. 1999). Heterocyclic amines and PAHs had been proposed earlier as possible causal agents in meat (Sugimura and Sato 1983); more recently, heme and nitrosation (Bingham et al. 2002; Cross et al. 2003) and O_6 carboxymethyl guanine have also been suggested (Lewin et al. 2006) as explanatory agents of the meat association. For related, detailed discussion of the genetics of carcinogen metabolism and risk, see Chap. 11.

Total Dietary Fat

Epidemiologic studies focused on individual behavior, in contrast to the international ecologic correlation studies, have failed to find evidence for an association with dietary fat. Most cohort studies found no association with total fat (WCRF Panel 1997), although total-fat consumption in the highest versus the lowest quintile of intake was associated with a twofold increased risk in the Nurses' Health Study (Willett et al. 1990). In contrast, the large majority of case-control studies reported increased risks in association with higher intakes of fat, with relative risks from 1.3 to 2.2 (WCRF Panel 1997). Nonetheless, in a combined analysis of 13 case-control studies across populations with different diets and cancer risks, dietary fat was not associated with risk after adjustment for total energy intake (Howe et al. 1997).

Animal or Saturated Fat

In the Nurses' Health Study, risk associated with the highest intakes of animal fat was almost twofold greater than that with the lowest intakes (Willett et al. 1990). A small increase in the risk of colon cancer was seen among women alone in the Netherlands cohort (Goldbohm et al. 1994), and no associations were seen in two other cohort studies (WCRF Panel 1997). The results of the case-control studies of animal fat or saturated fat are inconsistent (WCRF Panel 1997). Overall, about half of all studies, irrespective of design, show some evidence of elevated risk with higher intakes of saturated fat or animal fat, and no study shows the opposite, somewhat more consistent than the pattern for total fat. Giovannucci and Goldin (1997) concluded that the association with red meat does not appear to be explained by its fat content. The protein and iron content of red meat also are not strong candidates (WCRF Panel 1997). Reducing dietary fat in the Women's Health Initiative randomized trial did not reduce colorectal cancer rates (Beresford et al. 2006).

The data suggest a stronger association with meat than with any of its constituent nutrients; neither total fat nor total protein seems to play a major role. The public health recommendation has been made clearly: "If eaten at all, limit intake of red meat to less than three ounces daily" (WCRF Panel 1997).

Calcium and Vitamin D

Calcium and dairy foods have been studied for their association with colorectal neoplasia (Potter et al. 1993; WCRF Panel 1997). Most of these earlier studies suggested a reduced risk or no association. One large case-control study showed an adjusted relative risk for highest versus lowest quintile of dietary calcium of 0.6 (Kampman et al. 2000). Calcium was shown to reduce proliferation in the upper part of colonic crypts (Bostick et al. 1995), and observational data were consistent with a reduced risk of metachronous adenomas (Hyman et al. 1998).

A double-blind, randomized trial of 1,200 mg of elemental calcium per day versus placebo reported a statistically significant 15–20% reduction in metachronous colorectal adenomas (Baron et al. 1999). Subsequent analyses of these data revealed that the calcium-associated reduction of risk of metachronous adenoma was seen only among those with baseline 25-(OH) vitamin D levels in serum above the median. Further, those with high serum vitamin D levels had a reduced risk of adenoma only if they received calcium supplements (Grau et al. 2003). Thus, it is an interaction between vitamin D and calcium that alters adenoma risk. Other studies have added further data: two observational cohort studies within intervention studies showed that higher calcium and vitamin D (Hartman et al. 2005) or calcium alone (Peters et al. 2004) were associated with reduced risk of adenoma; a pooled analysis of cohort studies (Cho et al. 2004a) and three other cohort studies (Terry et al. 2002; Flood et al. 2005; Lin et al. 2005) each showed reduced risk of colorectal cancer with calcium; the WHI Calcium plus Vitamin D Supplementation Trial, with a low level of supplementation, showed no reduction in risk of colorectal cancer over 7 years (Wactawski-Wende et al. 2006). The majority of the calcium studies are reviewed in Chia and Newcomb (2004). Overall, the data are consistent with a reduced risk for, especially, the combination of a calcium source and vitamin D. More detail, as well as the role of genetics in modifying the protective effects of calcium and vitamin D, is presented in Chap. 12.

Physical Activity and Anthropometry

Physical activity is highly consistently associated with a reduced risk of colon cancer, as seen in studies of occupational activity, leisure activity, and total activity (Potter et al. 1993; WCRF Panel 1997). Of nine cohort studies, only two have reported no substantial association. Case-control studies are also very consistent. Individuals with high levels of activity throughout life are at lowest risk. Rectal cancer risk does not seem to be modified by physical activity. Two recent reviews (Lee 2003; Slattery 2004) and a meta-analysis (Samad et al. 2005) provide useful summaries.

Physical activity stimulates colon peristalsis, which should decrease the time that the fecal stream is in contact with the epithelium; however, transit time is not a risk factor for colon neoplasia. Exercise has both acute and persistent hormonal effects, as well as favorable effects on the immune system. Higher physical activity, especially in subjects with a low body mass, is associated with a metabolic milieu (lower insulin, other growth factors, glucose, and triacylglycerol levels), less favorable to the growth of cancer generally, and, perhaps, colon cancer in particular (McMichael and Potter 1980, 1985; McKeown-Eyssen 1994; Giovannucci 1995a; WCRF Panel 1997).

Obesity probably increases the risk of colon cancer, particularly in men, but, as with physical activity, probably not of rectal cancer. Most epidemiologic studies have found that obese men (the highest quintile for body mass) have as much as a twofold increased risk of colon cancer (Potter et al. 1993; WCRF Panel 1997; Singh and Fraser 1998); nonetheless, some studies have shown no association with body mass (Potter et al. 1993; WCRF Panel 1997). Data on women are much less consistent: two cohort studies found no association between body mass and colorectal cancer (WCRF Panel 1997); however, in the Iowa Women's Health Study, in contrast, subjects in the highest quintile of BMI had a statistically significant 40% higher risk than those in the lowest (Bostick et al. 1994). Three case-control studies reported inconsistent findings for women (WCRF Panel 1997), and a prospective study in Japan suggested that obesity and weight gain are associated with colon cancer in women, but not men (Tamakoshi et al. 2004).

Although body mass was not associated with elevated risk at high levels of long-term vigorous physical activity in a large case-control study, at lower levels of activity, risk was related both to total energy intake and to obesity. Those who, simultaneously, were the least active, had the highest energy intake, and had the largest BMI were at 3.4 times the risk of those at the opposite extremes, an association explained entirely by findings for men, in whom the odds ratio for a comparison of extremes was 7.2; there was little association in women (Slattery et al. 1997a). A high waist/hip ratio, a measure of intra-abdominal fat, has been reported to be associated with increased risk in men (Giovannucci et al. 1995a), but not women (Bostick et al. 1994). Framingham Study data showed that waist circumference was a better predictor of lifetime colon cancer risk than body mass, and, unlike that with BMI, this association was similar between the sexes (Moore et al. 2004).

Alcohol

Of five general population cohort studies of colon cancer, four have shown statistically significant elevated risk for alcohol consumption, as did each of the three studies that explored rectal cancer risk and two of the three studies that reported on colorectal cancer as a single entity (WCRF Panel 1997). A pooled cohort analysis showed an elevated risk for colon and rectum, a dose–response relationship, no heterogeneity by study or sex, and no differences by particular alcoholic beverage (Cho et al. 2004b). Alcohol has also been associated with an increased risk of colon and rectal cancer in about half of the case-control studies (WCRF Panel 1997), and almost no inverse associations have been reported. Acetaldehyde (a metabolite of alcohol) forms DNA adducts. Alcohol also inhibits DNA repair (Farinati et al. 1985). Finally, alcohol may also exert its effect through associated deficiencies in nutrients, particularly folate (Garro and Lieber 1990; Giovannucci et al. 1995c). The WCRF report concluded that "high alcohol consumption probably increases the risk of cancers of the colon and rectum" and that the association is likely to be "related to total ethanol intake, irrespective of the type of drink" (WCRF Panel 1997).

Tobacco

Tobacco was originally not associated with an elevated risk of colorectal cancer, although associations with cigar and pipe smoking have been reported (Wynder and

Shigematsu 1967; Slattery et al. 1990). Some more recent studies have described an increased risk of colon cancer with both an early onset and a long history of smoking cigarettes (Giovannucci et al. 1994a), although no associations were seen in two European studies (Tavani et al. 1998). The most recent data suggest that smoking is largely associated with microsatellite unstable (MSI-H) colorectal cancer (Slattery et al. 2000) and, relatedly, in tumors with loss of MLH1 expression (Lüchtenborg et al. 2005). Hyperplastic polyps and serrated adenomas, rather than *APC* truncation-mutation, probably define the progression pathway (Morimoto et al. 2002). See Chap. 5 for further discussion.

There are many carcinogens in tobacco smoke, including polycyclic hydrocarbons, heterocyclic amines, and nitrosamines, all plausible blood-borne carcinogens. Rat models show that heterocyclic amines cause specific *APC* mutations (Kakiuchi et al. 1995), although this is not seen in humans. NSAIDs generally reduce the risk of colorectal neoplasia (see below and Chapter 10), but they may not be effective in heavy smokers and especially against MSI-H tumors (Chia et al. 2006). For detailed discussion of the genetics of carcinogen metabolism in relation to risk, see Chap. 11.

Reproductive Factors

In 1969, Fraumeni et al. reported that, among nuns, there was an excess, not only of known reproductive and hormonal cancers, but also of colon cancer (Fraumeni et al. 1969). Case-control studies in the 1970s and 1980s noted a higher risk of colon cancer among the nulliparous. This association was initially postulated to be due to changes in lipids and bile acids that occur with changes in steroid hormone profiles (McMichael and Potter 1980). That estrogen-receptor expression in the colonic epithelium declines with age may also be important to the role of hormones in colon cancer (Issa et al. 1994; Potter et al. 1996). More than 20 epidemiologic studies have now reported on the relationships with reproductive history in women. Overall, age at first birth is not associated with colon cancer risk. The parity data seem also to be null, especially given that none of the cohort studies shows an association (Potter et al. 1993). Nevertheless, two caveats remain: the differences between the findings of the cohort and population-based case-control studies are not well explained and the original observations applied to excess risk among more elderly women (Slattery et al. 1994). See later for a discussion of the role of exogenous hormones in colon cancer, where associations are more markedly more consistent.

Infection

A few ecologic and case-control studies suggest that infection with *Schistosoma japonicum* is associated with increased risk of colorectal cancer, perhaps particularly rectal cancer (Xu and Su 1984). JC Virus has been associated with colorectal cancers (Laghi et al. 1999; Enam et al. 2002), but not consistently (Newcomb et al. 2004). It has been proposed that the infection triggers the classical chromosomal instability pathway (Niv et al. 2005). *Helicobacter pylori* has been proposed as a risk factor for colorectal adenoma (Breuer-Katschinski et al. 1999) and cancer (Talley et al. 1991); a recent meta-analysis suggested a small elevation in risk (Zumkeller et al. 2006). Mucosa-associated lymphoid tissue (MALT) lymphoma in the colon regresses with antibiotic therapy (Raderer et al. 2000) as does its gastric counterpart, where the causal agent is known to be *H. pylori*.

Occupation

Colon cancer risk is elevated in white-collar occupations, probably due to lower physical activity (Chow et al. 1994). Asbestos exposure has been observed intermittently as a risk factor (Kang et al. 1997). The Shanghai Textile Workers Cohort suggests that there may be an elevated risk in women exposed to dyes and metals but a reduced risk associated with exposure to cotton and cotton dust (De Roos et al. 2005).

Medical Conditions

Inflammatory Bowel Disease

Ulcerative colitis and Crohn disease increase risk of colorectal cancer (Ekbom et al. 1990a, b); up to 5% of all colorectal cancers in patients under the age of 50 occur in individuals with these inflammatory diseases. In those with ulcerative colitis, younger age of onset, longer duration, and greater extent of disease, as well as the additional complication of primary sclerosing cholangitis, each increases risk (Broome et al. 1992). In those with Crohn disease, involvement of the colon and younger age at onset are established risk factors for cancer. A recent decline in risk of colorectal cancer among inflammatory bowel disease (IBD) patients has been tentatively ascribed to greater use of anti-inflammatory agents to control the disease (Moody et al. 1996). The diseases may increase risk because of the loss of the intestinal brush border, thus bringing proliferating stem cells into contact with the fecal stream without requiring prior adenoma formation (Potter 1999a). Reactive oxygen and nitrogen species, as well as the processes associated with inflammation, may be involved (Munkholm 2003; Itzkowitz and Yio 2004).

Diabetes Mellitus

There is an elevated risk of colorectal cancer among those with diabetes mellitus (Hu et al. 1999). Both colon and rectal cancer incidence in men and women are increased by 30–40% with a comparable, if more heterogeneous, association with mortality (Larsson et al. 2005). See above for a related discussion on the role of obesity.

Cholecystectomy

Studies of risk associated with cholecystectomy (which alters bile flow into the intestinal tract) show both strong positive associations and null associations; most of the positive associations are seen in case-control studies, where selection bias among controls is of concern. The largest case-control study showed a 30% elevation in the risk of proximal colon cancer and a statistically nonsignificant inverse association with distal colon cancer; thus, overall risk is essentially unchanged (Todoroki et al. 1999). Two large cohort studies (Ekbom et al. 1993; Johansen et al. 1996) and two meta-analyses (Giovannucci et al. 1993a; Reid et al. 1996) showed a 10–30% increased risk of proximal colon cancer after 15 or more years. Another Swedish study (Lagergren et al. 2001) examined the risk of both small and large bowel cancer and reported a gradient of decreasing risk from duodenum to distal colon, consistent with the hypothesis that higher concentrations of bile acids are associated with a higher risk, which declines as bile acids become absorbed and diluted.

Medications

NSAIDs

Aspirin and other nonsteroidal anti-inflammatory drugs (NSAIDs) are consistently associated with a reduced risk of colorectal neoplasia. Most case-control studies (Kune et al. 1988; Rosenberg et al. 1991; Suh et al. 1993; Muscat et al. 1994; Peleg et al. 1994; La Vecchia et al. 1997; Rosenberg et al. 1998; Bigler et al. 2001) and cohort studies (Thun et al. 1991; Schreinemachers and Everson 1994; Giovannucci et al. 1994c, 1995b; Chan et al. 2005) of aspirin have reported a lower risk of colorectal cancer incidence or mortality. Similar findings have been reported for adenomas (Greenberg et al. 1993; Logan et al. 1993; Suh et al. 1993; Giovannucci et al. 1994c). Sulindac induces regression of adenomas in patients with FAP (Giardiello et al. 1993). There are a few null studies (Paganini-Hill et al. 1989; Gann et al. 1993; Stürmer et al. 1998). In rodent studies, aspirin (Craven and DeRubertis 1992), indomethacin (Pollard and Luckert 1980; Narisawa et al. 1981), sulindac (Moorghen et al. 1988), piroxicam (Reddy et al. 1987), and celecoxib (a specific COX-2 inhibitor) (Kawamori et al. 1998) inhibit carcinogenesis.

Metachronous adenomas occur at lower frequency in randomized trials of aspirin versus placebo in those who have had a prior cancer (Sandler et al. 2003) or adenoma (Baron et al. 2003). Specific COX-2 inhibitors are also effective in clinical trials (Arber et al. 2006; Bertagnolli et al. 2006); however, the increased heart-disease risk associated with COX inhibitors essentially precludes their use in cancer prevention other than in persons at very high risk of colorectal cancer (Psaty and Potter 2006).

A more detailed discussion of the role of NSAIDs, the cyclooxgenase enzymes, and the genetics of both metabolism and downstream signaling is presented in Chap. 10.

Postmenopausal Hormone Use

In 1981, the first investigation of postmenopausal hormones (PMH) and colorectal cancer found no association (Weiss et al. 1981). A statistically significantly lower risk of colon cancer was reported 2 years later in association with use of the high-estrogen oral contraceptives (OCs), but not with PMH (Potter and McMichael 1983). Findings across the many studies since this time are not entirely consistent, although, of the studies that provided separate data on colon cancer, about half showed a statistically significant lower risk with PMH or a less well-specified hormone variable (Chute et al. 1991; Gerhardsson de Verdier and London 1992; Jacobs et al. 1994; Calle et al. 1995; Newcomb and Storer 1995; Kampman et al. 1997; Fernandez et al. 1998), two showed a statistically nonsignificantly lower risk (Potter and McMichael 1983; Bostick et al. 1994), two were null (Peters et al. 1990; Risch and Howe 1995), and one showed an elevated risk (Wu-Williams et al. 1991). Consistent with some other observations, several recent investigations report an approximate halving of risk with recent PMH use (Newcomb and Storer 1995; Kampman et al. 1997), a degree of risk reduction that is maintained for about 10 years after cessation. Longer use is probably associated with lower risk. A similar pattern of association exists for adenomatous polyps of both colon and rectum (Jacobson et al. 1995; Potter et al. 1996).

In 2004, the WHI Estrogen plus Progestin (E + P) Intervention Trial showed that those on estrogen and progestin (but not on estrogen alone) had a considerably reduced risk of colorectal cancer than those on the placebo arm (Anderson et al. 2004; Chlebowski et al. 2004). The most recent data suggest that estrogen plus progestin is associated with MSI-low and microsatellite-stable colorectal cancer, but not MSI-H cancer (Newcomb et al. 2007).

Estrogen receptor hypermethylation increases with age and is a central feature of colon cancer, initially suggesting that declining levels of estrogen may be important (Issa et al. 1994); the protective role of PMH against both polyps and cancer may be a consequence of replacing these declining endogenous estrogen levels, thus reducing the likelihood that the estrogen-receptor gene will be silenced by methylation (Potter 1995). However, the WHI evidence and the case-control evidence (Newcomb et al. 2007), that progestin is crucial, complicate the picture further. Identifying the hormone-responsive targets that are involved in colorectal carcinogenesis is important to designing better targeted prevention modalities.

Conclusion

There are several risk factors for which evidence of a causal association with colorectal cancer risk is strong. Obesity among men and a sedentary lifestyle in both sexes are strongly implicated. The exact dietary constituents or patterns that increase risk are less clear, but higher red meat consumption and lower intake of plant foods and calcium are probably important. Aspirin and NSAIDs, as well as postmenopausal hormones, reduce risk. However, many of the known risk factors have not been explored in relation to molecularly defined cancers and, where they have, differences have emerged, smoking being the most obvious example. Because there is a known precursor lesion (the polyp), that is detectable at endoscopy, screening and early detection are feasible and effective. This knowledge and a much better recent understanding of the genetics, as described in the following chapters, make colorectal neoplasia one of the most preventable, screenable, detectable, and manageable of all cancers. However, colorectal cancer remains a formidable cause of morbidity and mortality.

References

ACS (2006). Cancer Facts and Figures 2006. Atlanta, American Cancer Society.

Alberts, D. S., M. E. Martinez, et al. (2000). "Lack of effect of a high-fiber cereal supplement on the recurrence of colorectal adenomas. Phoenix Colon Cancer Prevention Physicians' Network." *N Engl J Med* **342**(16): 1156–62.

Anderson, G. L., M. Limacher, et al. (2004). "Effects of conjugated equine estrogen in postmenopausal women with hysterectomy: the Women's Health Initiative Randomized Controlled Trial." *JAMA* **291**(14): 1701–12.

Arber, N., I. Racz, et al. (2006). Chemoprevention of colorectal adenomas with celecoxib in an international randomized, placebo-controlled, double-blind trial. 97th Annual Meeting of the American Association for Cancer Research, Washington, DC.

Augustsson, K., K. Skog, et al. (1999). "Dietary heterocyclic amines and cancer of the colon, rectum, bladder, and kidney: a population-based study [see comments]." *Lancet* **353**(9154): 703–7.

Baron, J. A., M. Beach, et al. (1999). "Calcium supplements for the prevention of colorectal adenomas." *N Engl J Med* **340**: 101–7.

Baron, J. A., B. F. Cole, et al. (2003). "A randomized trial of aspirin to prevent colorectal adenomas." *N Engl J Med* **348**(10): 891–9.

Beresford, S. A., K. C. Johnson, et al. (2006). "Low-fat dietary pattern and risk of colorectal cancer: the Women's Health Initiative Randomized Controlled Dietary Modification Trial." *JAMA* **295**(6): 643–54.

Bertagnolli, M. M., C. J. Eagle, et al. (2006). Celecoxib reduces sporadic colorectal adenomas: results from the Adenoma Prevention with Celecoxib (APC) trial. 97th Annual Meeting of the American Association for Cancer Research. Washington, DC.

Bigler, J., J. Whitton, et al. (2001). "CYP2C9 and UGT1A6 genotypes modulate the protective effect of aspirin on colon adenoma risk." *Cancer Res* **61**(9): 3566–9.

Bingham, S. A., R. Hughes, et al. (2002). "Effect of white versus red meat on endogenous N-nitrosation in the human colon and further evidence of a dose response." *J Nutr* **132** (11 Suppl): 3522S–5S.

Bingham, S. A., N. E. Day, et al. (2003). "Dietary fibre in food and protection against colorectal cancer in the European Prospective Investigation into Cancer and Nutrition (EPIC): an observational study." *Lancet* **361**(9368): 1496–501.

Bostick, R. M., J. D. Potter, et al. (1994). "Sugar, meat, and fat intake, and non-dietary risk factors for colon cancer incidence in Iowa women (United States)." *Cancer Causes Control* **5**: 38–52.

Bostick, R., L. Fosdick, et al. (1995). "Calcium and colorectoral epithelial cell proliferation in sporadic adenoma patients: a randomized, double-blinded, placebo-controlled clinical trial." *J Natl Cancer Inst* **87**: 1307–15.

Breuer-Katschinski, B., K. Nemes, et al. (1999). "Helicobacter pylori and the risk of colonic adenomas. Colorectal Adenoma Study Group." *Digestion* **60**(3): 210–5.

Broome, U., G. Lindberg, et al. (1992). "Primary sclerosing cholangitis in ulcerative colitis – a risk factor for the development of dysplasia and DNA aneuploidy?" *Gastroenterology* **102**(6): 1877–80.

Burkitt, D. P. (1969). "Related disease-related cause?" *Lancet* **1**: 1229–30.

Calle, E. E., H. L. Miracle-McMahill, et al. (1995). "Estrogen replacement therapy and risk of fatal colon cancer in a prospective cohort of postmenopausal women." *J Natl Cancer Inst* **87**: 517–23.

Chan, A. T., E. L. Giovannucci, et al. (2005). "Long-term use of aspirin and nonsteroidal anti-inflammatory drugs and risk of colorectal cancer." *JAMA* **294**(8): 914–23.

Chao, A., M. J. Thun, et al. (2005). "Meat consumption and risk of colorectal cancer." *JAMA* **293**(2): 172–82.

Chia, V. and P. A. Newcomb (2004). "Calcium and colorectal cancer: some questions remain." *Nutr Rev* **62**(3): 115–20.

Chia, V., P. Newcomb, et al. (2006). "Risk of MSI colorectal cancer is associated jointly with smoking and NSAID use." *Cancer Res* **66**: 6877–83.

Chlebowski, R. T., J. Wactawski-Wende, et al. (2004). "Estrogen plus progestin and colorectal cancer in postmenopausal women." *N Engl J Med* **350**(10): 991–1004.

Cho, E., S. A. Smith-Warner, et al. (2004a). "Dairy foods, calcium, and colorectal cancer: a pooled analysis of 10 cohort studies." *J Natl Cancer Inst* **96**(13): 1015–22.

Cho, E., S. A. Smith-Warner, et al. (2004b). "Alcohol intake and colorectal cancer: a pooled analysis of 8 cohort studies." *Ann Intern Med* **140**(8): 603–13.

Chow, W., H. Malker, et al. (1994). "Occupational risks for colon cancer in Sweden." *J Occup Med* **36**(6): 647–51.

Chute, C. G., W. C. Willett, et al. (1991). "A prospective study of reproductive history and exogenous estrogens on the risk of colorectal cancer in women." *Epidemiology* **2**: 201–7.

Cole, B. F., J. A. Baron, et al. (2007). "Folic acid for the prevention of colorectal adenomas: a randomized clinical trial." *JAMA* **297**(21): 2351–9.

Connelly-Frost, A., C. Poole, et al. (2006). "Selenium, apoptosis, and colorectal adenomas." *Cancer Epidemiol Biomarkers Prev* **15**(3): 486–93.

Craven, P. A. and F. R. DeRubertis (1992). "Effects of aspirin on 1,2-dimethylhydrazine-induced colonic carcinogenesis." *Carcinogenesis* **13**: 541–6.

Cross, A. J., J. R. Pollock, et al. (2003). "Haem, not protein or inorganic iron, is responsible for endogenous intestinal N-nitrosation arising from red meat." *Cancer Res* **63**(10): 2358–60.

De Roos, A. J., R. M. Ray, et al. (2005). "Colorectal cancer incidence among female textile workers in Shanghai, China: a case-cohort analysis of occupational exposures." *Cancer Causes Control* **16**(10): 1177–88.

Ekbom, A., C. Helmick, et al. (1990a). "Ulcerative colitis and colorectal cancer. A population-based study." *N Engl J Med* **323**(18): 1228–33.

Ekbom, A., C. Helmick, et al. (1990b). "Increased risk of large-bowel cancer in Crohn's disease with colonic involvement." *Lancet* **336**(8711): 357–9.

Ekbom, A., J. Yuen, et al. (1993). "Cholecystectomy and colorectal cancer." *Gastroenterology* **105**(1): 142–7.

Enam, S., L. Del Valle, et al. (2002). "Association of human polyomavirus JCV with colon cancer: evidence for interaction of viral T-antigen and beta-catenin." *Cancer Res* **62**(23): 7093–101.

English, D. R., R. J. MacInnis, et al. (2004). "Red meat, chicken, and fish consumption and risk of colorectal cancer." *Cancer Epidemiol Biomarkers Prev* **13**(9): 1509–14.

Farinati, F., N. Espina, et al. (1985). "In vivo inhibition by chronic ethanol exposure of methylguanine transferase activity and DNA repair." *Ital J Gastroenterol* **17**: 48–9.

Fearon, E. R. and B. Vogelstein (1990). "A genetic model for colorectal tumorigenesis." *Cell* **61**: 759–67.

Fernandez, E., C. La Vecchia, et al. (1998). "Hormone replacement therapy and risk of colon and rectal cancer." *Cancer Epidemiol Biomarkers Prev* **7**(4): 329–33.

Flood, A., U. Peters, et al. (2005). "Calcium from diet and supplements is associated with reduced risk of colorectal cancer in a prospective cohort of women." *Cancer Epidemiol Biomarkers Prev* **14**(1): 126–32.
Fraumeni, J. F., Jr., J. W. Lloyd, et al. (1969). "Cancer mortality among nuns: role of marital status in etiology of neoplastic disease in women." *J Natl Cancer Inst* **42**: 455–68.
Freudenheim, J. L., S. Graham, et al. (1991). "Folate intake and carcinogenesis of the colon and rectum." *Int J Epidemiol* **20**: 368–74.
Fuchs, C. S., G. A. Colditz, et al. (1999). "Dietary fiber and the risk of colorectal cancer and adenoma in women." *N Engl J Med* **340**: 169–76.
Gann, P. H., J. E. Manson, et al. (1993). "Low-dose aspirin and incidence of colorectal tumors in a randomized trial." *J Natl Cancer Inst* **85**: 1220–4.
Garro, A. J. and C. S. Lieber (1990). "Alcohol and cancer." *Annu Rev Pharmacol Toxicol* **30**: 219–49.
Gerhardsson de Verdier, M. and S. London (1992). "Reproductive factors, exogenous female hormones, and colorectal cancer by subsite." *Cancer Causes Control* **3**: 355–60.
Gerhardsson de Verdier, M., U. Hagman, et al. (1991). "Meat, cooking methods and colorectal cancer: a case-referent study in Stockholm." *Int J Cancer* **49**: 520–5.
Giardiello, F. M., S. R. Hamilton, et al. (1993). "Treatment of colonic and rectal adenomas with sulindac in familial adenomatous polyposis." *N Engl J Med* **328**: 1313–6.
Giovannucci, E. (1995). "Insulin and colon cancer." *Cancer Causes Control* **6**(2): 164–79.
Giovannucci, E. and B. Goldin (1997). "The role of fat, fatty acids, and total energy intake in the etiology of human colon cancer." *Am J Clin Nutr* **66**(6 Suppl): 1564S–71S.
Giovannucci, E., M. J. Stampfer, et al. (1992). "Relationship of diet to risk of colorectal adenoma in men." *J Natl Cancer Inst* **84**(2): 91–8.
Giovannucci, E., G. A. Colditz, et al. (1993a). "A meta-analysis of cholecystectomy and risk of colorectal cancer." *Gastroenterology* **105**(1): 130–41.
Giovannucci, E., M. J. Stampfer, et al. (1993b). "Folate, methionine, and alcohol intake and risk of colorectal adenoma." *J Natl Cancer Inst* **85**(11): 875–84.
Giovannucci, E., E. B. Rimm, et al. (1994a). "A prospective study of cigarette smoking and risk of colorectal adenoma and colorectal cancer in U.S. men." *J Natl Cancer Inst* **86**(3): 183–91.
Giovannucci, E., E. B. Rimm, et al. (1994b). "Intake of fat, meat, and fiber in relation to risk of colon cancer in men." *Cancer Res* **54**(9): 2390–7.
Giovannucci, E., E. B. Rimm, et al. (1994c). "Aspirin use and the risk for colorectal cancer and adenoma in male health professionals." *Ann Intern Med* **121**: 241–6.
Giovannucci, E., A. Ascherio, et al. (1995a). "Physical activity, obesity, and risk for colon cancer and adenoma in men." *Ann Intern Med* **122**(45): 327–34.
Giovannucci, E., K. M. Egan, et al. (1995b). "Aspirin and the risk of colorectal cancer in women." *N Engl J Med* **333**(10): 609–14.
Giovannucci, E., E. B. Rimm, et al. (1995c). "Alcohol, low-methionine – low-folate diets, and risk of colon cancer in men." *J Natl Cancer Inst* **87**(4): 265–73.
Giovannucci, E., M. J. Stampfer, et al. (1998). "Multivitamin use, folate, and colon cancer in women in the Nurses' Health Study." *Ann Intern Med* **129**(7): 517–24.
Goldbohm, R. A., P. A. van den Brandt, et al. (1994). "A prospective cohort study on the relation between meat consumption and the risk of colon cancer." *Cancer Res* **54**(3): 718–23.
Grau, M. V., J. A. Baron, et al. (2003). "Vitamin D, calcium supplementation, and colorectal adenomas: results of a randomized trial." *J Natl Cancer Inst* **95**(23): 1765–71.
Greenberg, E. R., J. A. Baron, et al. (1993). "Reduced risk of large-bowel adenomas among aspirin users." *J Natl Cancer Inst* **85**: 912–6.
Haenszel, W. (1961). "Cancer mortality among the foreign born in the United States." *J Natl Cancer Inst* **26**: 37–132.
Hartman, T. J., P. S. Albert, et al. (2005). "The association of calcium and vitamin D with risk of colorectal adenomas." *J Nutr* **135**(2): 252–9.
Hill, M. J., B. C. Morson, et al. (1978). "Aetiology of adenoma-carcinoma sequence in large bowel." *Lancet* **8058**: 245–7.

Howe, G. R., E. Benito, et al. (1992). "Dietary intake of fiber and decreased risk of cancers of the colon and rectum: evidence from the combined analysis of 13 case-control studies." *J Natl Cancer Inst* **84**(24): 1887–96.

Howe, G., K. Aronson, et al. (1997). "The relationship between dietary fat intake and risk of colorectal cancer: evidence from the combined analysis of 13 case-control studies." *Cancer Causes Control* **8**: 215–28.

Hu, F. B., J. E. Manson, et al. (1999). "Prospective study of adult onset diabetes mellitus (type 2) and risk of colorectal cancer in women." *J Natl Cancer Inst* **91**(6): 542–7.

Hyman, J., J. A. Baron, et al. (1998). "Dietary and supplemental calcium and the recurrence of colorectal adenomas." *Cancer Epidemiol* **7**: 291–5.

Issa, J. P., Y. L. Ottaviano, et al. (1994). "Methylation of the oestrogen receptor CpG island links ageing and neoplasia in human colon." *Nat Genet* **7**(4): 536–40.

Itzkowitz, S. H. and X. Yio (2004). "Inflammation and cancer IV. Colorectal cancer in inflammatory bowel disease: the role of inflammation." *Am J Physiol Gastrointest Liver Physiol* **287**(1): G7–17.

Jacobs, E. J., E. White, et al. (1994). "Exogenous hormones, reproductive history, and colon cancer (Seattle, Washington, USA)." *Cancer Causes Control* **5**: 359–66.

Jacobs, E. J., C. J. Connell, et al. (2003). "Multivitamin use and colorectal cancer incidence in a US cohort: does timing matter?" *Am J Epidemiol* **158**(7): 621–8.

Jacobson, J. S., A. I. Neugut, et al. (1995). "Reproductive risk factors for colorectal adenomatous polyps (New York City, NY, United States)." *Cancer Causes Control* **6**(6): 513–8.

Johansen, C., W. H. Chow, et al. (1996). "Risk of colorectal cancer and other cancers in patients with gall stones." *Gut* **39**(3): 439–43.

Kakiuchi, H., M. Watanabe, et al. (1995). "Specific 5′-GGGA-3′ → 5′-GGA-3′ mutation of the Apc gene in rat colon tumors induced by 2-amino-1-methyl-6-phenylimidazo[4,5-b]pyridine." *Proc Natl Acad Sci U S A* **92**(3): 910–4.

Kampman, E., J. D. Potter, et al. (1997). "Hormone replacement therapy, reproductive history, and colon cancer: a multicenter, case-control study in the United States." *Cancer Causes Control* **8**: 146–58.

Kampman, E., M. L. Slattery, et al. (1999). "Meat consumption, genetic susceptibility, and colon cancer risk: a United States multicenter case-control study." *Cancer Epidemiol Biomarkers Prev* **8**(1): 15–24.

Kampman, E., M. L. Slattery, et al. (2000). "Calcium, vitamin D, sunshine exposure, dairy products and colon cancer risk (United States)." *Cancer Causes Control* **11**(5): 459–66.

Kang, S., C. Burnett, et al. (1997). "Gastrointestinal cancer mortality of workers in occupations with high asbestos exposures." *Am J Ind Med* **31**(6): 713–8.

Kawamori, T., C. V. Rao, et al. (1998). "Chemopreventive activity of celecoxib, a specific cyclooxygenase-2 inhibitor, against colon carcinogenesis." *Cancer Res* **58**: 409–12.

Key, T. J., G. E. Fraser, et al. (1998). "Mortality in vegetarians and non-vegetarians: a collaborative analysis of 8300 deaths among 76,000 men and women in five prospective studies." *Public Health Nutr* **1**(1): 33–41.

Kune, G. A., S. Kune, et al. (1988). "Colorectal cancer risk, chronic illnesses, operations, and medications: case control results from the Melbourne Colorectal Cancer Study." *Cancer Res* **48**: 4399–404.

La Vecchia, C., E. Negri, et al. (1997). "Aspirin and colorectal cancer." *Br J Cancer* **76**: 675–7.

Lagergren, J., W. Ye, et al. (2001). "Intestinal cancer after cholecystectomy: is bile involved in carcinogenesis?" *Gastroenterology* **121**(3): 542–7.

Laghi, L., A. E. Randolph, et al. (1999). "JC virus DNA is present in the mucosa of the human colon and in colorectal cancers." *Proc Natl Acad Sci U S A* **96**(13): 7484–9.

Larsson, S. C., N. Orsini, et al. (2005). "Diabetes mellitus and risk of colorectal cancer: a meta-analysis." *J Natl Cancer Inst* **97**(22): 1679–87.

Lee, I. M. (2003). "Physical activity and cancer prevention – data from epidemiologic studies." *Med Sci Sports Exerc* **35**(11): 1823–7.

Lewin, M. H., N. Bailey, et al. (2006). "Red meat enhances the colonic formation of the DNA adduct O_6-carboxymethyl guanine: implications for colorectal cancer risk." *Cancer Res* **66**(3): 1859–65.

Lin, J., S. M. Zhang, et al. (2005). "Intakes of calcium and vitamin D and risk of colorectal cancer in women." *Am J Epidemiol* **161**(8): 755–64.

Logan, R. F., J. Little, et al. (1993). "Effect of aspirin and non-steroidal anti-inflammatory drugs on colorectal adenomas: case-control study of subjects participating in the Nottingham faecal occult blood screening programme." *BMJ* **307**: 285–9.

Lüchtenborg, M., M. P. Weijenberg, et al. (2005). "Cigarette smoking and colorectal cancer: APC mutations, hMLH1 expression, and GSTM1 and GSTT1 polymorphisms." *Am J Epidemiol* **161**(9): 806–15.

MacLennan, R., F. Macrae, et al. (1995). "Randomized trial of intake of fat, fiber, and beta carotene to prevent colorectal adenomas." *J Natl Cancer Inst* **87**: 1760–6.

Mathew, A., R. Sinha, et al. (2004). "Meat intake and the recurrence of colorectal adenomas." *Eur J Cancer Prev* **13**(3): 159–64.

McKeown-Eyssen, G. E. (1994). "Epidemiology of colorectal cancer revisited: are serum triglycerides and/or plasma glucose associated with risk?" *Cancer Epidemiol Biomarkers Prev* **3**(8): 687–95.

McMichael, A. J. and G. G. Giles (1988). "Cancer in migrants to Australia: extending the descriptive epidemiological data." *Cancer Res* **48**: 751–6.

McMichael, A. J. and J. D. Potter (1980). "Reproduction, endogenous and exogenous sex hormones, and colon cancer: a review and hypothesis." *J Natl Cancer Inst* **65**(6): 1201–7.

McMichael, A. J. and J. D. Potter (1985). "Host factors in carcinogenesis: certain bile-acid metabolic profiles that selectively increase the risk of proximal colon cancer." *J Natl Cancer Inst* **75**(2): 185–91.

Michels, K. B., E. Giovannucci, et al. (2000). "Prospective study of fruit and vegetable consumption and incidence of colon and rectal cancers." *J Natl Cancer Inst* **92**(21): 1740–52.

Michels, K. B., C. S. Fuchs, et al. (2005). "Fiber intake and incidence of colorectal cancer among 76,947 women and 47,279 men." *Cancer Epidemiol Biomarkers Prev* **14**(4): 842–9.

Michels, K. B., E. Giovannucci, et al. (2006). "Fruit and vegetable consumption and colorectal adenomas in the Nurses' Health Study." *Cancer Res* **66**(7): 3942–53.

Moody, G. A., V. Jayanthi, et al. (1996). "Long-term therapy with sulphasalazine protects against colorectal cancer in ulcerative colitis: a retrospective study of colorectal cancer risk and compliance with treatment in Leicestershire." *Eur J Gastroenterol Hepatol* **8**(12): 1179–83.

Moore, L. L., M. L. Bradlee, et al. (2004). "BMI and waist circumference as predictors of lifetime colon cancer risk in Framingham Study adults." *Int J Obes Relat Metab Disord* **28**(4): 559–67.

Moorghen, M., P. Ince, et al. (1988). "A protective effect of sulindac against chemically-induced primary colonic tumours in mice." *J Pathol* **156**: 341–7.

Morimoto, L. M., P. A. Newcomb, et al. (2002). "Risk factors for hyperplastic and adenomatous polyps: evidence for malignant potential?" *Cancer Epidemiol Biomarkers Prev* **11**(10 Pt 1): 1012–8.

Munkholm, P. (2003). "The incidence and prevalence of colorectal cancer in inflammatory bowel disease." *Aliment Pharmacol Ther* **18**(Suppl 2): 1–5.

Muscat, J. E., S. D. Stellman, et al. (1994). "Nonsteroidal antiinflammatory drugs and colorectal cancer." *Cancer* **74**: 1847–54.

Narisawa, T., M. Sato, et al. (1981). "Inhibition of development of methylnitrosourea-induced rat colon tumors by indomethacin treatment." *Cancer Res* **41**: 1954–7.

Newcomb, P. A. and B. E. Storer (1995). "Postmenopausal hormone use and risk of large-bowel cancer." *J Natl Cancer Inst* **87**: 1067–71.

Newcomb, P. A., A. C. Bush, et al. (2004). "No evidence of an association of JC virus and colon neoplasia." *Cancer Epidemiol Biomarkers Prev* **13**(4): 662–6.

Newcomb, P. A., Y. Zheng, et al. (2007). "Estrogen plus progestin use, microsatellite instability, and the risk of colorectal cancer in women." *Cancer Res* **67**(15): 7534–9.

Niv, Y., A. Goel, et al. (2005). "JC virus and colorectal cancer: a possible trigger in the chromosomal instability pathways." *Curr Opin Gastroenterol* **21**(1): 85–9.

Norat, T., S. Bingham, et al. (2005). "Meat, fish, and colorectal cancer risk: the European Prospective Investigation into Cancer and Nutrition." *J Natl Cancer Inst* **97**(12): 906–16.

Paganini-Hill, A., A. Chao, et al. (1989). "Aspirin use and chronic diseases: a cohort study of the elderly." *Br Med J* **299**: 1247–50.

Park, Y., D. J. Hunter, et al. (2005). "Dietary fiber intake and risk of colorectal cancer: a pooled analysis of prospective cohort studies." *JAMA* **294**(22): 2849–57.

Peleg, I. I., H. T. Maibach, et al. (1994). "Aspirin and nonsteroidal anti-inflammatory drug use and the risk of subsequent colorectal cancer." *Arch Intern Med* **154**: 394–9.

Peters, R. K., M. C. Pike, et al. (1990). "Reproductive factors and colon cancer." *Br J Cancer* **61**: 741–8.

Peters, U., R. Sinha, et al. (2003). "Dietary fibre and colorectal adenoma in a colorectal cancer early detection programme." *Lancet* **361**(9368): 1491–5.

Peters, U., N. Chatterjee, et al. (2004). "Calcium intake and colorectal adenoma in a US colorectal cancer early detection program." *Am J Clin Nutr* **80**(5): 1358–65.

Pollard, M. and P. H. Luckert (1980). "Indomethacin treatment of rats with dimethylhydrazine-induced intestinal tumors." *Cancer Treat Rep* **64**: 1323–7.

Potter, J. D. (1995). "Hormones and colon cancer." *J Natl Cancer Inst* **87**(14): 1039–40.

Potter, J. D. (1999a). "Colorectal cancer: molecules and populations." *J Natl Cancer Inst* **91**(11): 916–32.

Potter, J. D. (1999b). "Fiber and colorectal cancer – where to now?" *N Engl J Med* **340**(3): 223–4.

Potter, J. D. (2005). "Vegetables, fruit, and cancer." *Lancet* **366**: 527–30.

Potter, J. D. and A. J. McMichael (1983). "Large bowel cancer in women in relation to reproductive and hormonal factors: a case-control study." *J Natl Cancer Inst* **71**(4): 703–9.

Potter, J. D., M. L. Slattery, et al. (1993). "Colon cancer: a review of the epidemiology." *Epidemiol Rev* **15**(2): 499–545.

Potter, J. D., R. M. Bostick, et al. (1996). "Hormone replacement therapy is associated with lower risk of adenomatous polyps of the large bowel: the Minnesota Cancer Prevention Research Unit Case-Control Study." *Cancer Epidemiol Biomarkers Prev* **5**(10): 779–84.

Probst-Hensch, N. M., R. Sinha, et al. (1997). "Meat preparation and colorectal adenomas in a large sigmoidoscopy-based case-control study in California (United States)." *Cancer Causes Control* **8**(2): 175–83.

Psaty, B. M. and J. D. Potter (2006). "Risks and benefits of celecoxib to prevent recurrent adenomas." *New Engl J Med* **355**(9): 950–2.

Raderer, M., F. Pfeffel, et al. (2000). "Regression of colonic low grade B cell lymphoma of the mucosa associated lymphoid tissue type after eradication of Helicobacter pylori." *Gut* **46**(1): 133–5.

Reddy, B. S., H. Maruyama, et al. (1987). "Dose-related inhibition of colon carcinogenesis by dietary piroxicam, a nonsteroidal antiinflammatory drug, during different stages of rat colon tumor development." *Cancer Res* **47**: 5340–6.

Reid, F. D., P. M. Mercer, et al. (1996). "Cholecystectomy as a risk factor for colorectal cancer: a meta-analysis." *Scand J Gastroenterol* **31**(2): 160–9.

Risch, H. A. and G. R. Howe (1995). "Menopausal hormone use and colorectal cancer in Saskatchewan: a record linkage cohort study." *Cancer Epidemiol Biomarkers Prev* **4**: 21–8.

Robertson, D. J., R. S. Sandler, et al. (2005). "Fat, fiber, meat and the risk of colorectal adenomas." *Am J Gastroenterol* **100**(12): 2789–95.

Rosenberg, L., J. Palmer, et al. (1991). "A hypothesis: nonsteroidal anti-inflammatory drugs reduce the incidence of large-bowel cancer." *J Natl Cancer Inst* **83**: 355–8.

Rosenberg, L., C. Louik, et al. (1998). "Nonsteroidal antiinflammatory drug use and reduced risk of large bowel carcinoma." *Cancer* **82**(12): 2326–33.

Samad, A. K., R. S. Taylor, et al. (2005). "A meta-analysis of the association of physical activity with reduced risk of colorectal cancer." *Colorectal Dis* **7**(3): 204–13.

Sandler, R. S., S. Halabi, et al. (2003). "A randomized trial of aspirin to prevent colorectal adenomas in patients with previous colorectal cancer." *N Engl J Med* **348**(10): 883–90.

Sanjoaquin, M. A., N. Allen, et al. (2005). "Folate intake and colorectal cancer risk: a meta-analytical approach." *Int J Cancer* **113**(5): 825–8.

Satia-Abouta, J., J. A. Galanko, et al. (2003). "Associations of micronutrients with colon cancer risk in African Americans and whites: results from the North Carolina Colon Cancer Study." *Cancer Epidemiol Biomarkers Prev* **12**(8): 747–54.

Schatzkin, A., E. Lanza, et al. (2000). "Lack of effect of a low-fat, high-fiber diet on the recurrence of colorectal adenomas. Polyp Prevention Trial Study Group." *N Engl J Med* **342**(16): 1149–55.

Schiffman, M. H. and J. S. Felton (1990). "Re: "Fried foods and the risk of colon cancer"." *Am J Epidemiol* **131**(2): 376–8.

Schreinemachers, D. and R. Everson (1994). "Aspirin use and lung, colon, and breast cancer incidence in a prospective study." *Epidemiology* **5**: 138–46.

Singh, P. N. and G. E. Fraser (1998). "Dietary risk factors for colon cancer in a low-risk population." *Am J Epidemiol* **148**: 761–74.

Sinha, R., W. H. Chow, et al. (1999). "Well-done, grilled red meat increases the risk of colorectal adenomas." *Cancer Res* **59**: 4320–4.

Slattery, M. L. (2004). "Physical activity and colorectal cancer." *Sports Med* **34**(4): 239–52.

Slattery, M. L., D. W. West, et al. (1990). "Tobacco, alcohol, coffee, and caffeine as risk factors for colon cancer in a low-risk population." *Epidemiology* **1**(2): 141–5.

Slattery, M. L., J. D. Potter, et al. (1994). "Age and risk factors for colon cancer: are there implications for understanding differences in case control and cohort studies?" *Cancer Causes Control* **5**: 557–63.

Slattery, M. L., J. Potter, et al. (1997a). "Energy balance and colon cancer – beyond physical activity." *Cancer Res* **57**(1): 75–80.

Slattery, M. L., D. Schaffer, et al. (1997b). "Are dietary factors involved in DNA methylation associated with colon cancer?" *Nutr Cancer* **28**(1): 52–62.

Slattery, M. L., K. Curtin, et al. (2000). "Associations between cigarette smoking, lifestyle factors, and microsatellite instability in colon tumors." *J Natl Cancer Inst* **92**(22): 1831–6.

Steinmetz, K. A. and J. D. Potter (1991). "Vegetables, fruit, and cancer. II. Mechanisms." *Cancer Causes Control* **2**(6): 427–42.

Steinmetz, K. A. and J. D. Potter (1996). "Vegetables, fruit, and cancer prevention: a review." *J Am Diet Assoc* **96**(10): 1027–39.

Stewart, B. W. and P. Kleihues (2003). *World Cancer Report*. Lyon: IARC Press.

Stürmer, T., R. J. Glynn, et al. (1998). "Aspirin use and colorectal cancer: post-trial follow-up data from the Physicians' Health Study." *Ann Intern Med* **128**: 713–20.

Sugimura, T. and S. Sato (1983). "Mutagens-carcinogens in foods." *Cancer Res* **43**(5 Suppl): 2415s–21s.

Suh, O., C. Mettlin, et al. (1993). "Aspirin use, cancer, and polyps of the large bowel." *Cancer* **72**: 1171–7.

Talley, N. J., A. R. Zinsmeister, et al. (1991). "Gastric adenocarcinoma and helicobacter pylori infection." *J Natl Cancer Inst* **83**(23): 1734–9.

Tamakoshi, K., K. Wakai, et al. (2004). "A prospective study of body size and colon cancer mortality in Japan: The JACC Study." *Int J Obes Relat Metab Disord* **28**(4): 551–8.

Tavani, A., S. Gallus, et al. (1998). "Cigarette smoking and risk of cancers of the colon and rectum: a case-control study from Italy." *Eur J Epidemiol* **14**(7): 675–81.

Terry, P., E. Giovannucci, et al. (2001). "Fruit, vegetables, dietary fiber, and risk of colorectal cancer." *J Natl Cancer Inst* **93**(7): 525–33.

Terry, P., J. A. Baron, et al. (2002). "Dietary calcium and vitamin D intake and risk of colorectal cancer: a prospective cohort study in women." *Nutr Cancer* **43**(1): 39–46.

Thun, M. J., M. M. Namboodiri, et al. (1991). "Aspirin use and reduced risk of fatal colon cancer." *N Engl J Med* **325**: 1593–6.

Thun, M. J., E. E. Calle, et al. (1992). "Risk factors for fatal colon cancer in a large prospective study." *J Natl Cancer Inst* **84**: 1491–500.

Todoroki, I., G. D. Friedman, et al. (1999). "Cholecystectomy and the risk of colon cancer." *Am J Gastroenterol* **94**(1): 41–6.

Trock, B., E. Lanza, et al. (1990). "Dietary fiber, vegetables, and colon cancer: critical review and meta-analyses of the epidemiologic evidence." *J Natl Cancer Inst* **82**: 650–61.

Ulrich, C. M. and J. D. Potter (2007). "Folate and cancer – timing is everything." *JAMA* **297**(21): 2408–9.

Wactawski-Wende, J., J. M. Kotchen, et al. (2006). "Calcium plus vitamin D supplementation and the risk of colorectal cancer." *N Engl J Med* **354**(7): 684–96.

Wark, P. A., M. P. Weijenberg, et al. (2005). "Fruits, vegetables, and hMLH1 protein-deficient and -proficient colon cancer: The Netherlands cohort study." *Cancer Epidemiol Biomarkers Prev* **14**(7): 1619–25.

Wattenberg, L. W. (1978). "Inhibition of chemical carcinogenesis." *J Natl Cancer Inst* **60**(1): 11–8.

WCRF Panel (1997). *Food, Nutrition and the Prevention of Cancer: A Global Perspective*. Washington, DC: American Institute for Cancer Research.

Weiss, N. S., J. R. Daling, et al. (1981). "Incidence of cancer of the large bowel in women in relation to reproductive and hormonal factors." *J Natl Cancer Inst* **67**: 57–60.

Whelan, R. L., K. D. Horvath, et al. (1999). "Vitamin and calcium supplement use is associated with decreased adenoma recurrence in patients with a previous history of neoplasia." *Dis Colon Rectum* **42**(2): 212–7.

White, E., J. S. Shannon, et al. (1997). "Relationship between vitamin and calcium supplement use and colon cancer." *Cancer Epidemiol Biomarkers Prev* **6**(10): 769–74.

Willett, W. C., M. J. Stampfer, et al. (1990). "Relation of meat, fat, and fiber intake to the risk of colon cancer in a prospective study among women." *N Engl J Med* **323**(24): 1664–72.

Wu-Williams, A. H., M. Lee, et al. (1991). "Reproductive factors and colorectal cancer risk among Chinese females." *Cancer Res* **51**: 2307–11.

Wu, K., W. C. Willett, et al. (2002). "A prospective study on supplemental vitamin E intake and risk of colon cancer in women and men." *Cancer Epidemiol Biomarkers Prev* **11**(11): 1298–304.

Wynder, E. L. and T. Shigematsu (1967). "Environmental factors of cancer of the colon and rectum." *Cancer* **20**: 1520–61.

Xu, Z. and D. L. Su (1984). "Schistosoma japonicum and colorectum cancer: an epidemiological study in the People's Republic of China." *Int J Cancer* **34**(3): 315–8.

Zumkeller, N., H. Brenner, et al. (2006). "Helicobacter pylori infection and colorectal cancer risk: a meta-analysis." *Helicobacter* **11**(2): 75–80.

Chapter 2
Mouse Models of Intestinal Cancer

Erin M. Perchiniak and Joanna Groden

Introduction

Intestinal cancers are a category of heterogeneous tumors that occur sporadically or through inherited susceptibility, each characterized by genetic alterations affecting a number of molecular pathways. As a result of this complexity, numerous genetically engineered mice (GEM) have been generated to model different genetic, morphologic, or clinical features of intestinal cancer. Mouse models of intestinal cancer can be broadly divided into six groups based on the underlying signaling pathway disrupted or by the means with which tumors were induced: Wnt-related GEM; GEM associated with alterations in TGF-beta (β) signaling; mismatch repair-deficient GEM; immune-deficient mice; carcinogen-treated mice; and others that do not neatly fit into the aforementioned categories. Although differences have been noted in lesions arising in these broadly grouped genetic and other models, some characteristics are shared. Adenomas are the most common lesion in mouse models of intestinal cancer. Unlike humans, lesions can be present throughout the intestinal tract, with no predilection for the colon. Invasion and metastasis occur rarely. This chapter will summarize the findings from most of the available mouse models of intestinal cancer.

GEM and the Wnt Signaling Pathway

$^{Min/+}$ *and Related Mice*

The *APC* gene was initially identified by positional cloning as the disease gene for familial adenomatous polyposis coli (FAP) and was subsequently found to

E.M. Perchiniak and J. Groden (✉)
Department of Molecular Virology, Immunology and Medical Genetics
The Ohio State University College of Medicine
e-mail: joanna.groden@osumc.edu

J.D. Potter and N.M Lindor (eds.), *Genetics of Colorectal Cancer*,

DOI: 10.1007/978-0-387-09568-4_3,

be mutated in the majority of sporadic colorectal tumors (Groden et al. 1991; Kinzler et al. 1991). As a result of these initial findings, research has focused on understanding the cellular pathways in which APC participates and how deregulation of these pathways can lead to tumorigenesis. Although APC has many roles in the cell, regulation of the protein β-catenin is the primary function. APC is part of a protein complex that phosphorylates β-catenin, marking it for ubiquitination and proteolytic degradation. In the absence of APC or in the presence of a Wnt signal, β-catenin is stabilized and shuttled to the nucleus where it can transcriptionally alter the expression of downstream Wnt target genes. Therefore, when *APC* is disabled by mutations, the Wnt signaling pathway is constitutively activated, allowing for uncontrolled growth and tumor progression. Many *APC* mutants have been identified in persons with FAP, in whom genotype/phenotype correlations are well known (reviewed by Nieuwenhuis and Vasen 2007). *APC* mutations in the first or last third of the gene are associated with an attenuated polyposis, characterized by late onset and a small number of adenomas. Conversely, mutations in the central region of the gene correlate with a severe phenotype with thousands of adenomas developing at a young age. To investigate these observations more closely and to gain insight into the mechanism of disease onset, several mutant *Apc* mouse models have been created.

Perhaps the most widely used GEM model of gastrointestinal (GI) tumorigenesis is the $Apc^{Min/+}$ mouse. Thus, it has become the basis of comparison for other GI cancer mouse models (Moser et al. 1990). The Apc^{Min} (multiple intestinal neoplasia) allele carries an ethylnitrosourea (ENU)-induced nonsense mutation at codon 850 (Su et al. 1992), which leads to embryonic lethality in homozygote animals. Most studies use the heterozygote mice, $Apc^{Min/+}$, which typically live 4 months (Moser et al. 1990). Mice carrying the Apc^{Min} allele on the *C57BL/6* background develop an average of 24 polyps per mouse in the small intestine and five per mouse in the colon by 4 months of age. Most polyps are adenomas, with none progressing to invasive adenocarcinoma and, as expected for adenomas, tumors in the $Apc^{Min/+}$ have been found not to metastasize.

Given the advances in gene-knockout technologies, several other *Apc* mutant mice have been created. The importance of their study stems from the knowledge that several mutations have been detected within the *APC* gene in human tumor samples and in persons with FAP, which may underlie variations in disease progression among patients; the results from these subsequent mouse models indicate that, indeed, not all *Apc* mutations are equivalent. The precise location and the type of mutation within *Apc* dictate the degree of tumor susceptibility, which is probably the result of the multifunctional nature of Apc and its contribution to various cellular pathways.

$Apc^{716/+}$ mice harbor a truncating mutation at codon 716 and, like $Apc^{Min/+}$ mice, develop polyps mainly in the small intestine (Oshima et al. 1995); they develop an average of 300 polyps as early as 3 weeks of age (Oshima et al. 1995), and typically have a reduced lifespan compared to $Apc^{Min/+}$ mice, even on the same *C57BL/6* background. $Apc^{1309/+}$ mice have a truncating mutation at codon 1309 (Quesada et al. 1998). These mice typically develop an average of 34 adenomas by 14 weeks of age, a slightly higher incidence of polyp formation than the $Apc^{Min/+}$ mouse, and a lower incidence than the *Apc*$^{716/+}$ mouse. Again, these polyps are predominantly found in the small intestine. This GEM is particularly interesting because codon 1309 is the most frequently mutated residue in persons with FAP with severe polyposis (Nagase and Nakamura 1993). A 5-base-pair deletion results in truncation of APC three codons downstream from the mutation.

Apc$^{1638/+}$ mice carry an allele with a mutation at codon 1638 resulting in truncation of Apc; these mice also develop polyps mainly in the small intestine (Fodde et al. 1994; Oshima et al. 1995). However, *Apc*$^{1638/+}$ mice form only 3–5 tumors by 3.5 months of age and typically live 1 year (Yang et al. 1997). A modification to the *Apc*^{1638} allele design was engineered to produce a stable Apc protein (Smits et al. 1999). This allele, *Apc*^{1638T}, still encodes some of the β-catenin-binding motifs but lacks the C-terminal portion of Apc necessary for its interaction with other proteins important for growth control. On a mixed background, B6/129Ola, homozygosity for the *Apc*^{1638T} mutation does not result in embryonic lethality but leads to a number of phenotypic abnormalities in adult animals, including growth retardation and nipple-associated cysts. The same mutation on a B6 background leads to reduced postnatal survival. Heterozygous *Apc*^{1638T} mice are normal.

Given the predilection for intestinal tumors to form in the mouse models, conditional *Apc* mutant mice have been developed to investigate the initiation stage of intestinal adenoma formation (Shibata et al. 1997). *Apc*$^{580/+}$ mutant mice, on a mixed *129/BL6* background, carry an allele with *loxP* (or flox) sites flanking exon 14. Colonic introduction of a recombinant adenovirus expressing Cre recombinase, driven by the *SR-alpha* promoter into *Apc*$^{580/+}$ mice, induces a frameshift mutation at codon 580. Over 80% of homozygous animals (20 of 24 animals) have an average of 6.7 colonic adenomas 4 weeks after infection. No tumors were detected in either heterozygous or wild-type animals. Five of six homozygous mutants allowed to live after adenoviral infection survived over 1 year. Analysis of these animals showed invasion into the submucosal layer by tumor cells and hence progression to adenocarcinoma. Recent studies have used colon-specific promoters to drive Cre expression and generate colon tumors in the mouse, rather than the small intestinal distributions seen in the more established models.

Hypomorphic *Apc* mice were created in a study by Li et al. whereby the expression level of Apc was reduced to 10–20% of the wild-type *Apc* (Li et al. 2005). Polyp formation was reduced compared to the *Apc*$^{716/+}$ mice. These results argue that there is a threshold level (15% of wild type) of Apc expression that is required for proper growth control.

More recent studies have focused on conditionally inactivating *Apc* in order to understand the precise mechanism by which Wnt activation leads to polyps. Two mouse models have been generated, both making use of the *loxP* system. *Apc* is modified by an inducible *Cyp1A-Cre* transgene (Sansom et al. 2003) in one model, whereas the other uses a tamoxifen-regulated intestinal-specific *Villin-CreER* transgene (Andreu et al. 2005). Both studies reported that inactivation of *Apc* led to the rapid translocation of β-catenin to the nucleus and subsequent changes in the appearance of enterocytes and intestinal crypts. Following *Apc* loss, many of the epithelial cells along the crypt-villus axis enter S-phase. These studies establish that a single event, loss of *Apc*, is enough to promote early phenotypic changes in the crypt.

β-Catenin Transgenic Mice

β-Catenin is a multifunctional protein component of the Wnt signal transduction pathway (Sheng et al. 1998). It is also a mediator of cell adhesion through its interaction with cadherins. It is known that β-catenin rapidly translocates to the nucleus upon loss of APC, resulting in transcriptional alteration of downstream target genes involved with proliferation, apoptosis, and cell-cycle regulation. Therefore, overexpressed β-catenin is considered oncogenic, resulting from either a nonfunctional *APC* gene or a gain of function mutation within β-catenin. The finding of mutations in the β-catenin gene (*CTNNB1*) in human colon cancer cell lines, with no detectable mutations in *APC*, has led to the hypothesis that β-catenin acts as an oncogene in the development of intestinal neoplasia (Iwao et al. 1998; Morin et al. 1997; Sparks et al. 1998). Several groups have investigated the role of activated β-catenin using in vivo mouse models.

Wong et al. designed a transgenic mouse expressing a human β-catenin N-terminal truncation mutant (N89β-catenin) in the intestine driven by the fatty acid-binding protein gene (*Fabpl*) promoter (Wong et al. 1998). The absence of GSK-3β phosphorylation sites, normally targeting degradation of β-catenin, was associated with a longer half-life than wild-type β-catenin in cell culture studies (Aberle et al. 1997; Cadigan and Nusse 1997; Miller and Moon 1996; Munemitsu et al. 1996; Yost et al. 1996). The deletion of these amino acids did not affect the ability of β-catenin to interact with E-cadherin, α-catenin, or Tcf (Wong et al. 1998). Although there were some changes in the architecture of the villi and an increase in the rate of cell division within undifferentiated cells in the crypts of Liberkühn, no dysplasia was detected in the transgenic mice.

Romagnolo et al. generated a similar β-catenin transgenic mouse, but had dramatically different results (Romagnolo et al. 1999). This transgenic mouse expressed activated β-catenin in the epithelial cells of the intestine using a transgene with an N-terminal truncation, N131β-catenin, lacking both the GSK-3β phosphorylation site, important for protein stabilization, and the α-catenin-binding site, necessary for adhesive properties of β-catenin (Barth et al. 1997; Hulsken et al. 1994). A calbindin-D9K promoter and its regulatory sequences, active in differentiated epithelial cells of the villi and the kidney (Colnot et al. 1998; Romagnolo et al. 1996), and the enhancer of the adolase B gene were used to drive expression. Overexpression of N131β-catenin resulted in small intestine adenomas by 3–4 weeks of age. The intestines were characterized by multifocal dysplastic lesions and a 3- to 4-fold higher number of apoptotic cells than in nontransgenic mice. Further analysis of these animals was inhibited by mortality from polycystic kidney disease.

A third β-catenin GEM was generated in which exon 3 could be deleted by inducible homologous recombination using *loxP* sites (Harada et al. 1999). The loss of exon 3 does not alter the frame of the RNA. In this model, nearly 3,000 adenomas develop by 3 weeks of age, primarily in the duodenum and jejunum and with little involvement of the ileum, cecum, or colon. *Fabp1* regulatory regions were used to express Cre, resulting in a mutant β-catenin driven by its own enhancer and promoter. Differences in promoters, transgene copy numbers or locations, mouse strains, and or different types of dominant mutations may explain the dramatic differences in these three mouse models. Each, however, underscores the importance of the Wnt signaling pathway in mouse GI tumorigenesis.

Genes that Modify the Wnt Pathway

Cyclo-oxygenases (Cox) 1 and 2 are the key enzymes in prostanoid production and are the targets of nonsteroidal anti-inflammatory drugs (NSAIDs) and aspirin (Vane 1971, 1994). Both Cox-1 and Cox-2 enzymes convert arachidonic acid to prostaglandin G2 and then to prostaglandin H2 (DeWitt and Smith 1988; Hemler and Lands 1976; Miyamoto et al. 1976). Cox-1 is constitutively expressed in several mammalian tissues, whereas the distribution of Cox-2 expression is restricted to inflammatory cells such as monocytes and macrophages upon stimulation by cytokines, mitogens, serum, and endotoxins (Lee et al. 1992; Maier et al. 1990; O'Banion et al. 1992; O'Neill and Ford-Hutchinson 1993) Cyclo-oxygenase-2 (Cox-2) is expressed at early stages of adenoma formation, suggesting its importance as a therapeutic target. Cox-1 seems to work with Cox-2 in adenoma development by producing prostaglandin E2 (PGE_2) and stimulating angiogenesis (Takeda et al. 2003). Introduction of a *Cox-2* deletion onto the $Apc^{Min/+}$ background dramatically decreases tumor number (Oshima et al. 1996). The combination of *Cox-2* deletion with the $Apc^{716/+}$ mutation also leads to a dramatic decrease in the number and size of tumors. Not surprisingly, introduction of a *Cox-1* mutation to the $Apc^{Min/+}$ mouse reduces the number and size of tumors to about 80% of the reduction seen in *Cox-2;Apc* mutant mice (Chulada et al. 2000). As might be predicted, treatment of $Apc^{Min/+}$ mice with PGE_2 increases the number and size of intestinal adenomas (Wang et al. 2004). Clinical trials are ongoing to investigate Cox-2 inhibitors in FAP (Higuchi et al. 2003; Steinbach et al. 2000). See also Chap. 5.

To probe the arachidonic acid cascade for its contribution to intestinal tumorigenesis, several other compound mice were developed. Cytosolic phospholipase A_2 ($cPLA_2$) is one of the key enzymes responsible for cleavage of arachidonic acid, a substrate of Cox, from membrane phospholipids. Knockout of $cPla_2$ in $Apc^{716/+}$ mice reduces tumor number (Takaku et al. 2000). Additional studies have investigated the role of the G-protein coupled receptor Ep_2, which binds PGE_2, in tumor formation in $Apc^{Min/+}$ mice. Double heterozygotes displayed a marked reduction of tumor number (Sonoshita et al. 2001). PGE_2 indirectly transactivates peroxisome-proliferator activity receptor delta (PPARδ) through PI3K/Akt signaling. Deletion of *PPARδ* in $Apc^{Min/+}$ mice treated with PGE_2 negated the increase of intestinal adenomas found with treatment of PGE_2 alone (Wang et al. 2004).

Methylation contributes to the silencing of many genes which, in turn, leads to deleterious phenotypic changes depending on which genes have been affected. DNA methyltransferase 1 (DNMT1) is one of the enzymes responsible for methylating cytidine residues within genes. Mutations in the *Dnmt1* gene, in combination with an enzyme inhibitor, reduced the tumor number in $Apc^{Min/+}$ mice from one hundred to two or less (Laird et al. 1995). Mutation in the *Mbd2* gene, encoding a methyl-CpG-binding repressor, also reduced tumor numbers in $Apc^{Min/+}$ mice (Laird et al. 1995; Sansom et al. 2003). These results suggest a role for methylation in the development of intestinal polyposis.

Other modifier genes include the reqQ-like DNA helicase gene, *BLM*, which, when mutated, is responsible for the development of Bloom syndrome. When *Blm* heterozygous mice were bred to $Apc^{Min/+}$ mice, an increase in adenomas was observed as well as a change in the degree of tumor dysplasia (Goss et al. 2002; Luo et al. 2000). Mutation of the gene encoding the matrix metalloproteinases matrilysin (Mmp7), implicated in cancer invasion and metastasis, also reduces tumor number in $Apc^{Min/+}$ mice (Wilson et al. 1997).

GEM and the TGFβ Signaling Pathway

TGFβ1$^{-/-}$ *and Related Mice*

The transforming growth factor β (TGFβ) pathway plays an important role in both human and murine colon cancer. TGFβ controls cell growth, regulates epithelial cell differentiation and cell matrix interaction, and protects the epithelium from genetic damage caused by inflammatory cells (Brandes et al. 1991; Kulkarni et al. 1993; Roberts et al. 1992; Shull et al. 1992; Wahl et al. 1987). The multifunctional nature of the TGFβ family suggests several mechanisms by which defects in TGFβ signaling can lead to initiation, promotion, or progression of cancer. This hypothesis is supported by evidence from tumor-derived human colon cancer cell lines which are frequently resistant to the growth-inhibitory effects of TGFβ1 (Manning et al. 1991; Mulder et al. 1988). Mutations have been detected in *TGFβR2* in both sporadic and inherited colon cancers (Markowitz et al. 1995; Parsons et al. 1995). Additionally, inactivating mutations in *SMAD2* and *SMAD4*, two members of the family of intracellular proteins responsible for transducing signals from the activated TGFβ receptors, are present in many human colon cancers (Eppert et al. 1996; Takagi et al. 1996; Thiagalingam et al. 1996).

Inactivation of Tgfβ1 in mice results in autoimmune disease and death before 1 month of age. In order to study the role of Tgfβ1 in the development and progression of GI cancer, the *Tgfβ1*$^{-/-}$ mouse strain was crossed onto the immunodeficient *Rag2*$^{-/-}$ (Engle et al. 1999). Tgfβ1 deficiency (+/– or –/–) on the *Rag2*$^{-/-}$ background leads to cecal and colonic neoplasms (Engle et al. 1999). A marked increase in tumor incidence and severity was observed in the *Tgfβ1*$^{-/-}$ mice: adenomas are detectable at 2 months of age and carcinomas are detectable at 3–6 months with 100% penetrance. The carcinomas show no mutations of *Apc*, *Ras*, or *Ctnnb1*, which suggests that the tumor-suppressive function of Tgfβ is independent of other known signaling pathways disrupted in intestinal cancers. Notably, many of the tumors have a mucinous histopathology.

Smad$^{-/-}$ Mice

Signaling by Tgfβ family ligands is mediated by the Smad family of intracellular proteins (Graff et al. 1996). The Smad proteins are the core of the Tgfβ pathway through their translation of cellular signals into responses. There are eight Smad proteins encoded by the human and mouse genomes, five of which act as substrates for the Tgfβ family of receptors (Massague 1998). Smads 1, 2, 3, 5, and 8 are commonly referred to as receptor-regulated Smads (RSmads). Smad4, also called Co-Smad, serves as a common partner for all Smads. Smads 6 and 7 are inhibitory and serve as decoys by interfering with Smad-receptor and Smad-Smad interactions. Smads undergo a continuous nuclear-cytoplasmic shuttling cycle. Phosphorylation leads to nuclear accumulation by destabilizing the RSmad interaction with cytoplasmic anchors and increases their affinity for nuclear factors (Shi and Massague 2003; Xu and Massague 2004). This then allows Smads to transcriptionally regulate Tgfβ downstream targets. Dephosphorylation has the opposite effect, sequestering Smads to the cytoplasm (Inman et al. 2002). Because Tgfβ signaling affects cell division, differentiation, migration, adhesion, organization, and death, and because Smads are the translators of these signals, Smad deregulation could have many deleterious cellular affects. Therefore, several Smad GEM models have been generated, some of which have developed intestinal tract tumors.

Smad2 is 91% homologous to Smad3; however, it differs biologically. Unlike Smad3 and 4, Smad2 does not bind directly to DNA and has a unique thirty amino acid region absent from other Smad proteins (Dennler et al. 1998; Jonk et al. 1998; Kim et al. 1997; Labbe et al. 1998; Yingling et al. 1997; Zawel et al. 1998). Pertinent to human GI tumors, *SMAD2* is the only RSMAD for which mutations have been associated with colorectal cancer (Eppert et al. 1996). To investigate whether Smad2 can act as a tumor suppressor, knockout mice were generated. Homozygous deletion of *Smad2* results in embryonic lethality at day 8.5 (Heyer et al. 1999; Nomura and Li 1998; Waldrip et al. 1998; Weinstein et al. 1998). Heterozygous mice (*Smad2*$^{+/-}$) had no abnormalities when aged to 1.5 years. Hamamoto et al. generated double heterozygous mice that carried *Apc* and *Smad2* null alleles (Hamamoto et al. 2002). Inactivation of *Smad2* in heterozygous *Apc* mutant mice did not change the total number of intestinal tumors but decreased the time to death from intestinal obstruction due to extremely large tumors. Additionally, these mice developed multiple invasive cancers not observed in *Apc* heterozygotes. These results suggest that deletion of *Smad2* alone does not initiate tumor formation, but accelerates progression of tumors initiated by loss of *Apc*.

Unlike other Tgfβ-family null mice, *Smad3* null mice are viable and reasonably healthy. They develop intestinal adenomas that sometimes progress to adenocarcinoma (Zhu et al. 1998). The *Smad3* mutant allele was generated by homologous recombination and established in both *129/Sv* and *129/Sv C57BL/6* mixed background mice. Most tumors are mucinous. Metastatic spread (uncommon in mouse models) was detected in a small number of animals. There was great variability in the time-course of disease, but tumors were smaller and less aggressive in mixed background mice. These in vivo studies have defined a new role for Smad3 as a tumor suppressor protein in the intestine. *Smad3* mutant mice display many of the histopathological stages observed in human colon cancer progression; to date, no *SMAD3* mutations have been detected in human colorectal cancers.

SMAD4 was initially cloned as a tumor suppressor that is mutated in more than 50% of human pancreatic cancers (Hahn et al. 1996). *SMAD4* is also mutated in more than 30% of human sporadic colon cancers; germline mutations are associated with familial juvenile polyposis (Friedl et al. 1999; Nagatake et al. 1996). *Smad4* null mice die at embryonic day 6.5; therefore, *Smad4*$^{+/-}$ mice are often used for tumorigenesis studies (Sirard et al. 1998; Yang et al. 1998). Polyps can be detected in the fundus and antrum of the stomach of *Smad4*$^{+/-}$ mice; polyps found in the antrum can develop into adenocarcinoma with aging (Xu et al. 2000). Polyps can also be found in the duodenum and cecum, albeit at a lower frequency. From these studies, it seems reasonable to infer that Smad4 is particularly important for tumor suppression in the stomach. *Smad4*$^{+/-}$ mice have also been bred with *Apc*$^{+/-}$ mice; double heterozygotes develop intestinal adenocarcinomas that lack wild-type alleles at both loci (Takaku et al. 1998).

GEM and DNA Mismatch Repair

Individuals with Lynch Syndrome (see Chap. 6) carry heterozygous germline mutations in one of six DNA mismatch repair (MMR) genes. Tumors that arise have typically lost the wild-type copy of the gene through somatic events and are characterized by microsatellite instability (MSI). The mammalian MMR system detects and repairs base substitution or small nucleotide insertion/deletion mutations, sends apoptotic signals in response to DNA damage, and suppresses incorrect homologous recombination events. In eukaryotes, initiation of the repair process requires three different MutS yeast homologs: MSH2, MSH3, and MSH6. MSH2 and MSH6 form a heterodimeric complex that initiates base-base mispairing as well as single base insertion/deletion mispairs. The MSH2-MSH3 heterodimeric complex repairs larger insertion/deletion mispairs of 2–4 bases. Both complexes require interaction with eukaryotic MutL homologs to activate subsequent repair events. The four MutL homologs are: MLH1, PMS1, PMS2, and MLH3. Three heterodimeric complexes form: MLH1–PMS2 to provide the primary function for mitotic MMR, MLH1–PMS1, and MLH1–MLH3. The MLH1–PMS2 complex also interacts with the two MutS complexes. The majority of Lynch Syndrome mutations occur in three MMR genes, *MLH1*, *MSH2*, or *MSH6*, although in rare cases mutation in other MMR genes have been identified. Mouse lines carrying mutations all of the *MutS* and *MutL* genes have been generated, some of which have resulted in phenotypes similar to Lynch Syndrome.

Deletion of *Msh2*, *Msh6*, or *Mlh1* results in intestinal tumors, although there is great variation of the phenotypes. Given that Msh2 participates in two "MutS" complexes, it is not surprising that the *Msh2*$^{-/-}$ mice have a severe phenotype (de Wind et al. 1995; Reitmair et al. 1995). Fifty percent of *Msh2*$^{-/-}$ mice die by 6 months and all animals by 1 year. Mice develop adenomas of the small intestine and, after 6 months, adenocarcinoma (Reitmair et al. 1996b). *Msh3*$^{-/-}$ mice develop tumors very late in life, with an overall tumor spectrum somewhat similar to wild-type animals. This mild phenotype may be the result of only moderate repair defects being caused by deletion of *Msh3*, or by compensation by intact Msh2 and Msh6. These data are reminiscent of the absence of detectable *MSH3* mutations in Lynch Syndrome families. *Msh6*$^{-/-}$ mice develop a similar tumor spectrum of intestinal adenomas and adenocarcinomas as the *Msh2*$^{-/-}$ mice but with a delayed onset and subsequent increased survival (up to 16 months of age) (Edelmann et al. 1997). This delayed onset of tumor formation is attributed to the impairment of the repair of base-base mismatches, but retention of the 2- to 4-base-pair insertion/deletion repair. Also, as a result of this retention of 2- to 4-base-pair insertion/deletion repair, the MSI phenotype in tumors is absent (de Wind et al. 1999; Edelmann et al. 2000). Given the redundancy in function between MMR genes, compound-knockout mice have also been generated. Inactivation of both *Msh3* and *Msh6* in mice is associated with adenocarcinoma of the small intestine and decreased survival compared to the single-gene-inactivation controls. These phenotypes are more similar to *Msh2*$^{-/-}$ mice.

Mutation of *Mlh1* results in a severe phenotype and a markedly reduced lifespan (6 months) similar to *Msh2*$^{-/-}$ mice (Baker et al. 1996; Edelmann et al. 1996, 1999; Prolla et al. 1998). Intestinal adenocarcinoma, skin tumors, and T-cell lymphomas have also been detected. As a result of the complete ablation of repair mechanisms in *Mlh1*$^{-/-}$ mice, MSI is a characteristic of their tumors.

Because the lifespan of many homozygous MMR mice is markedly shortened by aggressive lymphomas, studies of spontaneous intestinal tumors are more complicated. To circumvent this, intestinal tumorigenesis can be accelerated by breeding homozygous mutant MMR mice to carry an *Apc* mutation. *Msh3*, *Msh6*, *Mlh1*, *Pms2*, and *Msh3/Msh6* deficient mice have all been bred with mutant *Apc* mice (Baker et al. 1998; Edelmann et al. 1999; Kuraguchi et al. 2001; Reitmair et al. 1996a; Wei et al. 2002). In each case, there is a significant increase in tumor number and a consequent decreased lifespan compared to controls.

More recent studies of the role of MMR genes in intestinal tumor formation have shifted to knock-in allele designs, to analyze individual Lynch Syndrome mutations. Often these are missense mutations, which have quite different outcomes than gene deletions. The first of the knock-in MMR mice, *Msh2*^{GA}, carries a mutation at codon 674 (glycine to alanine) in the *Msh2* coding region (Lin et al. 2004). This mutation affects a conserved ATPase domain of Msh2 that is crucial for initiation of repair by MutS homologs (Alani et al. 1997; Drotschmann et al. 1999;

Wu and Marinus 1994). Analysis of cells from *Msh2*$^{GA/GA}$ mice showed that, although apoptotic responses were comparable to wild-type cells, ATP-mediated mismatch release was impaired, similar to *Msh2*$^{-/-}$ cells. This repair defect results in cancer predisposition in vivo that is similar to *Msh2*$^{-/-}$ mice: all *Msh2*$^{GA/GA}$ mice succumb to lymphoid or intestinal tumors by 1 year. The delayed onset of cancer in *Msh*$^{GA/GA}$ mice compared to *Msh2*$^{-/-}$ mice indicates that the remaining functional apoptotic response can stall the onset of tumorigenesis.

Another knock-in mouse model carries a mutation at codon 1217 (threonine to aspartate) in the *Msh6* gene (Yang et al. 2004). The *Msh6*^{TD} mutation impairs ATP-binding or its processing steps in the repair process (Hess et al. 2002). Studies from mutant cell extracts found that the DNA damage response and mismatch-binding capacity was not impaired; however, cells were deficient in ATP-induced mismatch release. *Msh6*$^{TD/TD}$ cell extracts were deficient in repair of both base substitutions and dinucleotide insertion/deletion loops, in contrast to *Msh6*$^{-/-}$ cell extracts that were not. *Msh6*$^{TD/TD}$ mice had a cancer phenotype similar to *Msh6*$^{-/-}$ mice, although they were characterized by a delayed tumor onset.

Two additional genes involved with DNA MMR, Flap endonuclease 1 (Fen1) and exonuclease 1 (Exo1) have been studied to determine their potential contribution in GI tumors. Fen1 was found to promote tumor progression when combined with *Apc*$^{1638N/+}$ (Kucherlapati et al. 2002). *Exo1* in combination with *Apc*$^{1638N/+}$ showed a moderate increase in tumor incidence and multiplicity when compared to *Apc*$^{1638N/+}$ siblings (Kucherlapati et al. 2007). These mice have decreased median survival, which is due to infections resulting from an impaired immune response. Triple mutant mice *Apc*$^{1638N/+}$ *Exo1Fen1* mice survive longer and display invasive GI tumors with MSI.

Immune-Deficient GEM

Inflammatory bowel disease (IBD) in humans has been divided into two major forms, ulcerative colitis (UC) and Crohn's disease (Podolsky 1991). Although the underlying mechanisms of IBD development are not fully understood, it certainly involves an immune response to intestinal bacterial and subsequent inflammation. IBD very markedly increases the risk of GI cancer above that of the general population (Eaden et al. 2001; Itzkowitz 1997). The risk of colitis-associated colon cancer (CACC) among patients is related to the severity of colitis. Although the pathogenesis of CACC remains unclear, it is characterized by an increased rate of epithelial proliferation associated with repetitive cycles of inflammation, tissue damage, and regeneration. Various immune-deficient mouse models have been generated to model IBD and are commonly characterized by inflammation of the large bowel with proliferative lesions that occasionally progress to adenocarcinoma. Many of these models, when rederived in a germ-free (bacteria-free and virus-free) environment, have a less severe phenotype than those maintained under normal conditions, suggesting roles for both pathogens and inflammatory responses in tumor susceptibility.

Cytokine-Deficient Mice

Il-2 was initially believed critical for the proliferation of T-cells in vitro; however, in vivo studies indicate that this is not the case (Hatakeyama et al. 1989). More recent studies point to a newly defined role for Il-2 in the development and homeostasis of regulatory T-cells (Burchill et al. 2007). *Il-2*$^{-/-}$ mice develops symptoms of UC (Sadlack et al. 1993). Half of the mice die within 9 weeks from severe anemia while the rest die within 6 months due to wasting. None of these mice develop GI cancer. When *Il2*$^{-/-}$ mice are crossed with *β2-microglobulin* null mice, 32% develop colonic adenocarcinoma between 6 and 12 months of age (Simpson et al. 1995; Sohn et al. 2001). The late onset of adenocarcinoma suggests that prolonged chronic inflammation may be required for tumorigenesis. All tumors from these compound mice carry mutations in *Apc*; more than half carry *p53* mutations. *Il10*$^{-/-}$ mice develop symptoms characteristic of Crohn's disease; 60% of mice develop colonic adenocarcinoma (Berg et al. 1996; Kuhn et al. 1993). These adenocarcinomas are not associated with mutations in genes typical of GI cancer, such as *p53*, *Apc*, *Msh2*, or *K-ras*. Gαi2-knockout mice develop inflammation limited to the colon; 31% develop neoplasms throughout the colon anywhere from 15 to 36 weeks (Rudolph et al. 1995). A recent study by Edwards et al. (2008) found that the Gi2-α$^{-/-}$ colonic epithelium is hyperproliferative even before the onset of colitis and resistant to induction of apoptosis. They concluded from their study that Gi2α is a direct negative regulator of colonic epithelium. Seventy-five percent of these mice die by 28 weeks, preventing long-term studies. Recent work by Ko et al. (2008) investigated the effect of deletion of *IL-4Rα* gene on AOM-induced aberrant crypt foci number and size in Balb/c mice. IL-4Rα-dependent signaling was found to have a protective, anti-neoplastic role during the post-initiation phase of AOM-induced colorectal carcinogenesis in Balb/c mice. Deletion of the *IL-4Rα* gene led to high serum levels of IL-4. Additionally IL-13, which can signal through the IL-4Rα receptor normally, instead signals via the IL-13Rα2 receptor leading to induction of TGFβ, which has pro-tumororigenic activity at early stages of intestinal tumorigenesis.

Mucin-Deficient Mice

Mucins are highly glycosylated proteins that are the major component of the mucus that lubricates and protects underlying intestinal epithelium (Gendler and Spicer 1995). Alterations of mucin expression and glycosylation have been detected in human colon cancer, but their role in tumorigenesis is not well understood (Kim et al. 1996). MUC2 is the most abundant secreted gastrointestinal apomucin (Kim and Gum 1995; van Klinken et al. 1999). *Muc2*-deficient GEM were generated by replacing exons 2–4 of *Muc2* with a *PGK-neo* cassette (Velcich et al. 2002). The resultant *Muc2*$^{-/-}$ mice were characterized by the absence of recognizable goblet cells throughout the intestine. By 12 months, 65% of *Muc2*$^{-/-}$ mice had developed adenomas with an average of >1.5 tumors per mouse. Adenomas occurred in the small and large intestine, as well as the rectum. In older mice, adenomas spontaneously progressed to adenocarcinoma. The formation of rectal tumors distinguishes the *Muc2*$^{-/-}$ mouse from many of the other mice presented here and may reflect the disorganized inflammatory processes occurring in response to the loss of normal mucins. To understand the impact of the MUC2 and APC interaction on tumorigenesis, Yang et al. (2008) crossed *Muc2*$^{-/-}$ mice with both the *Apc*$^{1638N/+}$ and *Apc*$^{Min/+}$ mice respectively. They found that introduction of *Muc2* into *Apc*$^{1638N/+}$ and *Apc*$^{Min/+}$ greatly increased transformation induced by the *Apc* mutation and significantly shifted tumor development toward the colon as a function of *Muc2* gene dosage.

MUC1 is an epithelial cell glycoprotein overexpressed and hypoglycosylated in the majority of human adenocarcinomas; its expression is also increased in IBD (Vlad et al. 2004; Campbell et al. 2001; Rhodes 1996). *Il10*$^{-/-}$ mice display some of the characteristics of human IBD; however, this mouse model lacks Muc1 expression. To explore the importance of MUC1 in IBD, Beatty et al. (2007) introduced the human MUC1 molecule into the *Il10*$^{-/-}$ mouse model. These mice develop IBD, but the disease is characterized by an earlier age of onset, greater inflammation, and higher number of colon cancers than *Il10*$^{-/-}$ controls.

Carcinogen-Induced Models of Intestinal Tumorigenesis

Intestinal tumors can be induced in rodents by a number of carcinogens including *N*-methyl-*N*′-nitro-*N*-nitrosoguanidine (Schoental and Bensted 1969), *N*-ethyl-*N*′-nitro-*N*-nitrosoguanidine, 1,2-dimethylhydrazine (Colussi et al. 2001), 2-amino-3,4-dimethylimidazo[4,5-*f*]quinoline (Fujita et al. 1999), and *N*-methyl-*N*-nitrosourea (Qin et al. 2000). Azoxymethane (AOM), a metabolite of 1,2-dimethylhydrazine (DMH), is the most widely used compound and offers a number of advantages over the parent compound including enhanced potency and chemical stability. In AOM-treated rodents, most intestinal tumors arise in the colon and form grossly visible exophytic polypoid or plaque-like growths. The microscopic appearance of low-grade lesions in these models is similar to human colonic adenomas. There is also evidence that AOM-treated mice may be a useful model for studying metastatic colorectal cancer (Ochiai et al. 2001). Studies of AOM-treated mice have identified some of the molecular abnormalities associated with these tumors and suggest that in many ways they are indistinguishable from tumors initiated by activation of Wnt signaling (Perantoni and Rice 1999; Takahashi et al. 2000; Kaiser et al. 2007). The dramatic differences in tumor number and penetrance associated with AOM-treatment in different mouse strains also highlight the ability of the mouse to model the complexities of genetic background and possibly environment (e.g., intestinal bacteria) and their effects on tumor susceptibility and eventual response to therapy in the human.

Other GEM Models of Intestinal Cancer

$Rb^{MI/MI}$ Mice

In addition to mouse models engineered to perturb known pathways in the development of GI cancer, interesting findings have emerged from mouse models targeting pathways not associated with GI cancer. One of these is the $Rb^{MI/MI}$ mouse, which carries a knock-in mutation that eliminates the C-terminal caspase-cleavage site of the retinoblastoma (Rb) protein, a known regulator of cell proliferation and cell death (Chau et al. 2002). Apoptosis was attenuated in the intestine of the $Rb^{MI/MI}$ mice following endotoxic shock; embryo-derived fibroblasts were resistant to apoptosis induced by the type I receptor for tumor necrosis factor (TNFRI) (Chau et al. 2002). These results suggested that caspase cleavage of Rb is required for TNFRI-induced cell death and that the antiapoptotic function of the $Rb^{MI/MI}$ allele might promote tumor formation when tumor suppression function is altered. Borges et al. (2005) explored this hypothesis by combining the $Rb^{MI/MI}$ allele with a *p53*-null background. Introduction of $Rb^{MI/MI}$ statistically significantly increased the incidence of colonic adenomas as well as lymphoma. Colonic tumors are a rare phenotype in *p53*-null mice (Donehower et al. 1995; Jacks et al. 1994); 26% of $Rb^{MI/MI};p53^{-/-}$ mice developed colonic tumors versus 3% of $p53^{-/-}$ mice (Borges et al. 2005). In recent studies by Kucherlapati et al. (2008), mice were generated with an *Apc(1638N)* allele, *Rb(tm2brn)* floxed alleles, and a villin-cre transgene (RBVCA) to examine the role of Rb1 in GI tumors. RBVCA mice were found to have reduced median survival due to increased tumor incidence and multiplicity in the cecum and proximal colon. These results indicate that Rb1 may influence the location of the tumor within the GI tract, and that both cecal and duodenal tumors initiate through inactivation of Apc.

PI(3)K-Deficient Mice

Phosphoinositide-3-OH kinases (PI(3)Ks) constitute a family of evolutionarily conserved lipid kinases that regulate numerous fundamental cellular responses, including proliferation, transformation, differentiation, and protection from apoptosis (Leevers et al. 1999; Toker and Cantley 1997). Homozygous gene-targeted deletion of the $p110\gamma$ catalytic subunit of PI(3)K leads to the development of invasive colorectal adenocarcinomas in mice (Sasaki et al. 2000). Epithelial tumors were detected in the colon and represented all stages of histopathology, including tubular and villous adenomas and invasive adenocarcinoma. The large carcinomas demonstrated transmural, local invasion, and metastasis into the peritoneal cavity. No tumors were found in the small intestine, stomach, or other tissues.

$Cdx2^{-/-}$ Mice

Cdx2, one of the mouse homologs of the *Drosophila melanogaster* protein, caudal (Mlodzik and Gehring 1987), is a key transcription factor for intestinal development and differentiation (Beck et al. 1995; Lorentz et al. 1997; Traber and Silberg 1996). Homozygous knockout of the *Cdx2* gene in mice results in embryonic lethality (Chawengsaksophak et al. 1997; Tamai et al. 1999). Ninety percent of $Cdx2^{+/-}$ mice develop multiple (up to ten) intestinal adenomas by 3 months of age; these adenomas primarily occur in the proximal colon. To test whether reduced expression of Cdx2 may be responsible for colon tumor progression, the *Cdx2*-knockout allele was introduced into the $Apc^{716/+}$ background to generate double heterozygote mice, $Apc^{716/+}$;$Cdx2^{+/-}$ (Aoki et al. 2003). These mice develop colonic adenomas that are characterized by loss of heterozygosity (LOH) at the *Apc* locus. $Apc^{716/+}$; $Cdx2^{+/-}$ mice rarely survive more than 30 weeks, preventing the study of malignant progression.

Dominant Negative N-Cadherin Mice

Cadherins are transmembrane glycoproteins that mediate homophilic adhesive interactions between cells (Kemler 1993; Ranscht 1994). Their conserved cytoplasmic domains interact directly with β-catenin or plakoglobin and are essential for linkage to the actin cytoskeleton and for productive cell–cell adhesion (Hinck et al. 1994; Nathke et al. 1994). Control of cell adhesion is important during embryogenesis, and perturbations of cell adhesion are associated with tumor invasion and metastasis. To understand the role of cadherins in intestinal tumorigenesis, Hermiston and Gordon (1995) generated a transgenic mouse line on the *129SV/B6* background that expresses dominant negative N-cadherin in the crypt-villus epithelium of the small intestine using a *Fabp* promoter. By 3 months of age, the mice developed features of Crohn's disease; by 6 months, adenomas; this suggested relationships among the structural integrity of the intestinal epithelium, inflammatory responses, and, ultimately, tumor initiation.

Conclusions

This chapter highlights many of the mouse models currently in use that allow us to learn about the initiation and progression of intestinal cancers. It is important to highlight some considerations concerning mouse models while thinking about such studies. Species, strain, and sex of the mice may affect experimental outcomes. The same gene mutated in two mouse strains may lead to dramatically different phenotypes, with great variation in expressivity and penetrance. Male mice are more susceptible to gastric and hepatic cancers; therefore, studies without male mice may under-represent these tumors (Rogers and Fox 2004). Additionally, the environment in which mice are bred and housed can affect experimental outcomes. Microbial populations most certainly differ between facilities and perhaps even across rooms and cages and, as described earlier, can affect inflammatory responses and subsequent gastrointestinal disease. Dietary differences also affect tumor susceptibility. However, despite the variables affecting outcome in these long-term in vivo experiments, the ability to simulate the complex germline and somatic alterations that occur in intestinal tumor formation is very powerful. The effects of aging and environmental exposures can also be queried in these complex in vivo systems in order to model human cancer.

Acknowledgments The authors are supported by NIH awards CA-009338 (EMP), CA-98013 (JG), and CA-63517 (JG). For further information or ongoing updates, the NCI website (http://emice.nci.nih.gov/mouse_models/organ_models/gastro_models) on mouse models of cancer in an outstanding resource.

References

Aberle H, Bauer A, Stappert J, Kispert A, Kemler R (1997) beta-Catenin is a target for the ubiquitin-proteasome pathway. EMBO J 16:3797–3804

Alani E, Sokolsky T, Studamire B, Miret JJ, Lahue RS (1997) Genetic and biochemical analysis of Msh2p-Msh6p: role of ATP hydrolysis and Msh2p-Msh6p subunit interactions in mismatch base pair recognition. Mol Cell Biol 17:2436–2447

Andreu P, Colnot S, Godard C, Gad S, Chafey P, Niwa-Kawakita M, Laurent-Puig P, Kahn A, Robine S, Perret C, Romagnolo B (2005) Crypt-restricted proliferation and commitment to the Paneth cell lineage following Apc loss in the mouse intestine. Development 132: 1443–1451

Aoki K, Tamai Y, Horiike S, Oshima M, Taketo MM (2003) Colonic polyposis caused by mTOR-mediated chromosomal instability in Apc+/Delta716 Cdx2$^{+/-}$ compound mutant mice. Nat Genet 35:323–330

Baker SM, Plug AW, Prolla TA, Bronner CE, Harris AC, Yao X, Christie DM, Monell C, Arnheim N, Bradley A, Ashley T, Liskay RM (1996) Involvement of mouse Mlh1 in DNA mismatch repair and meiotic crossing over. Nat Genet 13:336–342

Baker SM, Harris AC, Tsao JL, Flath TJ, Bronner CE, Gordon M, Shibata D, Liskay RM (1998) Enhanced intestinal adenomatous polyp formation in Pms2$^{-/-}$;Min mice. Cancer Res 58:1087–1089

Barth AI, Nathke IS, Nelson WJ (1997) Cadherins, catenins and APC protein: interplay between cytoskeletal complexes and signaling pathways. Curr Opin Cell Biol 9:683–690

Beatty PL, Plevy SE, Sepulveda AR, Finn OJ (2007) Cutting edge: transgenic expression of human MUC1 in IL-10$^{-/-}$ mice accelerates inflammatory bowel disease and progression to colon cancer. J Immunol 179:735–739

Beck F, Erler T, Russell A, James R (1995) Expression of Cdx-2 in the mouse embryo and placenta: possible role in patterning of the extra-embryonic membranes. Dev Dyn 204:219–227

Berg DJ, Davidson N, Kuhn R, Muller W, Menon S, Holland G, Thompson-Snipes L, Leach MW, Rennick D (1996) Enterocolitis and colon cancer in interleukin-10-deficient mice are associated with aberrant cytokine production and CD4(+) TH1-like responses. J Clin Invest 98:1010–1020

Borges HL, Bird J, Wasson K, Cardiff RD, Varki N, Eckmann L, Wang JY (2005) Tumor promotion by caspase-resistant retinoblastoma protein. Proc Natl Acad Sci USA 102:15587–15592

Brandes ME, Mai UE, Ohura K, Wahl SM (1991) Type I transforming growth factor-beta receptors on neutrophils mediate chemotaxis to transforming growth factor-beta. J Immunol 147:1600–1606

Burchill MA, Yang J, Vang KB, Farrar MA (2007) Interleukin-2 receptor signaling in regulatory T cell development and homeostasis. Immunol Lett 114:1–8

Cadigan KM, Nusse R (1997) Wnt signaling: a common theme in animal development. Genes Dev 11:3286–3305

Campbell BJ, Yu LG, Rhodes JM (2001) Altered glycosylation in inflammatory bowel disease: a possible role in cancer development. Glycoconj J 18:851–858

Chau BN, Borges HL, Chen TT, Masselli A, Hunton IC, Wang JY (2002) Signal-dependent protection from apoptosis in mice expressing caspase-resistant Rb. Nat Cell Biol 4:757–765

Chawengsaksophak K, James R, Hammond VE, Kontgen F, Beck F (1997) Homeosis and intestinal tumours in Cdx2 mutant mice. Nature 386:84–87

Chulada PC, Thompson MB, Mahler JF, Doyle CM, Gaul BW, Lee C, Tiano HF, Morham SG, Smithies O, Langenbach R (2000) Genetic disruption of Ptgs-1, as well as Ptgs-2, reduces intestinal tumorigenesis in Min mice. Cancer Res 60:4705–4708

Colnot S, Romagnolo B, Lambert M, Cluzeaud F, Porteu A, Vandewalle A, Thomasset M, Kahn A, Perret C (1998) Intestinal expression of the calbindin-D9K gene in transgenic mice. Requirement for a Cdx2-binding site in a distal activator region. J Biol Chem 273:31939–31946

Colussi C, Fiumicino S, Giuliani A, Rosini S, Musiani P, Macri C, Potten CS, Crescenzi M, Bignami M (2001) 1,2-Dimethylhydrazine-induced colon carcinoma and lymphoma in msh2(–/–) mice. J Natl Cancer Inst 93:1534–1540

de Wind N, Dekker M, Berns A, Radman M, te Riele H (1995) Inactivation of the mouse Msh2 gene results in mismatch repair deficiency, methylation tolerance, hyperrecombination, and predisposition to cancer. Cell 82:321–330

de Wind N, Dekker M, Claij N, Jansen L, van Klink Y, Radman M, Riggins G, van der Valk M, van't Wout K, te Riele H (1999) HNPCC-like cancer predisposition in mice through simultaneous loss of Msh3 and Msh6 mismatch-repair protein functions. Nat Genet 23:359–362

Dennler S, Itoh S, Vivien D, ten Dijke P, Huet S, Gauthier JM (1998) Direct binding of Smad3 and Smad4 to critical TGF beta-inducible elements in the promoter of human plasminogen activator inhibitor-type 1 gene. EMBO J 17:3091–3100

DeWitt DL, Smith WL (1988) Primary structure of prostaglandin G/H synthase from sheep vesicular gland determined from the complementary DNA sequence. Proc Natl Acad Sci USA 85:1412–1416

Donehower LA, Harvey M, Vogel H, McArthur MJ, Montgomery CA, Jr., Park SH, Thompson T, Ford RJ, Bradley A (1995) Effects of genetic background on tumorigenesis in p53-deficient mice. Mol Carcinog 14:16–22

Drotschmann K, Clark AB, Tran HT, Resnick MA, Gordenin DA, Kunkel TA (1999) Mutator phenotypes of yeast strains heterozygous for mutations in the MSH2 gene. Proc Natl Acad Sci USA 96:2970–2975

Eaden JA, Abrams KR, Mayberry JF (2001) The risk of colorectal cancer in ulcerative colitis: a meta-analysis. Gut 48:526–535

Edelmann W, Cohen PE, Kane M, Lau K, Morrow B, Bennett S, Umar A, Kunkel T, Cattoretti G, Chaganti R, Pollard JW, Kolodner RD, Kucherlapati R (1996) Meiotic pachytene arrest in MLH1-deficient mice. Cell 85:1125–1134

Edelmann W, Yang K, Umar A, Heyer J, Lau K, Fan K, Liedtke W, Cohen PE, Kane MF, Lipford JR, Yu N, Crouse GF, Pollard JW, Kunkel T, Lipkin M, Kolodner R, Kucherlapati R (1997) Mutation in the mismatch repair gene Msh6 causes cancer susceptibility. Cell 91: 467–477

Edelmann W, Yang K, Kuraguchi M, Heyer J, Lia M, Kneitz B, Fan K, Brown AM, Lipkin M, Kucherlapati R (1999) Tumorigenesis in Mlh1 and Mlh1/Apc1638N mutant mice. Cancer Res 59:1301–1307

Edelmann W, Umar A, Yang K, Heyer J, Kucherlapati M, Lia M, Kneitz B, Avdievich E, Fan K, Wong E, Crouse G, Kunkel T, Lipkin M, Kolodner RD, Kucherlapati R (2000) The DNA mismatch repair genes Msh3 and Msh6 cooperate in intestinal tumor suppression. Cancer Res 60:803–807

Edwards RA, Wang K, Davis JS, Birnbaumer L (2008) Role for epithelial dysregulation in early-onset colitis-associated colon cancer in Gi2-alpha-/- mice. Inflamm Bowel Dis 14:898–907

Engle SJ, Hoying JB, Boivin GP, Ormsby I, Gartside PS, Doetschman T (1999) Transforming growth factor beta1 suppresses nonmetastatic colon cancer at an early stage of tumorigenesis. Cancer Res 59:3379–3386

Eppert K, Scherer SW, Ozcelik H, Pirone R, Hoodless P, Kim H, Tsui LC, Bapat B, Gallinger S, Andrulis IL, Thomsen GH, Wrana JL, Attisano L (1996) MADR2 maps to 18q21 and encodes a TGFbeta-regulated MAD-related protein that is functionally mutated in colorectal carcinoma. Cell 86:543–552

Fodde R, Edelmann W, Yang K, van Leeuwen C, Carlson C, Renault B, Breukel C, Alt E, Lipkin M, Khan PM, et al. (1994) A targeted chain-termination mutation in the mouse Apc gene results in multiple intestinal tumors. Proc Natl Acad Sci USA 91:8969–8973

Friedl W, Kruse R, Uhlhaas S, Stolte M, Schartmann B, Keller KM, Jungck M, Stern M, Loff S, Back W, Propping P, Jenne DE (1999) Frequent 4-bp deletion in exon 9 of the SMAD4/MADH4 gene in familial juvenile polyposis patients. Genes Chromosomes Cancer 25:403–406

Fujita H, Nagano K, Ochiai M, Ushijima T, Sugimura T, Nagao M, Matsushima T (1999) Difference in target organs in carcinogenesis with a heterocyclic amine, 2-amino-3,4-dimethylimidazo [4,5-f]quinoline, in different strains of mice. Jpn J Cancer Res 90:1203–1206

Gendler SJ, Spicer AP (1995) Epithelial mucin genes. Annu Rev Physiol 57:607–634

Goss KH, Risinger MA, Kordich JJ, Sanz MM, Straughen JE, Slovek LE, Capobianco AJ, German J, Boivin GP, Groden J (2002) Enhanced tumor formation in mice heterozygous for Blm mutation. Science 297:2051–2053

Graff JM, Bansal A, Melton DA (1996) Xenopus Mad proteins transduce distinct subsets of signals for the TGF beta superfamily. Cell 85:479–487

Groden J, Thliveris A, Samowitz W, Carlson M, Gelbert L, Albertsen H, Joslyn G, Stevens J, Spirio L, Robertson M, et al. (1991) Identification and characterization of the familial adenomatous polyposis coli gene. Cell 66:589–600

Hahn SA, Schutte M, Hoque AT, Moskaluk CA, da Costa LT, Rozenblum E, Weinstein CL, Fischer A, Yeo CJ, Hruban RH, Kern SE (1996) DPC4, a candidate tumor suppressor gene at human chromosome 18q21.1. Science 271:350–353

Hamamoto T, Beppu H, Okada H, Kawabata M, Kitamura T, Miyazono K, Kato M (2002) Compound disruption of smad2 accelerates malignant progression of intestinal tumors in apc knockout mice. Cancer Res 62:5955–5961

Harada N, Tamai Y, Ishikawa T, Sauer B, Takaku K, Oshima M, Taketo MM (1999) Intestinal polyposis in mice with a dominant stable mutation of the beta-catenin gene. EMBO J 18:5931–5942

Hatakeyama M, Tsudo M, Minamoto S, Kono T, Doi T, Miyata T, Miyasaka M, Taniguchi T (1989) Interleukin-2 receptor beta chain gene: generation of three receptor forms by cloned human alpha and beta chain cDNA's. Science 244:551–556

Hemler M, Lands WE (1976) Purification of the cyclooxygenase that forms prostaglandins. Demonstration of two forms of iron in the holoenzyme. J Biol Chem 251:5575–5579

Hermiston ML, Gordon JI (1995) Inflammatory bowel disease and adenomas in mice expressing a dominant negative N-cadherin. Science 270:1203–1207

Hess MT, Gupta RD, Kolodner RD (2002) Dominant *Saccharomyces cerevisiae* msh6 mutations cause increased mispair binding and decreased dissociation from mispairs by Msh2-Msh6 in the presence of ATP. J Biol Chem 277:25545–25553

Heyer J, Escalante-Alcalde D, Lia M, Boettinger E, Edelmann W, Stewart CL, Kucherlapati R (1999) Postgastrulation Smad2-deficient embryos show defects in embryo turning and anterior morphogenesis. Proc Natl Acad Sci USA 96:12595–12600

Higuchi T, Iwama T, Yoshinaga K, Toyooka M, Taketo MM, Sugihara K (2003) A randomized, double-blind, placebo-controlled trial of the effects of rofecoxib, a selective cyclooxygenase-2 inhibitor, on rectal polyps in familial adenomatous polyposis patients. Clin Cancer Res 9:4756–4760

Hinck L, Nathke IS, Papkoff J, Nelson WJ (1994) Dynamics of cadherin/catenin complex formation: novel protein interactions and pathways of complex assembly. J Cell Biol 125:1327–1340

Hulsken J, Birchmeier W, Behrens J (1994) E-Cadherin and APC compete for the interaction with beta-catenin and the cytoskeleton. J Cell Biol 127:2061–2069

Inman GJ, Nicolas FJ, Hill CS (2002) Nucleocytoplasmic shuttling of Smads 2, 3, and 4 permits sensing of TGF-beta receptor activity. Mol Cell 10:283–294

Itzkowitz SH (1997) Inflammatory bowel disease and cancer. Gastroenterol Clin North Am 26:129–139

Iwao K, Nakamori S, Kameyama M, Imaoka S, Kinoshita M, Fukui T, Ishiguro S, Nakamura Y, Miyoshi Y (1998) Activation of the beta-catenin gene by interstitial deletions involving exon 3 in primary colorectal carcinomas without adenomatous polyposis coli mutations. Cancer Res 58:1021–1026

Jacks T, Remington L, Williams BO, Schmitt EM, Halachmi S, Bronson RT, Weinberg RA (1994) Tumor spectrum analysis in p53-mutant mice. Curr Biol 4:1–7

Jonk LJ, Itoh S, Heldin CH, ten Dijke P, Kruijer W (1998) Identification and functional characterization of a Smad binding element (SBE) in the JunB promoter that acts as a transforming growth factor-beta, activin, and bone morphogenetic protein-inducible enhancer. J Biol Chem 273:21145–21152

Kaiser S, Park Y, Chen X, Freudenberg J, Halberg R, Haigis K, Jegga A, Kong S, Sakthivel B, Reichling T, Azhar M, Yeatman T, Doetschman T, Groden J, Threadgill D, Dove WF, Coffey RJ, Aronow B (2007) Recapitulation of developing colon transcriptional patterns in mouse and human colon cancer. Genome Physiol 8(7):R131

Kemler R (1993) From cadherins to catenins: cytoplasmic protein interactions and regulation of cell adhesion. Trends Genet 9:317–321

Kim YS, Gum JR, Jr. (1995) Diversity of mucin genes, structure, function, and expression. Gastroenterology 109:999–1001

Kim YS, Gum J, Jr., Brockhausen I (1996) Mucin glycoproteins in neoplasia. Glycoconj J 13: 693–707

Kim J, Johnson K, Chen HJ, Carroll S, Laughon A (1997) Drosophila Mad binds to DNA and directly mediates activation of vestigial by Decapentaplegic. Nature 388:304–308

Kinzler KW, Nilbert MC, Su LK, Vogelstein B, Bryan TM, Levy DB, Smith KJ, Preisinger AC, Hedge P, McKechnie D, et al. (1991) Identification of FAP locus genes from chromosome 5q21. Science 253:661–665

Ko CW, Cuthbert RJ, Orsi NM, Brooke DA, Perry SL, Markham AF, Coletta PL, Hull MA (2008) Lack of interleukin-4 receptor alpha chain-dependent signaling promotes azoxymethane-induced colorectal aberrant crypt focus formation in Balb/c mice. J Pathol 214:603–609

Kucherlapati M, Yang K, Kuraguchi M, Zhao J, Lia M, Heyer J, Kane MF, Fan K, Russell R, Brown AM, Kneitz B, Edelmann W, Kolodner RD, Lipkin M, Kucherlapati R (2002) Haploinsufficiency of Flap endonuclease (Fen1) leads to rapid tumor progression. PNAS 99:9924–9929

Kucherlapati MH, Yang K, Fan K, Kuraguchi M, Sonkin D, Rosulek A, Lipkin M, Bronson RT, Aronow BJ, Kucherlapati R (2008) Loss of Rb1 in the gastrointestinal tract of Apc1638N mice promotes tumors of the cecum and proximal colon. PNAS 105:15493–15498

Kucherlapati M, Nguyen A, Kuraguchi M, Yang K, Fan K, Bronson R, Wei K, Lipkin M, Edelmann W, Kucherlapati R (2007) Tumor progression in Apc(1638N) mice with Exo 1 and Fen1 deficiencies. Oncogene 26:6297–6306

Kuhn R, Lohler J, Rennick D, Rajewsky K, Muller W (1993) Interleukin-10-deficient mice develop chronic enterocolitis. Cell 75:263–274

Kulkarni AB, Huh CG, Becker D, Geiser A, Lyght M, Flanders KC, Roberts AB, Sporn MB, Ward JM, Karlsson S (1993) Transforming growth factor beta 1 null mutation in mice causes excessive inflammatory response and early death. Proc Natl Acad Sci USA 90:770–774

Kuraguchi M, Yang K, Wong E, Avdievich E, Fan K, Kolodner RD, Lipkin M, Brown AM, Kucherlapati R, Edelmann W (2001) The distinct spectra of tumor-associated Apc mutations in mismatch repair-deficient Apc1638N mice define the roles of MSH3 and MSH6 in DNA repair and intestinal tumorigenesis. Cancer Res 61:7934–7942

Labbe E, Silvestri C, Hoodless PA, Wrana JL, Attisano L (1998) Smad2 and Smad3 positively and negatively regulate TGF beta-dependent transcription through the forkhead DNA-binding protein FAST2. Mol Cell 2:109–120

Laird PW, Jackson-Grusby L, Fazeli A, Dickinson SL, Jung WE, Li E, Weinberg RA, Jaenisch R (1995) Suppression of intestinal neoplasia by DNA hypomethylation. Cell 81: 197–205

Lee SH, Soyoola E, Chanmugam P, Hart S, Sun W, Zhong H, Liou S, Simmons D, Hwang D (1992) Selective expression of mitogen-inducible cyclooxygenase in macrophages stimulated with lipopolysaccharide. J Biol Chem 267:25934–25938

Leevers SJ, Vanhaesebroeck B, Waterfield MD (1999) Signalling through phosphoinositide 3-kinases: the lipids take centre stage. Curr Opin Cell Biol 11:219–225

Li Q, Ishikawa TO, Oshima M, Taketo MM (2005) The threshold level of adenomatous polyposis coli protein for mouse intestinal tumorigenesis. Cancer Res 65:8622–8627

Lin DP, Wang Y, Scherer SJ, Clark AB, Yang K, Avdievich E, Jin B, Werling U, Parris T, Kurihara N, Umar A, Kucherlapati R, Lipkin M, Kunkel TA, Edelmann W (2004) An Msh2 point mutation uncouples DNA mismatch repair and apoptosis. Cancer Res 64:517–522

Lorentz O, Duluc I, Arcangelis AD, Simon-Assmann P, Kedinger M, Freund JN (1997) Key role of the Cdx2 homeobox gene in extracellular matrix-mediated intestinal cell differentiation. J Cell Biol 139:1553–1565

Luo G, Santoro IM, McDaniel LD, Nishijima I, Mills M, Youssoufian H, Vogel H, Schultz RA, Bradley A (2000) Cancer predisposition caused by elevated mitotic recombination in Bloom mice. Nat Genet 26:424–429

Maier JA, Hla T, Maciag T (1990) Cyclooxygenase is an immediate-early gene induced by interleukin-1 in human endothelial cells. J Biol Chem 265:10805–10808

Manning AM, Williams AC, Game SM, Paraskeva C (1991) Differential sensitivity of human colonic adenoma and carcinoma cells to transforming growth factor beta (TGF-beta): conversion of an adenoma cell line to a tumorigenic phenotype is accompanied by a reduced response to the inhibitory effects of TGF-beta. Oncogene 6:1471–1476

Markowitz S, Wang J, Myeroff L, Parsons R, Sun L, Lutterbaugh J, Fan RS, Zborowska E, Kinzler KW, Vogelstein B, et al. (1995) Inactivation of the type II TGF-beta receptor in colon cancer cells with microsatellite instability. Science 268:1336–1338

Massague J (1998) TGF-beta signal transduction. Annu Rev Biochem 67:753–791

Miller JR, Moon RT (1996) Signal transduction through beta-catenin and specification of cell fate during embryogenesis. Genes Dev 10:2527–2539

Miyamoto T, Ogino N, Yamamoto S, Hayaishi O (1976) Purification of prostaglandin endoperoxide synthetase from bovine vesicular gland microsomes. J Biol Chem 251:2629–2636

Mlodzik M, Gehring WJ (1987) Expression of the caudal gene in the germ line of Drosophila: formation of an RNA and protein gradient during early embryogenesis. Cell 48:465–478

Morin PJ, Sparks AB, Korinek V, Barker N, Clevers H, Vogelstein B, Kinzler KW (1997) Activation of beta-catenin-Tcf signaling in colon cancer by mutations in beta-catenin or APC. Science 275:1787–1790

Moser AR, Pitot HC, Dove WF (1990) A dominant mutation that predisposes to multiple intestinal neoplasia in the mouse. Science 247:322–324

Mulder KM, Ramey MK, Hoosein NM, Levine AE, Hinshaw XH, Brattain DE, Brattain MG (1988) Characterization of transforming growth factor-beta-resistant subclones isolated from a transforming growth factor-beta-sensitive human colon carcinoma cell line. Cancer Res 48:7120–7125

Munemitsu S, Albert I, Rubinfeld B, Polakis P (1996) Deletion of an amino-terminal sequence beta-catenin in vivo and promotes hyperphosporylation of the adenomatous polyposis coli tumor suppressor protein. Mol Cell Biol 16:4088–4094

Nagase H, Nakamura Y (1993) Mutations of the APC (adenomatous polyposis coli) gene. Hum Mutat 2:425–434

Nagatake M, Takagi Y, Osada H, Uchida K, Mitsudomi T, Saji S, Shimokata K, Takahashi T, Takahashi T (1996) Somatic in vivo alterations of the DPC4 gene at 18q21 in human lung cancers. Cancer Res 56:2718–2720

Nathke IS, Hinck L, Swedlow JR, Papkoff J, Nelson WJ (1994) Defining interactions and distributions of cadherin and catenin complexes in polarized epithelial cells. J Cell Biol 125:1341–1352

Nieuwenhuis MH. Vasen HF (2007). Correlations between mutation site in APC and phenotype of familial adenomatous polyposis (FAP): a review of the literature. Crit Rev Oncol Hematol 61(2):153–161

Nomura M, Li E (1998) Smad2 role in mesoderm formation, left-right patterning and craniofacial development. Nature 393:786–790

O'Banion MK, Winn VD, Young DA (1992) cDNA cloning and functional activity of a glucocorticoid-regulated inflammatory cyclooxygenase. Proc Natl Acad Sci USA 89:4888–4892

O'Neill GP, Ford-Hutchinson AW (1993) Expression of mRNA for cyclooxygenase-1 and cyclooxygenase-2 in human tissues. FEBS Lett 330:156–160

Ochiai M, Ubagai T, Kawamori T, Imai H, Sugimura T, Nakagama H (2001) High susceptibility of Scid mice to colon carcinogenesis induced by azoxymethane indicates a possible caretaker role for DNA-dependent protein kinase. Carcinogenesis 22:1551–1555

Oshima M, Oshima H, Kitagawa K, Kobayashi M, Itakura C, Taketo M (1995) Loss of Apc heterozygosity and abnormal tissue building in nascent intestinal polyps in mice carrying a truncated Apc gene. Proc Natl Acad Sci USA 92:4482–4486

Oshima M, Dinchuk JE, Kargman SL, Oshima H, Hancock B, Kwong E, Trzaskos JM, Evans JF, Taketo MM (1996) Suppression of intestinal polyposis in Apc delta716 knockout mice by inhibition of cyclooxygenase 2 (COX-2). Cell 87:803–809

Parsons R, Myeroff LL, Liu B, Willson JK, Markowitz SD, Kinzler KW, Vogelstein B (1995) Microsatellite instability and mutations of the transforming growth factor beta type II receptor gene in colorectal cancer. Cancer Res 55:5548–5550

Perantoni AO, Rice JM (1999) Mutation patterns in non-ras oncogenes and tumour suppressor genes in experimentally induced tumours. IARC Sci Publ (146):87–122

Podolsky DK (1991) Inflammatory bowel disease (1). N Engl J Med 325:928–937

Prolla TA, Baker SM, Harris AC, Tsao JL, Yao X, Bronner CE, Zheng B, Gordon M, Reneker J, Arnheim N, Shibata D, Bradley A, Liskay RM (1998) Tumour susceptibility and spontaneous mutation in mice deficient in Mlh1, Pms1 and Pms2 DNA mismatch repair. Nat Genet 18:276–279

Qin X, Shibata D, Gerson SL (2000) Heterozygous DNA mismatch repair gene PMS2-knockout mice are susceptible to intestinal tumor induction with *N*-methyl-*N*-nitrosourea. Carcinogenesis 21:833–838

Quesada CF, Kimata H, Mori M, Nishimura M, Tsuneyoshi T, Baba S (1998) Piroxicam and acarbose as chemopreventive agents for spontaneous intestinal adenomas in APC gene 1309 knockout mice. Jpn J Cancer Res 89:392–396

Ranscht B (1994) Cadherins and catenins: interactions and functions in embryonic development. Curr Opin Cell Biol 6:740–746

Reitmair AH, Schmits R, Ewel A, Bapat B, Redston M, Mitri A, Waterhouse P, Mittrucker HW, Wakeham A, Liu B, et al. (1995) MSH2 deficient mice are viable and susceptible to lymphoid tumours. Nat Genet 11:64–70

Reitmair AH, Cai JC, Bjerknes M, Redston M, Cheng H, Pind MT, Hay K, Mitri A, Bapat BV, Mak TW, Gallinger S (1996a) MSH2 deficiency contributes to accelerated APC-mediated intestinal tumorigenesis. Cancer Res 56:2922–2926

Reitmair AH, Redston M, Cai JC, Chuang TC, Bjerknes M, Cheng H, Hay K, Gallinger S, Bapat B, Mak TW (1996b) Spontaneous intestinal carcinomas and skin neoplasms in Msh2-deficient mice. Cancer Res 56:3842–3849

Rhodes JM (1996) Unifying hypothesis for inflammatory bowel disease and associated colon cancer: sticking the pieces together with sugar. Lancet 347:40–44

Roberts AB, McCune BK, Sporn MB (1992) TGF-beta: regulation of extracellular matrix. Kidney Int 41:557–559

Rogers AB, Fox JG (2004) Inflammation and cancer. I. Rodent models of infectious gastrointestinal and liver cancer. Am J Physiol Gastrointest Liver Physiol 286:G361–G366

Romagnolo B, Cluzeaud F, Lambert M, Colnot S, Porteu A, Molina T, Tomasset M, Vandewalle A, Kahn A, Perret C (1996) Tissue-specific and hormonal regulation of calbindin-D9K fusion genes in transgenic mice. J Biol Chem 271:16820–16826

Romagnolo B, Berrebi D, Saadi-Keddoucci S, Porteu A, Pichard AL, Peuchmaur M, Vandewalle A, Kahn A, Perret C (1999) Intestinal dysplasia and adenoma in transgenic mice after overexpression of an activated beta-catenin. Cancer Res 59:3875–3879

Rudolph U, Finegold MJ, Rich SS, Harriman GR, Srinivasan Y, Brabet P, Boulay G, Bradley A, Birnbaumer L (1995) Ulcerative colitis and adenocarcinoma of the colon in G alpha i2-deficient mice. Nat Genet 10:143–150

Sadlack B, Merz H, Schorle H, Schimpl A, Feller AC, Horak I (1993) Ulcerative colitis-like disease in mice with a disrupted interleukin-2 gene. Cell 75:253–261

Sansom OJ, Berger J, Bishop SM, Hendrich B, Bird A, Clarke AR (2003) Deficiency of Mbd2 suppresses intestinal tumorigenesis. Nat Genet 34:145–147

Sasaki T, Irie-Sasaki J, Horie Y, Bachmaier K, Fata JE, Li M, Suzuki A, Bouchard D, Ho A, Redston M, Gallinger S, Khokha R, Mak TW, Hawkins PT, Stephens L, Scherer SW, Tsao M, Penninger JM (2000) Colorectal carcinomas in mice lacking the catalytic subunit of PI(3)Kγ. Nature 406:897–902

Schoental R, Bensted JP (1969) Gastro-intestinal tumours in rats and mice following various routes of administration of *N*-methyl-*N*-nitroso-*N*′-nitroguanidine and *N*-ethyl-*N*-nitroso-*N*′-nitroguanidine. Br J Cancer 23:757–764

Sheng H, Shao J, Williams CS, Pereira MA, Taketo MM, Oshima M, Reynolds AB, Washington MK, DuBois RN, Beauchamp RD (1998) Nuclear translocation of beta-catenin in hereditary and carcinogen-induced intestinal adenomas. Carcinogenesis 19:543–549

Shi Y, Massague J (2003) Mechanisms of TGF-beta signaling from cell membrane to the nucleus. Cell 113:685–700

Shibata H, Toyama K, Shioya H, Ito M, Hirota M, Hasegawa S, Matsumoto H, Takano H, Akiyama T, Toyoshima K, Kanamaru R, Kanegae Y, Saito I, Nakamura Y, Shiba K, Noda T (1997) Rapid colorectal adenoma formation initiated by conditional targeting of the Apc gene. Science 278:120–123

Shull MM, Ormsby I, Kier AB, Pawlowski S, Diebold RJ, Yin M, Allen R, Sidman C, Proetzel G, Calvin D, et al. (1992) Targeted disruption of the mouse transforming growth factor-beta 1 gene results in multifocal inflammatory disease. Nature 359:693–699

Simpson SJ, Mizoguchi E, Allen D, Bhan AK, Terhorst C (1995) Evidence that CD4+, but not CD8+ T cells are responsible for murine interleukin-2-deficient colitis. Eur J Immunol 25:2618–2625

Sirard C, de la Pompa JL, Elia A, Itie A, Mirtsos C, Cheung A, Hahn S, Wakeham A, Schwartz L, Kern SE, Rossant J, Mak TW (1998) The tumor suppressor gene Smad4/Dpc4 is required for gastrulation and later for anterior development of the mouse embryo. Genes Dev 12:107–119

Smits R, Kielman MF, Breukel C, Zurcher C, Neufeld K, Jagmohan-Changur S, Hofland N, van Dijk J, White R, Edelmann W, Kucherlapati R, Khan PM, Fodde R (1999) Apc1638T: a mouse model delineating critical domains of the adenomatous polyposis coli protein involved in tumorigenesis and development. Genes Dev 13:1309–1321

Sohn KJ, Shah SA, Reid S, Choi M, Carrier J, Comiskey M, Terhorst C, Kim YI (2001) Molecular genetics of ulcerative colitis-associated colon cancer in the interleukin 2- and beta(2)-microglobulin-deficient mouse. Cancer Res 61:6912–6917

Sonoshita M, Takaku K, Sasaki N, Sugimoto Y, Ushikubi F, Narumiya S, Oshima M, Taketo MM (2001) Acceleration of intestinal polyposis through prostaglandin receptor EP2 in Apc(Delta 716) knockout mice. Nat Med 7:1048–1051

Sparks AB, Morin PJ, Vogelstein B, Kinzler KW (1998) Mutational analysis of the APC/beta-catenin/Tcf pathway in colorectal cancer. Cancer Res 58:1130–1134

Steinbach G, Lynch PM, Phillips RK, Wallace MH, Hawk E, Gordon GB, Wakabayashi N, Saunders B, Shen Y, Fujimura T, Su LK, Levin B (2000) The effect of celecoxib, a cyclooxygenase-2 inhibitor, in familial adenomatous polyposis. N Engl J Med 342:1946–1952

Su LK, Kinzler KW, Vogelstein B, Preisinger AC, Moser AR, Luongo C, Gould KA, Dove WF (1992) Multiple intestinal neoplasia caused by a mutation in the murine homolog of the APC gene. Science 256:668–670

Takagi Y, Kohmura H, Futamura M, Kida H, Tanemura H, Shimokawa K, Saji S (1996) Somatic alterations of the DPC4 gene in human colorectal cancers in vivo. Gastroenterology 111:1369–1372

Takahashi M, Nakatsugi S, Sugimura T, Wakabayashi K (2000) Frequent mutations of the beta-catenin gene in mouse colon tumors induced by azoxymethane. Carcinogenesis 21:1117–1120

Takaku K, Oshima M, Miyoshi H, Matsui M, Seldin MF, Taketo MM (1998) Intestinal tumorigenesis in compound mutant mice of both Dpc4 (Smad4) and Apc genes. Cell 92:645–656

Takaku K, Sonoshita M, Sasaki N, Uozumi N, Doi Y, Shimizu T, Taketo MM (2000) Suppression of intestinal polyposis in Apc(delta 716) knockout mice by an additional mutation in the cytosolic phospholipase A(2) gene. J Biol Chem 275:34013–34016

Takeda H, Sonoshita M, Oshima H, Sugihara K, Chulada PC, Langenbach R, Oshima M, Taketo MM (2003) Cooperation of cyclooxygenase 1 and cyclooxygenase 2 in intestinal polyposis. Cancer Res 63:4872–4877

Tamai Y, Nakajima R, Ishikawa T, Takaku K, Seldin MF, Taketo MM (1999) Colonic hamartoma development by anomalous duplication in Cdx2 knockout mice. Cancer Res 59:2965–2970

Thiagalingam S, Lengauer C, Leach FS, Schutte M, Hahn SA, Overhauser J, Willson JK, Markowitz S, Hamilton SR, Kern SE, Kinzler KW, Vogelstein B (1996) Evaluation of candidate tumour suppressor genes on chromosome 18 in colorectal cancers. Nat Genet 13:343–346

Toker A, Cantley LC (1997) Signalling through the lipid products of phosphoinositide-3-OH kinase. Nature 387:673–676

Traber PG, Silberg DG (1996) Intestine-specific gene transcription. Annu Rev Physiol 58:275–297

van Klinken BJ, Einerhand AW, Duits LA, Makkink MK, Tytgat KM, Renes IB, Verburg M, Buller HA, Dekker J (1999) Gastrointestinal expression and partial cDNA cloning of murine Muc2. Am J Physiol 276:G115–G124

Vane JR (1971) Inhibition of prostaglandin synthesis as a mechanism of action for aspirin-like drugs. Nat New Biol 231:232–235

Vane J (1994) Towards a better aspirin. Nature 367:215–216

Velcich A, Yang W, Heyer J, Fragale A, Nicholas C, Viani S, Kucherlapati R, Lipkin M, Yang K, Augenlicht L (2002) Colorectal cancer in mice genetically deficient in the mucin Muc2. Science 295:1726–1729

Vlad AM, Kettel JC, Alajez NM, Carlos CA, Finn OJ (2004) MUC1 immunobiology: from discovery to clinical applications. Adv Immunol 82:249–293

Wahl SM, Hunt DA, Wakefield LM, McCartney-Francis N, Wahl LM, Roberts AB, Sporn MB (1987) Transforming growth factor type beta induces monocyte chemotaxis and growth factor production. Proc Natl Acad Sci USA 84:5788–5792

Waldrip WR, Bikoff EK, Hoodless PA, Wrana JL, Robertson EJ (1998) Smad2 signaling in extraembryonic tissues determines anterior-posterior polarity of the early mouse embryo. Cell 92:797–808

Wang D, Wang H, Shi Q, Katkuri S, Walhi W, Desvergne B, Das SK, Dey SK, DuBois RN (2004) Prostaglandin E(2) promotes colorectal adenoma growth via transactivation of the nuclear peroxisome proliferator-activated receptor delta. Cancer Cell 6:285–295

Wei K, Kucherlapati R, Edelmann W (2002) Mouse models for human DNA mismatch-repair gene defects. Trends Mol Med 8:346–353

Weinstein M, Yang X, Li C, Xu X, Gotay J, Deng CX (1998) Failure of egg cylinder elongation and mesoderm induction in mouse embryos lacking the tumor suppressor smad2. Proc Natl Acad Sci USA 95:9378–9383

Wilson CL, Heppner KJ, Labosky PA, Hogan BL, Matrisian LM (1997) Intestinal tumorigenesis is suppressed in mice lacking the metalloproteinase matrilysin. Proc Natl Acad Sci USA 94:1402–1407

Wong MH, Rubinfeld B, Gordon JI (1998) Effects of forced expression of an NH2-terminal truncated beta-Catenin on mouse intestinal epithelial homeostasis. J Cell Biol 141:765–777

Wu TH, Marinus MG (1994) Dominant negative mutator mutations in the mutS gene of Escherichia coli. J Bacteriol 176:5393–5400

Xu L, Massague J (2004) Nucleocytoplasmic shuttling of signal transducers. Nat Rev Mol Cell Biol 5:209–219

Xu X, Brodie SG, Yang X, Im YH, Parks WT, Chen L, Zhou YX, Weinstein M, Kim SJ, Deng CX (2000) Haploid loss of the tumor suppressor Smad4/Dpc4 initiates gastric polyposis and cancer in mice. Oncogene 19:1868–1874

Yang K, Edelmann W, Fan K, Lau K, Kolli VR, Fodde R, Khan PM, Kucherlapati R, Lipkin M (1997) A mouse model of human familial adenomatous polyposis. J Exp Zool 277:245–254

Yang X, Li C, Xu X, Deng C (1998) The tumor suppressor SMAD4/DPC4 is essential for epiblast proliferation and mesoderm induction in mice. Proc Natl Acad Sci USA 95:3667–3672

Yang G, Scherer SJ, Shell SS, Yang K, Kim M, Lipkin M, Kucherlapati R, Kolodner RD, Edelmann W (2004) Dominant effects of an Msh6 missense mutation on DNA repair and cancer susceptibility. Cancer Cell 6:139–150

Yang K, Popova NV, Yang WC, Lozonschi I, Tadesse S, Kent S, Bancroft L, Matise I, Cormier RT, Scherer SJ, Edelmann W, Lipkin M, Augenlicht L, Velcich A (2008) Interaction of Muc2 and Apc on Wnt signaling and in intestinal tumorigenesis: potential role of chronic inflammation. Cancer Res 68:7313–7322

Yingling JM, Datto MB, Wong C, Frederick JP, Liberati NT, Wang XF (1997) Tumor suppressor Smad4 is a transforming growth factor beta-inducible DNA binding protein. Mol Cell Biol 17:7019–7028

Yost C, Torres M, Miller JR, Huang E, Kimelman D, Moon RT (1996) The axis-inducing activity, stability, and subcellular distribution of beta-catenin is regulated in Xenopus embryos by glycogen synthase kinase 3. Genes Dev 10:1443–1454

Zawel L, Dai JL, Buckhaults P, Zhou S, Kinzler KW, Vogelstein B, Kern SE (1998) Human Smad3 and Smad4 are sequence-specific transcription activators. Mol Cell 1:611–617

Zhu Y, Richardson JA, Parada LF, Graff JM (1998) Smad3 mutant mice develop metastatic colorectal cancer. Cell 94:703–714

Section II
Pathways to Colorectal Cancer

Chapter 3
The Chromosomal-Instability Pathway and *APC* Gene Mutation in Colorectal Cancer

Robert Gryfe

Introduction

Cancer is fundamentally a disease in which the clonal accumulation of genetic alterations by the cell allows uncontrolled growth, evasion of cell death, local invasiveness, and metastatic potential (Fearon and Vogelstein 1990; Nowell 1976; Vogelstein and Kinzler 2004; Vogelstein et al. 1988). No cancer better exemplifies our current knowledge of the molecular genetic basis of neoplasia than cancer of colon and rectum. The progressive accumulation of point mutations in genes such as *APC*, *K-Ras*, and *p53*, in addition to larger genetic losses in chromosome arms 5q, 17p, and 18q not only elucidates specifically the adenoma to carcinoma pathway of colorectal cancer (Vogelstein et al. 1988), but also serves as a model for the generalized cancer concepts of genomic instability and the somatic evolution of neoplasia. In this chapter, we will discuss one form of proposed genomic instability observed in colorectal cancer, chromosomal instability, with specific emphasis on the relationship of Adenomatous Polyposis Coli (*APC*) gene mutation and function with this instability pathway.

It is widely accepted that a significant number of genetic alterations are required for cancer initiation and progression (Delattre et al. 1989; Fearon and Vogelstein 1990; Vogelstein et al. 1988). In a general sense, genetic alterations in cancer have been observed to occur "macroscopically" as alterations in chromosome number and structure (Boveri 1914; Law et al. 1988; Vogelstein et al. 1989) and "microscopically" as nucleotide changes involving individual genes (Bos et al. 1987; Forrester et al. 1987). Similarly, both macro- and microepigenetic alterations have been observed in human cancers (Goelz et al. 1985; Greger et al. 1989). Basal mutation rates appear to be insufficient to account for the 6,000–11,000 somatic alterations experimentally estimated to be present in a colon-cancer cell genome (Stoler et al. 1999; Wang et al. 2002) and has prompted the hypothesis that widespread genomic

R. Gryfe
Department of Surgery, Samuel Lunenfeld Research Institute, Mount Sinai Hospital, University of Toronto
e-mail: RGryfe@mtsinai.on.ca

J.D. Potter and N.M Lindor (eds.), *Genetics of Colorectal Cancer*,
DOI: 10.1007/978-0-387-09568-4_4,

(or epigenomic) instability is an essential early step in carcinogenesis (Loeb 1991; Loeb et al. 1974). The proposed inherent defect that makes cancer cells susceptible to genomic instability is often referred to as the mutator phenotype. There now appear to be at least three distinct mutator-phenotype pathways in colorectal and other cancers – the microsatellite instability (MSI) and CpG island methylator (CIMP) pathways, covered in other chapters, and the chromosomal-instability pathway, reviewed here.

Evidence for the Existence of a Chromosomal-Instability Pathway

Chromosomal instability is defined as an increased rate of loss or gain of large portions of chromosomes, or whole chromosomes in cancer (Rajagopalan et al. 2003). The majority of colorectal cancers are aneuploid, consistent with a chromosomal-instability pathway (Goh and Jass 1986). Colorectal cancers have long been known to harbor both widespread and frequent allelic losses at numerous chromosomal arms, most notably 5q, 17p, and 18q (Fearon and Vogelstein 1990; Vogelstein et al. 1988, 1989). Although abnormal chromosome number or content may be observed as the end result of chromosomal instability, the frequent observation of cancer-cell aneuploidy itself does not prove the existence of a chromosomal-instability pathway. Aneuploidy could theoretically arise by mechanisms other than chromosomal instability. In contrast to a dynamic, rate defined, chromosomal-instability mechanism, aneuploidy could arise from clonal selection and expansion of cells with a normal baseline rate of chromosomal changes, but an increased rate of replication (Rajagopalan et al. 2003) or, alternatively, as a result of exposure of cells to either an endogenous or exogenous force that creates a stable, but abnormal chromosomal content at a single point in time (Li et al. 2000). Furthermore, aneuploidy could result from a basal rate of chromosomal alteration that, in a normal cell, leads to cell death but is tolerated and clonally expanded in a cancer cell (Rajagopalan et al. 2003).

To date, only a limited number of studies have documented an increased dynamic rate of chromosomal alteration in aneuploid human colon-cancer cells (Lengauer et al. 1997; Phear et al. 1996). For the *APRT* locus on chromosome 16q, Phear et al. (1996) observed loss of heterozygosity (LOH) at a rate ten times higher ($\sim 6 \times 10^{-6}$) in the aneuploid SW460 human colon-cancer cell line than that in the near-diploid DLD1 colon-cancer cell line ($\sim 6 \times 10^{-7}$) (Phear et al. 1996). Furthermore, new LOH events observed in aneuploid SW460 cells involved most, or all, of chromosome 16q compared to smaller losses of 16q heterozygosity in DLD1 cells. Similarly, using fluorescent in situ hybridization (FISH) detection of centrosome probes on ten different chromosomes, Lengauer et al. (1997) established a chromosomal gain or loss rate of 0.01 per chromosome per generation in four aneuploid human colon-cancer cell lines (HT29, SW480, SW837, LoVo), whereas the rate of chromosomal instability in four near-diploid human colorectal-cancer cell lines (HCT116, DLD1, RKO, SW48) was too low to be measured accurately (Lengauer et al.

1997). Fusion of aneuploid HT29 cells with near-diploid DLD1 colorectal-cancer cells corrected the rate of microsatellite instability inherent in the DLD1 cells, but not the chromosomal instability inherent in the HT29 cells. Thus, in comparison to a recessive cellular predisposition to microsatellite instability, the underlying cause of chromosomal instability appeared to be dominant. Furthermore, fusion of two near-diploid colorectal-cancer cell lines (i.e., HCT116 × HCT116, DLD1 × DLD1 and HCT116 × DLD1) did not produce chromosomal instability. Similarly, transfection of an extra chromosome 3, alone, into HCT116 cells did not produce chromosomal instability, implying that abnormal chromosomal content by itself is not the cause of chromosomal instability (Lengauer et al. 1997).

Chromosomal instability is thought to arise early in colorectal neoplastic progression. Consistent with this hypothesis, LOH has been observed in dysplastic human colorectal aberrant crypt foci (ACF) and minute adenomatous polyps (Luo et al. 2006; Vogelstein et al. 1988). In older studies using flow cytometry, 6–27% of colorectal adenomas were classified as aneuploid (Goh and Jass 1986; Quirke et al. 1986; van den Ingh et al. 1985). In more recent studies, using the combination of more sensitive molecular techniques and microdissected or laser-captured specimens, a number of investigators have demonstrated that more than 85% of adenomatous polyps display insertions and deletions of genetic material ranging in size from hundreds of bases to entire chromosomal arms (Cardoso et al. 2006; Shih et al. 2001; Stoler et al. 1999). In a study by Shih et al. (2001), 88% of 1- to 3-mm sporadic adenomas with low-grade dysplasia showed allelic imbalance at 1–2 of five chromosome arms (5q, 1p, 8p, 15q, 18q) analyzed, consistent with chromosomal instability as a very early event in tumorigenesis (Shih et al. 2001). After excluding gains or losses surrounding the APC gene (on 5q), 66% of adenomas displayed allelic imbalance involving other loci. Similarly, Cardoso et al. (2006) used genome-wide detection with array comparative genomic hybridization (CGH) in analyzing 80 adenomas with low-grade dysplasia retrieved from 8 individuals with familial adenomatous polyposis (FAP) and 5 patients with MYH-associated polyposis (MAP) (Cardoso et al. 2006): genomic imbalances were observed in 53% of adenomas from patients with FAP and 92% from those with MAP. FAP adenomas were observed to have an average of 8.2 chromosomal losses or gains, including 2.7 complete chromosomal arm aberrations, whereas MAP adenomas displayed an average of 13.4 chromosomal alterations, including 5.9 whole-arm events. Chromosomal instability was also detected by array CGH in a small number of histologically normal colonic epithelial samples adjacent to adenomas, but not in nonadjacent, normal samples.

Although aneuploidy was detected in a substantial number of early adenomas in some studies (Goh and Jass 1986; Quirke et al. 1986; van den Ingh et al. 1985), not all recent analyses have demonstrated similar results (Haigis et al. 2002; Sieber et al. 2002). In a study by Sieber et al. (2002) evidence for chromosomal instability was sought in 55 adenomas from 18 patients with FAP using a combination of flow cytometry, LOH microsatellite marker analysis, and/or fluorescent CGH (Sieber et al. 2002). Whereas chromosome 5q LOH was detected in 60% of samples, other forms of chromosomal losses or gains were observed in only a small proportion of adenomas – 3/20 by flow cytometry, 2/49 by chromosome 15q LOH, 1/20 by chromosome 1p LOH, and 0/5 by CGH. Similarly, Haigis et al.

(2002) did not detect tumor chromosomal instability using FISH analysis of two chromosomes (chromosomes 7 and 18) in six human adenomas with paired normal tissue (Haigis et al. 2002). It is important to note that, compared to studies in which chromosomal alterations were detected in small adenomas (Cardoso et al. 2006; Shih et al. 2001; Stoler et al. 1999), neither of these studies (Haigis et al. 2002; Sieber et al. 2002) studied microdissected tissues, and thus the presence of normal cells could have hindered detection of aneuploidy.

The use of mathematical models that include mutational data provides further theoretical evidence for the existence of a chromosomal-instability pathway (Komarova and Wodarz 2004; Michor et al. 2005; Nowak et al. 2002). Allelic loss not only affords a cancer cell a potentially advantageous mechanism for losing a copy of a tumor-suppressor gene, but may lead to cell death by loss of genetic content essential to cell viability. Models that considered both beneficial and deleterious possibilities of chromosomal loss have concluded that:

1. The observed chromosomal gain or loss rate of 0.01 per chromosome per generation (Lengauer et al. 1997) closely mirrors the optimal theoretical allelic loss rate of tumor-suppressor genes (Komarova and Wodarz 2004).
2. Chromosomal instability probably follows point mutation inactivation of the first allele of a cancer-initiating tumor-suppressor gene, such as *APC*, and leads to loss of the second allele if there is significant selective cost to the chromosomal-instability pathway (Nowak et al. 2002).
3. Chromosomal instability may precede inactivation of either copy of a cancer-initiating tumor-suppressor gene, such as *APC*, if the chromosomal-instability pathway is effectively neutral or beneficial for cell viability (Nowak et al. 2002).
4. Chromosomal instability is probably required to initiate carcinogenesis in most circumstances unless mutation of a *single* copy of an initiating tumor-suppressor gene, such as *APC*, is sufficient to increase cellular proliferation (Michor et al. 2005).
5. Chromosomal instability is probably required to initiate carcinogenesis if allelic loss of two or more tumor-suppressor loci such as 5q, 17p, and 18q are rate limiting in cancer formation (Michor et al. 2005).

The Genetic Basis of the Chromosomal-Instability Pathway

In comparison to the microsatellite instability pathway, where a deficiency in DNA mismatch repair has been firmly established as an underlying mechanism (Gryfe 2006) (and see Chap. 6), the cause of chromosomal instability in colorectal cancer remains enigmatic. Analyses of both inherited cancer syndromes and nonfamilial cancers have been undertaken to investigate the genetic basis of chromosomal instability. Numerous mechanisms could theoretically contribute to chromosomal instability, including deregulation of: mitotic and cell-cycle checkpoints, telomere shortening and telomerase expression, centrosome number, double-strand break repair, kinetochore function, and chromatid separation (Lengauer et al. 1997; Wang et al. 2004).

Table 3.1 Inherited cancer syndromes with chromosomal instability (Adapted from Eyfjord and Bodvarsdottir 2005; Heinen et al. 2002)

Syndrome	Gene	Repair deficiency	Primary cancer predisposition
Ataxia telangectasia	ATM	Double-strand break repair	Lymphoma, leukemia
Bloom	BLM	Homologous recombination, sister-chromatid exchange	Many, including colorectal
Familial breast cancer	BRCA2	Homologous recombination, repair of crosslinks	Breast
Familial breast-ovarian cancer	BRCA1	Homologous recombination	Breast, ovary
Fanconi anemia	FANC-A, -C, -G	Repair of crosslinks, homologous recombination	Leukemia
Li Fraumeni	p53	Multiple DNA damage responses	Sarcoma, breast
Nijmegen breakage	NBS1	Double-strand break repair	Lymphoma
Werner	WRN	Homologous recombination, sister chromatid exchange	Many

Table 3.2 Proposed genetic causes of colorectal cancer chromosomal instability (see text for details)

Gene	Function
BUB1	Mitotic-spindle checkpoint
MAD2	Mitotic-spindle checkpoint
CDC4	cyclin E regulator
Aurora-A	Mitotic-spindle checkpoint
APC	Mitotic-spindle assembly, mitotic-spindle checkpoint

The molecular basis of a number of inherited cancer syndromes associated with chromosomal instability has been identified (Table 3.1) (Eyfjord and Bodvarsdottir 2005; Heinen et al. 2002). However, with the possible exception of Bloom syndrome (where there is an impairment in sister-chromatid exchange and chromosome breakage resulting from mutations in the BLM gene encoding a rec-Q DNA helicase), colorectal neoplasia is not a feature of inherited chromosomal-instability cancer syndromes (Eyfjord and Bodvarsdottir 2005; Lowy et al. 2001). Furthermore, with the exception of the familial breast cancer associated with inherited truncation mutations in *BRCA1* and *BRCA2* (both double-strand break repair genes), these syndromes are very rare and have not yet served to elucidate a common mechanism of chromosomal instability in nonfamilial cancer.

As described later, there is now increasing evidence that somatic mutations present in colorectal cancers may play a causative role in chromosomal instability (Table 3.2). The basis for most plausible causes of colorectal cancer chromosomal instability appears to involve direct disruption of regulation of the mitotic spindle. The mitotic spindle is part of the eukaryotic cell cytoskeleton that aligns and separates replicated chromosomes (sister chromatids) into daughter cells during

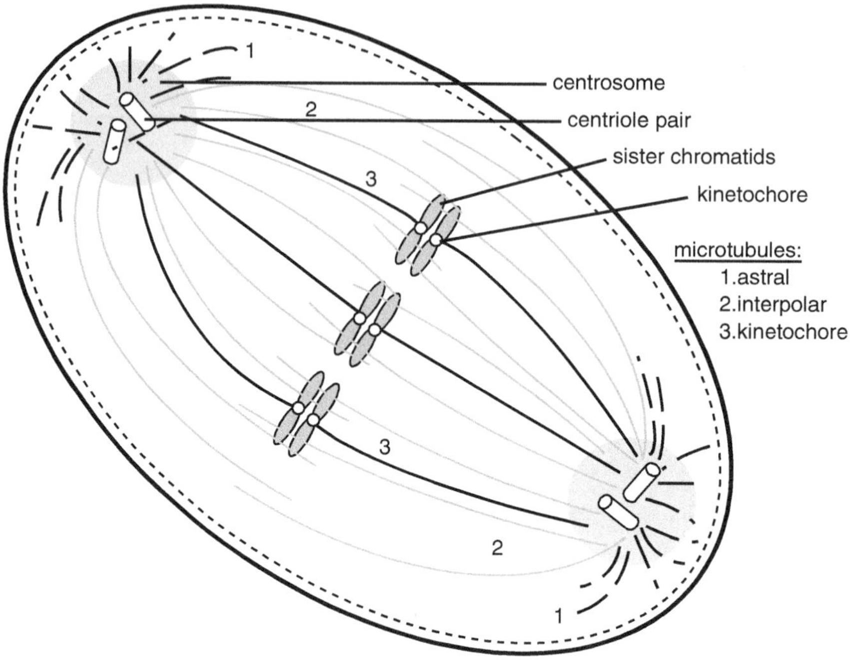

Fig. 3.1 The mitotic spindle

mitosis (Fig. 3.1). Mitotic-spindle arrest, due to abnormal chromosomal alignment, is dependent on the activity of a number of kinetochore-associated proteins including BUB1, BUB3, and MAD2 (Orr-Weaver and Weinberg 1998). Nocodazole is a microtubule-polymerizing agent that leads to microtubule disruption, subsequent chromosomal misalignment, and ultimately mitotic-spindle arrest. Nocodazole-treated aneuploid colon-cancer cell lines have been observed to be defective in this mitotic-spindle checkpoint arrest, whereas near-diploid cancer cell lines arrest appropriately (Cahill et al. 1998). Correspondingly, heterozygous splice-site mutations of the mitotic-spindle assembly checkpoint gene, *BUB1*, have been observed in a small number (2/19) of aneuploid colorectal-cancer cell lines and result in expression of a truncated, as well as a wild-type, protein (Cahill et al. 1998). *BUBR1*, a *BUB1* homolog, was observed to be mutated in an additional 2/19 colorectal cancers with chromosomal instability. Expression of either of the two identified *BUB1* mutants, in near-diploid colon-cancer cells (HCT116 or DLD1) that had wild-type *BUB1* expression, disrupted mitotic checkpoint arrest, consistent with a dominant effect. Similar involvement of somatic *BUB1* defects in a small proportion of colon cancers has been observed by others (Shichiri et al. 2002).

Similar to *BUB1*, loss of the *MAD2* mitotic-spindle checkpoint gene has been shown experimentally to cause chromosomal instability in colon-cancer cells (Michel et al. 2001). Michel et al. (2001) generated $MAD2^{+/-}$ HCT116 colon-cancer cells and observed that they did not undergo mitotic-spindle checkpoint arrest with nocodazole treatment. Haploinsufficient *MAD2*$^{+/-}$ HCT116 cells showed an 80% increase in aneuploid metaphases and a 100% increase in chromosomal loss rate compared to wild-type cells. Similar results were observed for murine embryonic fibroblasts derived from *Mad2*$^{+/-}$ mice and these mice developed lung cancers. However, despite these results, the role of MAD2 in colorectal cancer remains ambiguous as no mutations of this gene have been observed in colorectal cancer and expression appears to be significantly *increased*, not decreased, in many colorectal cancers (Cahill et al. 1999; Li et al. 2003).

The cyclin E regulator, *CDC4* (also known as Fbw7) gene, has been observed to be somatically mutated in cancer (Akhoondi et al. 2007; Rajagopalan et al. 2004). Normally, CDC4 participates in ubiquitin-mediated proteolysis of cyclin E and regulation of the G1-S cell-cycle checkpoint. *CDC4* is somatically mutated in 6–10% of colorectal cancers, a similar proportion of adenomas (Akhoondi et al. 2007; Kemp et al. 2005; Rajagopalan et al. 2004), and a number of other malignancies, including T-cell acute lymphocytic leukemia (31%) and cancers of the bile duct (35%), stomach (15%), pancreas (9%), and endometrium (9%) (Akhoondi et al. 2007). Furthermore, the majority of *CDC4* mutations identified to date have involved specific hotspots (Arg$^{465,\ 479,\ 224,\ 278,\ \text{and}\ 393}$ and Ser^{582}). These commonly mutated hotspot amino acids are key to normal CDC4 function. The majority of these mutations are C to T, or G to A transitions, the same CpG island sites that are prone to methylation. Frequent methylation of Arg^{479} has been observed in cancer, raising the possibility that epigenetic disruption of CDC4 function may also contribute to chromosomal instability (Akhoondi et al. 2007).

From a functional standpoint, biallelic *CDC4* knockout experiments on near-diploid HCT116 and DLD1 colon-cancer cells resulted in accumulation of cyclin E, nuclear atypia (micronuclei), and aneuploidy (Rajagopalan et al. 2004). Concurrent cyclin E knockdown in *CDC4*$^{-/-}$ cells abrogated chromosomal instability, whereas overexpression of cyclin E in the presence of normal *CDC4* recapitulated the *CDC4*$^{-/-}$ chromosomal-instability phenotype, implying an essential role for cyclin E in CDC4-deficient chromosomal instability. However, some authors have challenged the role of CDC4 as an important cause of chromosomal instability as the majority of mutations reported to date appear to involve only a single *CDC4* allele (Kemp et al. 2005), whereas experimental evidence appears to require biallelic inactivation of this gene (Rajagopalan et al. 2004). Nonetheless, recent experiments may support a functionally dominant role for *CDC4* mutation as coexpression of both mutant and wild-type *CDC4* in chromosomally stable HCT116 colon-cancer cells resulted in marked accumulation of cyclin E compared to wild-type cells (Akhoondi et al. 2007). While the effects of *CDC4* mutation on cyclin E accumulation appeared to act dominantly in these experiments, the authors did not report any direct evidence that they had generated chromosomal instability in these HCT116 cells with both mutant and wild-type *CDC4* expression.

Aurora-A (also known as *Aurora2* and *STK15*), another mitotic-spindle checkpoint gene, has been observed to be amplified in 30–50% of colorectal cancers as well as in other neoplasms such as cancers of breast, ovary, pancreas, prostate, head and neck, and cervix and chondrosarcoma (Bischoff et al. 1998; Nishida et al. 2007; Zhou et al. 1998). Amplification of the Aurora-A gene has been observed to be associated with overexpression of both m-RNA and protein (Bischoff et al. 1998; Zhou et al. 1998). By immunofluorescence microscopy, Aurora-A was seen to localize to the mitotic-spindle centrosome in HeLa cells (Zhou et al. 1998) and overexpression of human *Aurora-A* in Rat1 (rat embryo) and NIH 3T3 (mouse embryonic) fibroblasts caused malignant transformation as assessed by growth in soft agar and nude mice (Bischoff et al. 1998; Zhou et al. 1998). Transient *Aurora-A* overexpression in near-diploid MCF10A human breast-cancer cells leads to centromere-number and -distribution abnormalities as well as aneuploidy (Zhou et al. 1998). Although amplification of the Aurora-A gene is common in colorectal and other cancers and appears to lead to chromosomal instability, the underlying mechanism that accounts for genetic amplification in cancer remains unclear.

In addition to these examples, mutational analyses of a large number of putative instability genes characterized in model systems have revealed other possible human colorectal-cancer chromosomal-instability genes. Wang et al. (2004) sequenced the open reading frame of 100 candidate genes in 24 colorectal cancers and matched normal tissue (Wang et al. 2004). Genes found to harbor somatic mutations were analyzed in an additional 168 colorectal cancers. The DNA double-strand-break gene, *MRE11*, and a putative anaphase inhibitor gene, *Ding*, were mutated in approximately 4% of colon cancers each, whereas mutations in three putative spindle checkpoint genes (*ZW10*, *ZWILCH*, and *ROD*) were observed in a total of 2% of samples. Further studies will be required to establish the functional importance of these observations in colorectal-cancer chromosomal instability.

The p53 transcription factor has been called the "guardian of the genome," and inactivation of p53 would appear to be an excellent candidate as the cause of chromosomal instability in colorectal and other cancers (Lane 1992). *p53* is the most frequently mutated tumor-suppressor gene in all cancers, including colorectal cancers (Vogelstein and Kinzler 2004). Inactivation of p53 leads to cell-cycle checkpoint failure and evasion of apoptosis in the presence of DNA damage (Duensing and Duensing 2005). However, whereas *p53* mutations are common in aneuploid cancers, loss of p53 probably plays an important role in tolerating (as opposed to generating) DNA damage, including chromosomal instability (Duensing and Duensing 2005). Furthermore, *p53* mutation and chromosome 17p allelic loss have consistently been observed to be later events in the colorectal adenoma-to-carcinoma sequence and are therefore unlikely to be the primary cause of chromosomal instability (Fearon and Vogelstein 1990; Vogelstein et al. 1988).

APC Mutation and Chromosomal Instability

Truncating germline mutations of the *APC* tumor-suppressor gene on chromosome 5q are responsible for FAP (Groden et al. 1991; Kinzler et al. 1991a), and somatic mutations of this gene are believed to initiate the majority of sporadic colorectal adenomas and carcinomas (Kinzler et al. 1991b; Miyaki et al. 1994; Miyoshi et al. 1992; Powell et al. 1992). This has led researchers to dub APC as the "gatekeeper" of colorectal neoplasia (Kinzler and Vogelstein 1996). Whereas the most firmly established role of *APC* mutation in colorectal neoplasia relates to its involvement in β-catenin stabilization and upregulation of canonical WNT signaling (reviewed in (Polakis 1997, 2007)), recent work has intriguingly pointed to *APC* as a "caretaker" gene probably playing a central mechanistic role in chromosomal instability (Pellman 2001).

Both indirect and direct scientific evidence have emerged to link *APC* mutation with chromosomal instability in colorectal cancer. As described later, indirect evidence for this association can be drawn from studies elucidating the functional domains of the APC protein in addition to mutational analyses of both *APC* and *CTNNB1* (which encodes β-catenin). More recently, reverse-genetic approaches, in both human and model organism cell systems, have provided more direct evidence for *APC* mutation as a causative factor in colorectal cancer chromosomal instability.

Early clues to possible APC involvement in chromosomal instability came from elucidating the many functional domains of the *APC* gene. The C terminus of APC possesses three cytoskeletal-interacting domains (Fig. 3.2):

1. MT (microtubule)-binding domain (Munemitsu et al. 1994; Smith et al. 1994)
2. EB1 (microtubule end-binding protein)-binding domain (Su et al. 1995)
3. DLG (discs large gene)-binding domain (Matsumine et al. 1996)

The vast majority of *APC* mutations, in both FAP and nonfamilial neoplasia, are predicted to be truncating and cluster between codons 1286 and 1513 and thus lead to loss of these C terminus cytoskeletal-interacting domains (Miyaki et al. 1994; Miyoshi et al. 1992). Given the importance of microtubules and other cytoskeletal structures in normal mitosis and chromosomal integrity maintenance, loss of the MT-, EB1-, or DLG-binding domains all serve as reasonable candidates for chromosomal instability from an ontological standpoint. In early experiments,

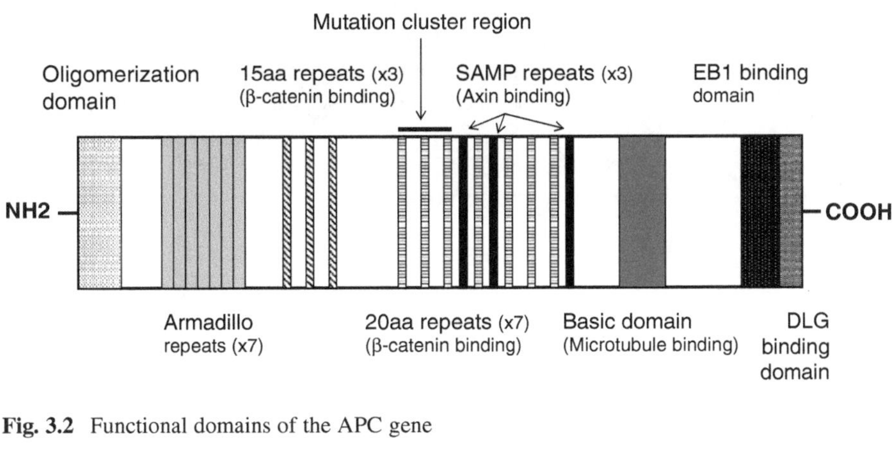

Fig. 3.2 Functional domains of the APC gene

wild-type, but not truncated, APC was observed to colocalize with microtubules to form a filamentous network with cell cytoplasm in vivo (Munemitsu et al. 1994; Smith et al. 1994). Furthermore, inhibition of microtubules by nocodazole resulted in wild-type APC becoming diffusely cytoplasmic (Smith et al. 1994). Furthermore, C terminus, but not N terminus APC protein fragments, promoted microtubule assembly in vitro (Munemitsu et al. 1994). Further, early experiments established the APC C terminus binding partner, EB1, as a mitotic-spindle checkpoint gene in yeast (Muhua et al. 1998). EB1 was shown to localize to spindle microtubules, and mutation of *EB1* resulted in failure of cell-cycle arrest in the presence of a misaligned mitotic spindles.

In accordance with Knudson's two-hit hypothesis, the majority of early colorectal neoplasms have been observed to harbor inactivating mutations of both *APC* alleles (Albuquerque et al. 2002; Lamlum et al. 1999; Miyaki et al. 1994; Miyoshi et al. 1992; Rowan et al. 2000). According to Knudson's hypothesis, these two events should be independent of one another with the end result being loss of tumor-suppressor function. However, data from both FAP and sporadic tumors indicate that APC genetic alterations are nonrandom in nature (Albuquerque et al. 2002; Lamlum et al. 1999; Rowan et al. 2000; Smits et al. 2000). The nonrandom nature of these genetic alterations appears to be related to loss or retention of specific numbers of the seven 20-amino acid (aa) repeats that act as β-catenin-binding domains in the APC protein (Fig. 3.2). These 20-aa repeats are critical to wild-type APC-mediated degradation of β-catenin (Munemitsu et al. 1995). When the 20-aa repeats are lost with typical truncating *APC* mutations, β-catenin stabilization and accumulation occurs. Furthermore, the nature of the first *APC* mutation appears to dictate both the nature and the type of second *APC* "hit" (Albuquerque et al. 2002; Lamlum et al. 1999; Rowan et al. 2000): specifically, truncating *APC* mutations between codons 1,194 and 1,392 have been associated with loss of the entire second *APC* allele (LOH), whereas mutations that are either 5' or 3' to these codons were accompanied by frameshift mutations that incorporated specific numbers of 20-aa repeats. This observed *APC* mutation scheme has been referred to as the "just-right" signaling model and is summarized in Table 3.3.

Much of the focus of this "just right" hypothesis is based on the plausible role of the specific *APC* truncation plus retention of one 20-aa repeat optimizing canonical WNT signaling for neoplastic initiation and/or progression (Albuquerque et al. 2002). However, because mutations near codon 1300 are specifically and nonrandomly accompanied by second-allele loss rather than second-allele mutation, it remains entirely plausible that specific *APC* mutation plays a critical role in initiating chromosomal instability in colorectal tumors.

Table 3.3 "Just-right" signaling: the link between the first and second *APC* hits

APC mutation	Second *APC* hit
Retention of none of the 20-aa repeats	Mutation with retention of one 20-aa repeat
Retention of one 20-aa repeats	Loss of heterozygosity
Retention of two or more 20-aa repeats	Mutation removing all 20-aa repeats

Similar to the nonrandom relationship of first and second hits in *APC*, the nonrandom association of WNT pathway deregulation with the mutator phenotype serves to provide a further potential link between *APC* and chromosomal instability. Approximately 85% of colorectal neoplasms harbor truncating *APC* mutations, and approximately half of the remainder display oncogenic exon 3 *CTNNB1* gene mutations that results in β-catenin stabilization and constitutive overexpression in the absence of an *APC* loss-of-function mutation (Huang et al. 1996; Sparks et al. 1998). *APC* and *CTNNB1* mutations appear to be mutually exclusive and, furthermore, there are significant correlations between *APC* mutations and aneuploidy, whereas *CTNNB1* mutations appear to be exclusively associated with near-diploid cancer status (Gayet et al. 2001; Sparks et al. 1998). Furthermore, the *APC* mutations identified in a subset of near-diploid colorectal cancers are genetically distinct from those observed in aneuploid cancers and therefore may also be functionally distinct (Huang et al. 1996). Taken together, these results provide circumstantial evidence that canonical WNT pathway overexpression, through either *APC* or *CTNNB1* mutation, acts as the gatekeeper to colorectal neoplasia, whereas *APC* mutation with retention of one 20-aa repeat specifically plays a caretaker role in initiating chromosomal instability.

Although earlier experiments had raised the possibility, clearer realization that *APC* mutation was mechanistically linked to chromosomal instability came from two laboratories in 2001, a decade after *APC* was initially cloned (Fodde et al. 2001; Kaplan et al. 2001). Both Kaplan et al. (2001) and Fodde et al. (2001) observed a high rate of karyotypic abnormalities in Min ($Apc^{+/-}$) mouse embryonic stem (ES) cells, but not in wild-type $Apc^{+/+}$ cells (Fodde et al. 2001; Kaplan et al. 2001). Similarly, using human HeLa, mouse ES, or marsupial PtK cells, both groups showed that wild-type *APC* associated with the plus-end of microtubules and colocalized with the mitotic-spindle kinetochore and, further, that this association was disrupted by nocodazole or colcemid, indicating that microtubules were required for APC-kinetochore interaction.

In relation to mitotic-checkpoint dynamics, Kaplan et al. (2001) observed that wild-type APC-microtubule colocalization occurred adjacent to the mitotic-checkpoint protein, Bub3, and APC coimmunoprecipitated with Bub1 and Bub3 in mitotically arrested cells (Kaplan et al. 2001). Moreover, recombinant experiments provided evidence that Bub1–Bub3 kinase complexes specifically phosphorylated APC in vitro and that APC, phosphorylated by its protein kinase partner, GSK3β, provided a better substrate for Bub1–Bub3 phosphorylation than unphosphorylated APC.

In contrast to this association of wild-type APC with mitotic-spindle microtubules, Fodde et al. (2001) observed that in $Apc^{+/-}$ ES mouse cells, mutant APC no longer localized to the kinetochore and that staining with antibodies against tubulin or EB1 demonstrated randomly projected microtubules (Fodde et al. 2001). Similar to earlier experiments establishing the dominant phenotype of chromosomal-instability experiments (Akhoondi et al. 2007; Cahill et al. 1998; Lengauer et al. 1997; Michel et al. 2001), Fodde et al. (2001) stably transfected the near-diploid, *APC* wild-type, HCT116 human colorectal-cancer cell line with an inducible, truncated *APC* construct and observed that induction caused a 2.5- to 5-fold increase in numerical chromosomal aberrations (Fodde et al. 2001). Because HCT116 colon-cancer cells harbor an oncogenic *CTNNB1* alteration and not an *APC* mutation (Sparks et al. 1998), these results provide genetic evidence that truncating *APC* mutations specifically, but not canonical WNT overexpression in general, lead to chromosomal instability. Taken together, these studies by Kaplan et al. (2001) and Fodde et al. (2001) established an association between wild-type APC, plus-end microtubules of the mitotic-spindle kinetochore, EB1, and the mitotic checkpoint proteins BUB1 and BUB3. Truncating *APC* mutations appeared to lead to interruption of mitotic-spindle dynamics and resulted in chromosomal abnormalities (Fodde et al. 2001; Kaplan et al. 2001).

Following initial observations of APC-mitotic spindle interactions, several more recent experiments have provided further insight into the relationship of *APC* mutation, the mitotic spindle, and chromosomal instability. Dikovskaya et al. (2004) studied mitotic-spindle formation in Xenopus (frog) egg extracts (Dikovskaya et al. 2004). Cytostatic factor (CSF) Xenopus egg extracts are arrested in metaphase of meiosis II. APC depletion in CSF Xenopus egg extracts resulted in decreased mitotic microtubule density and an abnormal distribution of microtubules compared to identical non-APC-depleted Xenopus egg extracts, suggesting that spindles formed in the absence of APC contain decreased amounts of inappropriately distributed tubulin. Similar to previous experiments, wild-type APC localized to the kinetochore in Xenopus egg extracts; results of this study suggested that APC specifically controlled centrosome-directed spindle formation. The abnormal spindle phenotype of APC-depleted CSF Xenopus egg extracts could be rescued by the endogenous expression, or the exogenous addition, of full-length APC, but not N-terminally truncated APC fragments that lacked the microtubule (MT)-binding site, indicating that this APC domain is required for correct spindle formation.

Using quantitative immunofluorescent microscopy, Green and Kaplan carefully characterized the mitotic spindle in *APC*-mutant colon-cancer cells with chromosomal instability (SW480, HT29, Caco, LoVo), and *APC*-wild-type colon-cancer cells without chromosomal instability (HCT116, RKO) (Green and Kaplan 2003). Wild-type APC normally localized with the plus-end of microtubules; however, when chromosomally stable 293 (also known as HEK; human embryonic kidney) cells with wild-type *APC* were transfected with an N terminus *APC* (N-APC$^{1-1450}$ encoding *APC* codons 1–1450) fragment expression vector, there was direct and dominant interference with mitotic spindle and kinetochore microtubule plus-end attachments. The 293 cells expressing the truncated N-APC$^{1-1450}$ quantitatively resembled colon-cancer cells with chromosomal instability in a variety of mitotic-spindle assays: increased collapsed mitotic spindles, decreased mitotic spindle pole-to-pole length, increased chromosomal width to height (congression index), and increased kinetochore localization of BubR1 (indicative of an aberrant mitotic-spindle checkpoint). This occurred despite the presence of equal or excess wild-type APC expression in these cells, implying a dominant role for N-APC$^{1-1450}$. In contrast to these results, transfection of 293 cells with expression vectors for either wild-type APC or a C terminus APC$^{2560-2843}$ fragment (that retains the EB1 microtubule-binding domain) did not cause interference of mitotic-spindle microtubule plus-end attachments. Although the effects of N-APC$^{1-1450}$ on the mitotic spindle appeared to be dominant and led to chromosomal positioning errors, expression of N-APC$^{1-1450}$ in 293 cells did not lead to measurable chromosomal instability. However, 293 cells are normally hypotriploid and have a markedly compromised mitotic-spindle checkpoint (Tighe et al. 2004).

Similar to the previously described experiments in human 293 cells (Green and Kaplan 2003), Tighe et al. expressed an N terminus fragment of APC (N-APC$^{1-750}$) in near-diploid HCT116 cells that normally express wild-type APC (Tighe et al. 2004). Compared to *APC*$^{+/+}$ HCT116 cells, N-APC$^{1-750}$ HCT116 cells were observed to have a defective mitotic-spindle checkpoints with nocadazole treatment. Following wash-out of the nocodazole at 48 h, long-term surviving N-APC$^{1-750}$ HCT116 cells became highly aneuploid. In contrast, the surviving population of similarly treated wild-type *APC* HCT116 cells remained near diploid. Similar to N-APC$^{1-1450}$ 293 cells immunofluorescent confocal microscopy results (Green and Kaplan 2003), N-APC$^{1-750}$ HCT116 cells were observed to have weakened mitotic-spindle kinetochore-microtubule interactions more typical of other colon-cancer cells with chromosomal instability than chromosomally stable, wild-type HCT116 (Tighe et al. 2004).

Recent work has further elucidated the functional dynamics of truncating *APC* mutations, *EB1*, and chromosomal instability (Green et al. 2005). Using human 293 cells, Green et al. (2005) observed that siRNA inhibition of either *APC* or *EB1* recapitulated the mitotic-spindle phenotype observed in their earlier experiments with N-APC expression. Inhibitory effects of APC and EB1 were distinct from inhibition of other microtubule dynamic-regulating (+TIPs) proteins. Furthermore, inhibition of APC and EB1 was nonadditive suggesting that these proteins work together maintaining mitotic-spindle function.

Using multiple N-APC-deletion constructs, experiments by Green et al. (2005) have genetically dissected the dominant-negative effects of *APC* truncation (Green et al. 2005). Removal of the first 58 amino acids responsible for APC oligomerization (N-APC$^{58-1450}$) eliminated the mitotic-spindle phenotype observed in 293 cells expressing N-APC$^{1-1450}$, strongly suggesting that APC oligomerization is critical for dominant negative activity. As expected from these results, N-APC$^{1-1450}$, but not N-APC$^{58-1450}$, was observed to associate with full-length APC. Further deletions of the C-terminal region of N-APC$^{1-1450}$ (N-APC$^{1-1309}$, N-APC$^{1-1020}$, N-APC$^{1-850}$) progressively reduced mitotic-spindle defects in 293 cells and the mitotic-spindle function of N-APC$^{1-768}$ 293 cells was indistinguishable from wild type. EB1 was observed to copurify with full-length APC; however, the expression N-APC$^{1-1450}$ directly eliminated the interaction between EB1 and wild-type APC. Confocal immunofluorescent microscopy indicated that the effect of N-APC$^{1-1450}$ on EB1 was to increase the number of pausing events in growing mitotic-spindle microtubules, suggesting that wild-type APC normally stimulates the antipause activity of EB1 on mitotic-spindle microtubule polymerization. These effects of N-APC$^{1-1450}$ on EB1-associated microtubule pausing were not affected by nocodazole and were partially rescued by microtubule-independent C-APC$^{2560-2843}$ (which does not contain a microtubule-binding domain), suggesting that APC normally regulates EB1 function in the cytosol, prior to association with microtubule plus-ends.

Recent studies have further established an association between *APC* deficiency and a number of other mitosis-related genes (Abal et al. 2007; Hadjihannas et al. 2006). Using FISH analysis Hadjihannas et al. (2006) observed that 60% of colorectal cancers with chromosomal instability expressed high levels of Conductin compared to 7% of colorectal cancers without chromosomal instability (Hadjihannas et al. 2006). When Conductin was expressed in near-diploid, *APC*-wild-type, HCT116 cells, 14–46% of cells exhibited chromosomal gains or losses compared to 2–7% of HCT116 with low, basal levels of Conductin expression. siRNA depletion of APC in HCT116 cells leads to upregulation of Conductin, and 12–20% of cells were observed to contain chromosomal gains or losses. Conductin levels were cell-cycle dependent and coimmunoprecipitated specifically with the known mitotic regulator PLK1, but not with other mitotic regulators (Mad2, Emi1, cdcd27, cyclin B1, Cdc2). Both Conductin and PLK1 localized to mitotic centrosomes and spindles. Expression of Conductin in HCT116, DLD1, and SW480 colon-cancer cells consistently led to significant impairment of mitotic-spindle checkpoint with nocodazole treatment (20–30% arrest in Conductin-overexpressing cells compared to >80% arrest in cells with basal Conductin expression). Conductin inhibition in chromosomally unstable SW480 cells by RNAi restored nocodazole-induced mitotic-spindle arrest in these cells.

Abal et al. (2007) performed videomicroscopy and transcriptome analysis on diploid adenomas from transgenic pVillin-KRAS^{V12G} mice expressing oncogenic human K-Ras and on adenomas from compound Apc$^{+/1638N}$/pVillin-KRAS^{V12G} mice with dominant truncating *Apc* mutations (Abal et al. 2007). Videomicroscopy revealed mitotic defects in the Apc$^{+/1638N}$/pVillin-KRAS^{V12G}, but not pVillin-KRAS^{V12G} polyps. Transcriptome analysis revealed that *Apc* mutation was statistically significantly associated with upregulation of *MAD2L1*, *BUB1B*, and *STMN1*. Both MAD2L1 and BUB1B are components of the mitotic-spindle checkpoint and STMN1 regulates microtubule polymerization. All three genes were observed to be overexpressed in nonfamilial human adenomas, carcinomas, and FAP adenomas compared to normal mucosa. Furthermore, transfection of wild-type APC in *APC*-mutant SW480 colon-cancer cells moderately reduced MAD2L1, BUB1B, and STMN1 expression as did *APC* silencing by siRNA in 293 cells, further linking APC derangement with additional important mitotic-spindle proteins.

In addition to direct effects on the mitotic spindle which appear to give rise to chromosomal instability, APC mutation has long been linked to apoptosis resistance as inducible APC expression in colon-cancer cells with an *APC* mutation triggers apoptosis (Morin et al. 1996). There is an obvious benefit for cancer cells in linking mitotic-spindle deregulation that gives rise to chromosomal misalignment with cellular resistance to the identification (which triggers cell death) of chromosomal derangements. Dikovskaya et al. (2007) have recently investigated the association of APC with both mitotic spindle and apoptotic function (Dikovskaya et al. 2007). RNAi APC-depleted U2OS human osteosarcoma cells displayed mild metaphase kinetochore tension abnormalities. Although these abnormalities were expected to lead to the accumulation of spindle-checkpoint proteins, decreased Bub1, as well as BubR1 association with kinetochores, was observed in these cells, as was an accelerated rate of mitotic progression. These APC-deficient U2OS cells displayed greater resistance to nocodazole- and taxol-induced mitotic-checkpoint arrest than APC-sufficient control cells. Inappropriate mitotic exit of APC-depleted U2OS cells leads to mitotic slippage and an increased population of tetraploid cells that had exited mitosis (as distinct from cells in M phase that normally possess double chromosome content). The number of tetraploid cells that had exited mitosis was increased even further with nocodazole or taxol treatment indicating that APC deficiency led to both mitotic-spindle-assembly abnormalities and mitotic-spindle-checkpoint abnormalities. Apoptosis was directly assessed by detection of cleaved Caspase 3. Although APC-depleted cells did undergo appropriate apoptosis after staurosporine treatment, basal levels of apoptosis and apoptotic responses to nocodazole and taxol were statistically significantly reduced in APC-depleted cells compared to controls. From these experiments, it thus appears that APC loss (depletion, as distinct from truncation) leads to a combination of defects compromising the mitotic spindle and mitotic checkpoint as well as apoptosis.

Although much of the information presented in this chapter appears to support a role for APC in chromosomal instability that is separate from its role in the canonical WNT pathway, recent evidence suggests that there may be a greater overlap in these roles of APC than has been previously appreciated (Tighe et al. 2007). Tighe et al. (2007) inhibited GSK-3β kinase activity in HeLa and DLD1 human cancer cells using a panel of small molecule inhibitors and analyzed the effects on mitosis. GSK-3β is a kinase that, in concert with APC and Axin, targets β-catenin for proteolysis in canonical WNT signaling. Compared to controls, cancer cells treated with GSK-3β inhibitors displayed delayed mitotic entry and exit, delayed chromosomal alignment, disoriented chromosomal alignment, altered mitotic spindle morphology, and, ultimately, elevated levels of chromosome missegregation. GSK-3β inhibition of DLD1 cells with siRNA similarly affected spindle morphology and chromosome alignment and segregation. Whether these mitotic-spindle-related roles of GSK3β are separate from its role in canonical WNT signaling was not addressed in this study. Furthermore, although *GSK3β* mutation does not appear to play a role in colorectal carcinogenesis and thus is unlikely to contribute to chromosomal instability, the results of this study raise concern over the safety of GSK3β inhibitors currently under investigation for diabetes and neurodegenerative disorders.

Conclusions

A number of apparently distinct pathways of genomic instability including chromosomal, microsatellite, and CpG-island-methylation instabilities appear to be critical in human carcinogenesis. The most common form of this genetic instability in colorectal cancer is chromosomal instability. Chromosomal instability is characterized by an increased rate of loss or gain of large portions of chromosomes or whole chromosomes. Approximately 85% of colorectal cancers are aneuploid and appear to be both genetically and clinically distinct from cancers that do not display this hallmark of chromosomal instability. Chromosomal instability appears to be a genetically dominant phenotype, and the underlying cause of this mutator pathway has, until recently, been largely enigmatic. A number of mitotic-spindle and cell-cycle genes, such as *BUB1*, *MAD2*, *Aurora-A*, and *CDC4*, appear to play a causative role in chromosomal instability in a subset of colorectal cancers with aneuploidy. In addition, it now appears that specific truncating mutations of the colorectal cancer gatekeeper gene, *APC*, play a critical role in establishing chromosomal instability in the majority of colorectal adenomas and carcinomas. Evidence for this relationship initially came indirectly from studies of *APC* functional domains and mutational analyses. Recently, our understanding of *APC* mutation in chromosomal instability has been advanced through both forward- and reverse-genetic experiments establishing APC as a key regulator of mitotic-spindle assembly and the mitotic-spindle checkpoint. A better understanding of this apparent causal role of *APC* mutation in colorectal cancer chromosomal instability may one day play a critical role in the development of effective molecular-based chemoprevention, screening, and therapy.

References

Abal, M., A. Obrador-Hevia, K.P. Janssen, L. Casadome, M. Menendez, S. Carpentier, E. Barillot, M. Wagner, W. Ansorge, G. Moeslein, H. Fsihi, V. Bezrookove, J. Reventos, D. Louvard, G. Capella, and S. Robine. 2007. APC inactivation associates with abnormal mitosis completion and concomitant BUB1B/MAD2L1 up-regulation. Gastroenterology 132:2448–58.

Akhoondi, S., D. Sun, N. von der Lehr, S. Apostolidou, K. Klotz, A. Maljukova, D. Cepeda, H. Fiegl, D. Dofou, C. Marth, E. Mueller-Holzner, M. Corcoran, M. Dagnell, S.Z. Nejad, B.N. Nayer, M.R. Zali, J. Hansson, S. Egyhazi, F. Petersson, P. Sangfelt, H. Nordgren, D. Grander, S.I. Reed, M. Widschwendter, O. Sangfelt, and C. Spruck. 2007. FBXW7/hCDC4 is a general tumor suppressor in human cancer. Cancer Res 67:9006–12.

Albuquerque, C., C. Breukel, R. van der Luijt, P. Fidalgo, P. Lage, F.J. Slors, C.N. Leitao, R. Fodde, and R. Smits. 2002. The 'just-right' signaling model: APC somatic mutations are selected based on a specific level of activation of the beta-catenin signaling cascade. Hum Mol Genet 11:1549–60.

Bischoff, J.R., L. Anderson, Y. Zhu, K. Mossie, L. Ng, B. Souza, B. Schryver, P. Flanagan, F. Clairvoyant, C. Ginther, C.S. Chan, M. Novotny, D.J. Slamon, and G.D. Plowman. 1998. A homologue of Drosophila aurora kinase is oncogenic and amplified in human colorectal cancers. EMBO J 17:3052–65.

Bos, J.L., E.R. Fearon, S.R. Hamilton, M. Verlaan-de Vries, J.H. van Boom, A.J. van der Eb, and B. Vogelstein. 1987. Prevalence of ras gene mutations in human colorectal cancers. Nature 327:293–7.

Boveri, T. 1914. Zur Frage der Entstehung maligner Tumoren Gustav Fischer Verlag, Jena, Germany.

Cahill, D.P., C. Lengauer, J. Yu, G.J. Riggins, J.K. Willson, S.D. Markowitz, K.W. Kinzler, and B. Vogelstein. 1998. Mutations of mitotic checkpoint genes in human cancers. Nature 392:300–3.

Cahill, D.P., L.T. da Costa, E.B. Carson-Walter, K.W. Kinzler, B. Vogelstein, and C. Lengauer. 1999. Characterization of MAD2B and other mitotic spindle checkpoint genes. Genomics 58:181–7.

Cardoso, J., L. Molenaar, R.X. de Menezes, M. van Leerdam, C. Rosenberg, G. Moslein, J. Sampson, H. Morreau, J.M. Boer, and R. Fodde. 2006. Chromosomal instability in MYH- and APC-mutant adenomatous polyps. Cancer Res 66:2514–9.

Delattre, O., S. Olschwang, D.J. Law, T. Melot, Y. Remvikos, R.J. Salmon, X. Sastre, P. Validire, A.P. Feinberg, and G. Thomas. 1989. Multiple genetic alterations in distal and proximal colorectal cancer. Lancet 2:353–6.

Dikovskaya, D., I.P. Newton, and I.S. Nathke. 2004. The adenomatous polyposis coli protein is required for the formation of robust spindles formed in CSF Xenopus extracts. Mol Biol Cell 15:2978–91.

Dikovskaya, D., D. Schiffmann, I.P. Newton, A. Oakley, K. Kroboth, O. Sansom, T.J. Jamieson, V. Meniel, A. Clarke, and I.S. Nathke. 2007. Loss of APC induces polyploidy as a result of a combination of defects in mitosis and apoptosis. J Cell Biol 176:183–95.

Duensing, A., and S. Duensing. 2005. Guilt by association? p53 and the development of aneuploidy in cancer. Biochem Biophys Res Commun 331:694–700.

Eyfjord, J.E., and S.K. Bodvarsdottir. 2005. Genomic instability and cancer: networks involved in response to DNA damage. Mutat Res 592:18–28.

Fearon, E.R., and B. Vogelstein. 1990. A genetic model for colorectal tumorigenesis. Cell 61:759–67.

Fodde, R., J. Kuipers, C. Rosenberg, R. Smits, M. Kielman, C. Gaspar, J.H. van Es, C. Breukel, J. Wiegant, R.H. Giles, and H. Clevers. 2001. Mutations in the APC tumour suppressor gene cause chromosomal instability. Nat Cell Biol 3:433–8.

Forrester, K., C. Almoguera, K. Han, W.E. Grizzle, and M. Perucho. 1987. Detection of high incidence of K-ras oncogenes during human colon tumorigenesis. Nature 327:298–303.

Gayet, J., X.P. Zhou, A. Duval, S. Rolland, J.M. Hoang, P. Cottu, and R. Hamelin. 2001. Extensive characterization of genetic alterations in a series of human colorectal cancer cell lines. Oncogene 20:5025–32.

Goelz, S.E., B. Vogelstein, S.R. Hamilton, and A.P. Feinberg. 1985. Hypomethylation of DNA from benign and malignant human colon neoplasms. Science 228:187–90.

Goh, H.S., and J.R. Jass. 1986. DNA content and the adenoma-carcinoma sequence in the colorectum. J Clin Pathol 39:387–92.

Green, R.A., and K.B. Kaplan. 2003. Chromosome instability in colorectal tumor cells is associated with defects in microtubule plus-end attachments caused by a dominant mutation in APC. J Cell Biol 163:949–61.

Green, R.A., R. Wollman, and K.B. Kaplan. 2005. APC and EB1 function together in mitosis to regulate spindle dynamics and chromosome alignment. Mol Biol Cell 16:4609–22.

Greger, V., E. Passarge, W. Hopping, E. Messmer, and B. Horsthemke. 1989. Epigenetic changes may contribute to the formation and spontaneous regression of retinoblastoma. Hum Genet 83:155–8.

Groden, J., A. Thliveris, W. Samowitz, M. Carlson, L. Gelbert, H. Albertsen, G. Joslyn, J. Stevens, L. Spirio, M. Robertson, et al. 1991. Identification and characterization of the familial adenomatous polyposis coli gene. Cell 66:589–600.

Gryfe, R. 2006. Clinical implications of our advancing knowledge of colorectal cancer genetics: inherited syndromes, prognosis, prevention, screening and therapeutics. Surg Clin North Am 86:787–817.

Hadjihannas, M.V., M. Bruckner, B. Jerchow, W. Birchmeier, W. Dietmaier, and J. Behrens. 2006. Aberrant Wnt/beta-catenin signaling can induce chromosomal instability in colon cancer. Proc Natl Acad Sci USA 103:10747–52.

Haigis, K.M., J.G. Caya, M. Reichelderfer, and W.F. Dove. 2002. Intestinal adenomas can develop with a stable karyotype and stable microsatellites. Proc Natl Acad Sci USA 99:8927–31.

Heinen, C.D., C. Schmutte, and R. Fishel. 2002. DNA repair and tumorigenesis: lessons from hereditary cancer syndromes. Cancer Biol Ther 1:477–85.

Huang, J., N. Papadopoulos, A.J. McKinley, S.M. Farrington, L.J. Curtis, A.H. Wyllie, S. Zheng, J.K. Willson, S.D. Markowitz, P. Morin, K.W. Kinzler, B. Vogelstein, and M.G. Dunlop. 1996. APC mutations in colorectal tumors with mismatch repair deficiency. Proc Natl Acad Sci USA 93:9049–54.

Kaplan, K.B., A.A. Burds, J.R. Swedlow, S.S. Bekir, P.K. Sorger, and I.S. Nathke. 2001. A role for the adenomatous polyposis coli protein in chromosome segregation. Nat Cell Biol 3:429–32.

Kemp, Z., A. Rowan, W. Chambers, N. Wortham, S. Halford, O. Sieber, N. Mortensen, A. von Herbay, T. Gunther, M. Ilyas, and I. Tomlinson. 2005. CDC4 mutations occur in a subset of colorectal cancers but are not predicted to cause loss of function and are not associated with chromosomal instability. Cancer Res 65:11361–6.

Kinzler, K.W., and B. Vogelstein. 1996. Lessons from hereditary colorectal cancer. Cell 87:159–70.

Kinzler, K.W., M.C. Nilbert, L.K. Su, B. Vogelstein, T.M. Bryan, D.B. Levy, K.J. Smith, A.C. Preisinger, P. Hedge, D. McKechnie, et al. 1991a. Identification of FAP locus genes from chromosome 5q21. Science 253:661–5.

Kinzler, K.W., M.C. Nilbert, B. Vogelstein, T.M. Bryan, D.B. Levy, K.J. Smith, A.C. Preisinger, S.R. Hamilton, P. Hedge, A. Markham, et al. 1991b. Identification of a gene located at chromosome 5q21 that is mutated in colorectal cancers. Science 251:1366–70.

Komarova, N.L., and D. Wodarz. 2004. The optimal rate of chromosome loss for the inactivation of tumor suppressor genes in cancer. Proc Natl Acad Sci USA 101:7017–21.

Lamlum, H., M. Ilyas, A. Rowan, S. Clark, V. Johnson, J. Bell, I. Frayling, J. Efstathiou, K. Pack, S. Payne, R. Roylance, P. Gorman, D. Sheer, K. Neale, R. Phillips, I. Talbot, W. Bodmer, and I. Tomlinson. 1999. The type of somatic mutation at APC in familial adenomatous polyposis is determined by the site of the germline mutation: a new facet to Knudson's 'two-hit' hypothesis. Nat Med 5:1071–5.

Lane, D.P. 1992. Cancer. p53, guardian of the genome. Nature 358:15–6.

Law, D.J., S. Olschwang, J.P. Monpezat, D. Lefrancois, D. Jagelman, N.J. Petrelli, G. Thomas, and A.P. Feinberg. 1988. Concerted nonsyntenic allelic loss in human colorectal carcinoma. Science 241:961–5.

Lengauer, C., K.W. Kinzler, and B. Vogelstein. 1997. Genetic instability in colorectal cancers. Nature 386:623–7.

Li, G.Q., H. Li, and H.F. Zhang. 2003. Mad2 and p53 expression profiles in colorectal cancer and its clinical significance. World J Gastroenterol 9:1972–5.

Li, R., A. Sonik, R. Stindl, D. Rasnick, and P. Duesberg. 2000. Aneuploidy vs. gene mutation hypothesis of cancer: recent study claims mutation but is found to support aneuploidy. Proc Natl Acad Sci USA 97:3236–41.

Loeb, L.A. 1991. Mutator phenotype may be required for multistage carcinogenesis. Cancer Res 51:3075–9.

Loeb, L.A., C.F. Springgate, and N. Battula. 1974. Errors in DNA replication as a basis of malignant changes. Cancer Res 34:2311–21.

Lowy, A.M., J.J. Kordich, V. Gismondi, L. Varesco, R.I. Blough, and J. Groden. 2001. Numerous colonic adenomas in an individual with Bloom's syndrome. Gastroenterology 121:435–9.

Luo, L., G.Q. Shen, K.A. Stiffler, Q.K. Wang, T.G. Pretlow, and T.P. Pretlow. 2006. Loss of heterozygosity in human aberrant crypt foci (ACF), a putative precursor of colon cancer. Carcinogenesis 27:1153–9.

Matsumine, A., A. Ogai, T. Senda, N. Okumura, K. Satoh, G.H. Baeg, T. Kawahara, S. Kobayashi, M. Okada, K. Toyoshima, and T. Akiyama. 1996. Binding of APC to the human homolog of the Drosophila discs large tumor suppressor protein. Science 272:1020–3.

Michel, L.S., V. Liberal, A. Chatterjee, R. Kirchwegger, B. Pasche, W. Gerald, M. Dobles, P.K. Sorger, V.V. Murty, and R. Benezra. 2001. MAD2 haplo-insufficiency causes premature anaphase and chromosome instability in mammalian cells. Nature 409:355–9.

Michor, F., Y. Iwasa, B. Vogelstein, C. Lengauer, and M.A. Nowak. 2005. Can chromosomal instability initiate tumorigenesis? Semin Cancer Biol 15:43–9.

Miyaki, M., M. Konishi, R. Kikuchi-Yanoshita, M. Enomoto, T. Igari, K. Tanaka, M. Muraoka, H. Takahashi, Y. Amada, M. Fukayama, et al. 1994. Characteristics of somatic mutation of the adenomatous polyposis coli gene in colorectal tumors. Cancer Res 54:3011–20.

Miyoshi, Y., H. Nagase, H. Ando, A. Horii, S. Ichii, S. Nakatsuru, T. Aoki, Y. Miki, T. Mori, and Y. Nakamura. 1992. Somatic mutations of the APC gene in colorectal tumors: mutation cluster region in the APC gene. Hum Mol Genet 1:229–33.

Morin, P.J., B. Vogelstein, and K.W. Kinzler. 1996. Apoptosis and APC in colorectal tumorigenesis. Proc Natl Acad Sci USA 93:7950–4.

Muhua, L., N.R. Adames, M.D. Murphy, C.R. Shields, and J.A. Cooper. 1998. A cytokinesis checkpoint requiring the yeast homologue of an APC-binding protein. Nature 393:487–91.

Munemitsu, S., B. Souza, O. Muller, I. Albert, B. Rubinfeld, and P. Polakis. 1994. The APC gene product associates with microtubules in vivo and promotes their assembly in vitro. Cancer Res 54:3676–81.

Munemitsu, S., I. Albert, B. Souza, B. Rubinfeld, and P. Polakis. 1995. Regulation of intracellular beta-catenin levels by the adenomatous polyposis coli (APC) tumor-suppressor protein. Proc Natl Acad Sci USA 92:3046–50.

Nishida, N., T. Nagasaka, K. Kashiwagi, C.R. Boland, and A. Goel. 2007. High copy amplification of the Aurora-A gene is associated with chromosomal instability phenotype in human colorectal cancers. Cancer Biol Ther 6:525–33.

Nowak, M.A., N.L. Komarova, A. Sengupta, P.V. Jallepalli, M. Shih Ie, B. Vogelstein, and C. Lengauer. 2002. The role of chromosomal instability in tumor initiation. Proc Natl Acad Sci USA 99:16226–31.

Nowell, P.C. 1976. The clonal evolution of tumor cell populations. Science 194:23–8.

Orr-Weaver, T.L., and R.A. Weinberg. 1998. A checkpoint on the road to cancer. Nature 392:223–4.

Pellman, D. 2001. Cancer. A CINtillating new job for the APC tumor suppressor. Science 291:2555–6.

Phear, G., N.P. Bhattacharyya, and M. Meuth. 1996. Loss of heterozygosity and base substitution at the APRT locus in mismatch-repair-proficient and -deficient colorectal carcinoma cell lines. Mol Cell Biol 16:6516–23.

Polakis, P. 1997. The adenomatous polyposis coli (APC) tumor suppressor. Biochim Biophys Acta 1332:F127–47.

Polakis, P. 2007. The many ways of Wnt in cancer. Curr Opin Genet Dev 17:45–51.

Powell, S.M., N. Zilz, Y. Beazer-Barclay, T.M. Bryan, S.R. Hamilton, S.N. Thibodeau, B. Vogelstein, and K.W. Kinzler. 1992. APC mutations occur early during colorectal tumorigenesis. Nature 359:235–7.

Quirke, P., J.B. Fozard, M.F. Dixon, J.E. Dyson, G.R. Giles, and C.C. Bird. 1986. DNA aneuploidy in colorectal adenomas. Br J Cancer 53:477–81.

Rajagopalan, H., M.A. Nowak, B. Vogelstein, and C. Lengauer. 2003. The significance of unstable chromosomes in colorectal cancer. Nat Rev Cancer 3:695–701.

Rajagopalan, H., P.V. Jallepalli, C. Rago, V.E. Velculescu, K.W. Kinzler, B. Vogelstein, and C. Lengauer. 2004. Inactivation of hCDC4 can cause chromosomal instability. Nature 428:77–81.

Rowan, A.J., H. Lamlum, M. Ilyas, J. Wheeler, J. Straub, A. Papadopoulou, D. Bicknell, W.F. Bodmer, and I.P. Tomlinson. 2000. APC mutations in sporadic colorectal tumors: a mutational "hotspot" and interdependence of the "two hits". Proc Natl Acad Sci USA 97:3352–7.

Shichiri, M., K. Yoshinaga, H. Hisatomi, K. Sugihara, and Y. Hirata. 2002. Genetic and epigenetic inactivation of mitotic checkpoint genes hBUB1 and hBUBR1 and their relationship to survival. Cancer Res 62:13–7.

Shih, I.M., W. Zhou, S.N. Goodman, C. Lengauer, K.W. Kinzler, and B. Vogelstein. 2001. Evidence that genetic instability occurs at an early stage of colorectal tumorigenesis. Cancer Res 61:818–22.

Sieber, O.M., K. Heinimann, P. Gorman, H. Lamlum, M. Crabtree, C.A. Simpson, D. Davies, K. Neale, S.V. Hodgson, R.R. Roylance, R.K. Phillips, W.F. Bodmer, and I.P. Tomlinson. 2002. Analysis of chromosomal instability in human colorectal adenomas with two mutational hits at APC. Proc Natl Acad Sci USA 99:16910–5.

Smith, K.J., D.B. Levy, P. Maupin, T.D. Pollard, B. Vogelstein, and K.W. Kinzler. 1994. Wild-type but not mutant APC associates with the microtubule cytoskeleton. Cancer Res 54:3672–5.

Smits, R., N. Hofland, W. Edelmann, M. Geugien, S. Jagmohan-Changur, C. Albuquerque, C. Breukel, R. Kucherlapati, M.F. Kielman, and R. Fodde. 2000. Somatic Apc mutations are selected upon their capacity to inactivate the beta-catenin downregulating activity. Genes Chromosomes Cancer 29:229–39.

Sparks, A.B., P.J. Morin, B. Vogelstein, and K.W. Kinzler. 1998. Mutational analysis of the APC/beta-catenin/Tcf pathway in colorectal cancer. Cancer Res 58:1130–4.

Stoler, D.L., N. Chen, M. Basik, M.S. Kahlenberg, M.A. Rodriguez-Bigas, N.J. Petrelli, and G.R. Anderson. 1999. The onset and extent of genomic instability in sporadic colorectal tumor progression. Proc Natl Acad Sci USA 96:15121–6.

Su, L.K., M. Burrell, D.E. Hill, J. Gyuris, R. Brent, R. Wiltshire, J. Trent, B. Vogelstein, and K.W. Kinzler. 1995. APC binds to the novel protein EB1. Cancer Res 55:2972–7.

Tighe, A., V.L. Johnson, and S.S. Taylor. 2004. Truncating APC mutations have dominant effects on proliferation, spindle checkpoint control, survival and chromosome stability. J Cell Sci 117:6339–53.

Tighe, A., A. Ray-Sinha, O.D. Staples, and S.S. Taylor. 2007. GSK-3 inhibitors induce chromosome instability. BMC Cell Biol 8:34.

van den Ingh, H.F., G. Griffioen, and C.J. Cornelisse. 1985. Flow cytometric detection of aneuploidy in colorectal adenomas. Cancer Res 45:3392–7.

Vogelstein, B., and K.W. Kinzler. 2004. Cancer genes and the pathways they control. Nat Med 10:789–99.

Vogelstein, B., E.R. Fearon, S.R. Hamilton, S.E. Kern, A.C. Preisinger, M. Leppert, Y. Nakamura, R. White, A.M. Smits, and J.L. Bos. 1988. Genetic alterations during colorectal-tumor development. N Engl J Med 319:525–32.

Vogelstein, B., E.R. Fearon, S.E. Kern, S.R. Hamilton, A.C. Preisinger, Y. Nakamura, and R. White. 1989. Allelotype of colorectal carcinomas. Science 244:207–11.

Wang, T.L., C. Rago, N. Silliman, J. Ptak, S. Markowitz, J.K. Willson, G. Parmigiani, K.W. Kinzler, B. Vogelstein, and V.E. Velculescu. 2002. Prevalence of somatic alterations in the colorectal cancer cell genome. Proc Natl Acad Sci USA 99:3076–80.

Wang, Z., J.M. Cummins, D. Shen, D.P. Cahill, P.V. Jallepalli, T.L. Wang, D.W. Parsons, G. Traverso, M. Awad, N. Silliman, J. Ptak, S. Szabo, J.K. Willson, S.D . Markowitz, M.L. Goldberg, R. Karess, K.W. Kinzler, B. Vogelstein, V.E. Velculescu, and C. Lengauer. 2004. Three classes of genes mutated in colorectal cancers with chromosomal instability. Cancer Res 64:2998–3001.

Zhou, H., J. Kuang, L. Zhong, W.L. Kuo, J.W. Gray, A. Sahin, B.R. Brinkley, and S. Sen. 1998. Tumour amplified kinase STK15/BTAK induces centrosome amplification, aneuploidy and transformation. Nat Genet 20:189–93.

Chapter 4
DNA Methylation in Colorectal Cancer: Multiple Facets of Tumorigenesis

Joanne P. Young and Peter W. Laird

Introduction

Epigenetic mechanisms of gene regulation result in stable cellular phenotypes that are passed on at cell division but are not explained by alterations in the primary structure of DNA; they contribute to both phenotypic diversity and disease (Jones et al. 1999; Serman et al. 2006). Epigenetic control relies on multiple interrelated processes. Perhaps foremost among these are the post-translational modifications to the N-terminal tails of the histone proteins that constitute the nucleosome packaging of genomic DNA. The type and distribution of these modifications can influence the degree of chromatin compaction and can govern interactions with other chromatin proteins and trans-acting factors. In higher eukaryotes, an additional, very stable epigenetic silencing mechanism is mediated by C-5 methylation of cytosine residues; in mammals, this is in the context of CpG dinucleotides (Boyer et al. 2006; Lee et al. 2006). There is also evidence for a role in transcriptional control by subsets of RNA molecules (Lippman et al. 2004). Interactions between the DNA and protein complexes provide a means whereby transcription is controlled by altering the three-dimensional structure of chromatin within the nucleus of a cell and by the selective recruitment or exclusion of transcription factors.

Cytosine-5 DNA methylation at CpG dinucleotides plays a key role in long-term silencing of parasitic DNA elements and of other types of repetitive elements, and is essential for maintenance of: silenced genes on the inactive X-chromosome in females; imprinted genes; and some developmentally controlled genes (Serman et al. 2006). Complex organ systems in humans require a reservoir both of stem-like and proliferating cells for tissue building and renewal, and of differentiated cells which give rise to the phenotypic features of a tissue or organ. Epigenetic mechanisms contribute significantly to the coordinated gene expression that underlies this process (Jaenisch et al. 2003). The pathogenesis of human disease via inherited disorders, inflammation and aging, response to environmental agents

J.P. Young (✉) and P.W. Laird
Cancer Council, Queensland Institute of Medical Research
e-mail: Joanne.Young@qimr.edu.au

J.D. Potter and N.M Lindor (eds.), *Genetics of Colorectal Cancer*,
DOI: 10.1007/978-0-387-09568-4_5,

(Sathyanarayana et al. 2007), and the action of infectious organisms all result in altered gene expression, accompanied or mediated by DNA methylation changes (Santos-Reboucas et al. 2007).

The target CpG dinucleotide sequence for DNA methylation in mammals is distributed unevenly across the genome. Much of the genome is depleted of most CpGs, interrupted by occasional stretches of about 500 bp of normal CpG density, referred to as CpG islands (Bird 1986). About half of all mammalian promoters have a CpG island spanning the transcription start site, but a significant minority of CpG islands can be found in downstream areas of genes or even in intergenic regions. CpG islands and CpG-poor areas of the genome display quite different epigenetic behaviors, so it is important to keep this distinction in mind as we discuss epigenetic alterations in colorectal cancer. CpG islands are usually not methylated, whether or not the gene is expressed (Issa 2004). However, when they become methylated in a normally expressed gene, there is a reciprocal relationship between the density of methylated cytosine residues in the promoter region of a gene and the level of transcription (Bird 1986). It is not certain whether this is a direct physical-sequence-dependent phenomenon or whether it is mediated indirectly through the deacetylation of histones with resulting changes in the spatial configuration of the DNA.

Congenital disorders involving abnormal imprinting are seen in the fetal and postnatal overgrowth conditions such as Beckwith–Wiedemann and Silver–Russell syndromes (Delaval et al. 2006). Aging is associated with increasing DNA methylation of some CpG islands in the colon (Ahuja et al. 1998, 2000; Issa 1999, 2000; Chan et al. 2002a, b) and in other tissues (Santos-Reboucas et al. 2007). Environmental exposures such as tobacco smoke (Samowitz et al. 2006; Slattery et al. 2007) and perhaps toxins (Shen et al. 2002) have been shown to be associated with increased CpG-island DNA methylation in cancer. The presence of infectious agents has also been associated with promoter DNA hypermethylation in premalignant and malignant lesions of the epithelia and hemopoietic tissues. Oncogenic viruses, in particular Epstein–Barr virus (EBV) in gastric cancer (Chang et al. 2006) and human papilloma virus (HPV) in cervical cancer (Feng et al. 2007; Kang et al. 2007), and bacteria, such as *H. pylori* in gastric cancer, are associated with frequent promoter DNA hypermethylation (Perri et al. 2007). The exact mechanism is unknown and, though there is some evidence for suppression of host immune response, and host suppression of integrated sequences, DNA methylation is thought to be a result of chronic inflammatory processes, possibly associated with increased cell turnover. Chronic inflammation in the absence of known infectious agents is seen in Barrett metaplasia in the esophagus and inflammatory bowel diseases, two conditions where DNA methylation is frequent (Issa et al. 2001; Baumann et al. 2006; Clement et al. 2006; Hamilton et al. 2006). Hence, it may be the inflammation and associated cell turnover, rather than the agents themselves, that induce promoter DNA hypermethylation (Shames et al. 2007). Importantly, DNA methylation plays a fundamental role in the establishment of neoplasia in a wide variety of human tumors. Neoplastic transformation in high-cell-turnover tissues, such as the colorectum, is associated with changes in chromatin architecture and epigenetic alterations at multiple levels, creating an environment dominated by disordered cell growth (Shames et al. 2007).

DNA Methylation and Neoplasia

The alteration of DNA methylation patterns in cancer has been recognized for several decades, but the causal relevance of these changes to the cancer process has only recently gained acceptance (Jones et al. 1999). In the 1970s, hybridization studies of malignant teratomas with normal murine cells implicated epigenetic factors in the resulting nonmalignant phenotype (Mintz et al. 1975). More recently, Jaenisch and colleagues have demonstrated that the passage of melanoma nuclei through an embryo, a process known to remove and re-establish DNA methylation patterns, resulted in diminution or abrogation of some of the malignant behaviors of the melanoma (Hochedlinger et al. 2004). Alterations of DNA methylation in human cancer cells were first reported in 1982 as a global decrease of DNA methylation content in human cancer cell lines (Diala et al. 1982, 1983), and then as localized (Feinberg and Vogelstein 1983a, b; Gama-Sosa et al. 1983) and global (Flatau et al. 1983; Gama-Sosa et al. 1983) DNA hypomethylation in primary tumors. Tumor-specific CpG-island hypermethylation was first reported in 1986 for the calcitonin gene in human lung tumors and lymphomas (Baylin et al. 1986). These early reports were followed by extensive documentation of these two kinds of DNA methylation changes in cancer genomes: widespread loss of global DNA methylation content, largely occurring in CpG-poor parts of the genome, and localized increased or de novo methylation of CpG islands, which are often unmethylated in normal tissues. This highly localized CpG-island hypermethylation does not affect CpG dinucleotides in sufficient numbers to offset the genome-wide reduction in DNA methylation content. Nevertheless, CpG-island hypermethylation has important phenotypic consequences, because many gene promoters are covered by CpG islands and become silenced by abnormal hypermethylation in cancer. Therefore, promoter CpG-island hypermethylation has received, by far, the most attention in cancer epigenetics research. Although global reductions in DNA methylation content are thought to be associated with increased genomic instability, relatively little is known about the phenotypic effects of localized hypomethylation of specific gene promoters. In recent years, it has become apparent that various mechanisms underlie global or localized hypomethylation and different types of localized hypermethylation.

An interesting mechanistic example of a cancer-associated epigenetic defect that involves both regional DNA hypo- and hypermethylation is loss of genomic imprinting. Imprinted loci are expressed monoallelically in a parent-of-origin-dependent manner, with the nonexpressed allele generally displaying promoter CpG-island hypermethylation. However, regulation of these regions can be complex, and can involve competition for enhancers and hypomethylation of regulatory regions near the silent allele. Therefore, loss of genomic imprinting can involve both examples of conversion of monoallelic DNA methylation to biallelic DNA methylation and monoallelic DNA methylation to biallelic DNA hypomethylation.

Two clearly distinct mechanisms of CpG-island hypermethylation in colorectal cancer have emerged. One is the concordant hypermethylation of a specific group of CpG islands in a subset of colorectal cancers, referred to as CpG-island methylator phenotype (CIMP). It is not clear what mechanism underlies CIMP-associated CpG-island hypermethylation, but it is strongly associated with mutation of the *BRAF* oncogene. The other class of CpG-island hypermethylation found in colorectal cancer is the promoter methylation of stem-cell polycomb targets. Polycomb proteins occupy promoters of genes encoding master regulators of differentiation and development in stem cells, and keep these genes in a lightly repressed state, poised for activation. These polycomb targets appear to be predisposed to CpG-island hypermethylation in cancer. It is likely that further analysis will yield additional subgroups of concordant CpG-island hypermethylation in colorectal cancer, providing clues to specific underlying mechanisms. Not all CpG islands that undergo cancer-specific DNA methylation are located in the promoters of tumor suppressor genes or even in genes that are expressed in normal tissue. However, this passenger-style DNA methylation is probably reflective of an increased rate of epigenetic inactivation affecting other key targets which contribute to tumorigenesis (Jones et al. 1999).

In cancer of the colorectum, and a diversity of other malignancies of epithelial and hemopoietic origin, promoter-DNA hypermethylation is a frequent occurrence (Ducasse et al. 2006; Hayslip et al. 2006; Wu et al. 2006; Yasui et al. 2006; Costa et al. 2007; Feinberg 2007; Li et al. 2007). It is present early in the establishment of neoplasia (Chan et al. 2002a, b) and is cumulative over the course of multistep carcinogenesis (Baylin et al. 2006). It is important to note that promoter DNA hypermethylation shows tissue specificity, is associated with expression silencing of the adjacent coding gene (Jones et al. 1999), and is considered to be an alternative inactivation mechanism for tumor suppressor genes in cancers. The association between promoter DNA hypermethylation and loss of expression was first mooted in 1986 during studies of the calcitonin gene, and demonstrated with some degree of certainty with subsequent studies of the *CDKN2A* gene (Baylin et al. 1986; Herman et al. 1995; Merlo et al. 1995).

The genes affected by promoter-DNA hypermethylation-induced silencing are involved in transcriptional regulation, DNA repair, cell-cycle control, and growth-factor pathway control and, fundamental to the establishment of neoplasia, include antiproliferative and proapoptotic genes. Promoter-DNA hypermethylation can sometimes silence multiple members of the same gene family, suggesting a targeted specificity (Akiyama et al. 2003). Further, multiple adjacent genes in a particular region of a chromosome may also be silenced, reflecting changes in chromatin structure, and with implications analogous to loss of heterozygosity (Frigola et al. 2006). Finally, DNA hypermethylation of particular genes shows tissue-specific differences in frequency, suggestive of functional relevance in cancer (Suzuki et al. 2006). Though some promoter-DNA hypermethylation changes may be the result of stochastic events, there is compelling evidence that methylation-induced silencing of particular genes drives critical events in multistep carcinogenesis. For example, in a subset of colorectal cancers, epigenetic inactivation of the DNA mismatch-repair gene *MLH1* is likely to be the event that effects the malignant transformation of benign serrated polyps (Jass et al. 2000).

Other alterations observed in colorectal cancer where DNA methylation plays a conspicuous role include the presence of aberrant gene-specific promoter-DNA methylation in the constitutive genome which has become known as germline epimutation.

The mechanisms governing epigenetic processes and promoter-DNA hypermethy lation, in the colorectum in particular, remain elusive (Shames et al. 2007). Promoter-DNA hypermethylation does not affect all genes with equal probability. The presence of a large CpG island in a promoter does not dictate whether a gene will be methylated in cancer. Well-documented differential methylation in cancer includes the mismatch-repair genes: *MLH1* is frequently silenced by methylation, but *MSH2* and *PMS2* almost never, even though all three have CpG islands within their promoters and play important roles in the genesis of colorectal cancer. Similarly, *CDKN2A* and *RB* vary markedly by organ site in their methylation status, despite their universal involvement in tumors (Shames et al. 2007). These observations suggest that DNA methylation of promoters in neoplastic progression is unlikely to be random and, like microsatellite instability (Simms et al. 1997) and allelic loss, is a directed mechanism in tumorigenesis. Recent evidence for such an instructive mechanism has come from work published by Keshet and coworkers (2006), where genes that underwent apparent de novo DNA methylation in colorectal-cancer cells were derived from functionally distinct gene groups, had promoter regions featuring common sequence motifs (Das et al. 2006), and consistent with the findings of Frigola and colleagues (2006), were found in clusters on chromosomes.

Epigenetic Changes in Colorectal Neoplasia

Genome-Wide Hypomethylation of DNA

Hypomethylation of DNA sequences has been recorded in the very early stages of colonic neoplasia (Goelz et al. 1985; Sharrard et al. 1992), prior to malignant transformation (Bariol et al. 2003). It is observed in noncoding regions (Jaenisch et al. 2003) and has been associated with genomic instability (Matsuzaki et al. 2005), loss of imprinting (Cui et al. 2002), and the induction of expression of oncofetal genes and repetitive sequences (Yoder et al. 1997; Walsh et al. 1998). DNA hypomethylation is observed most often in repetitive sequences (Jaenisch et al. 2003). The human genome contains vast amounts of repetitive DNA. Some of these elements represent the entire coding sequence of retroviruses embedded into the genome, with associated mutations and DNA methylation (Englander et al. 1993; Ostertag et al. 2005), and it has been postulated that the primary function of DNA hypermethylation is the repression of such repetitive elements, on the basis that expression could lead to insertional mutagenesis (Walsh et al. 1998). In the human genome, many repetitive elements, including endogenous retroviruses, LINES, SINES, and *Alu* repeats, have associated promoters. The notion that demethylation of these potentially mobile elements is associated with their mobilization, and possible role in insertional mutagenic events has also been proposed (Jackson-Grusby et al. 2001).

Although it has often been hypothesized, there is as yet little concrete evidence for a role of DNA hypomethylation in facilitating or encouraging the upregulation of genes that are not expressed in normal adult cells, such as those for embryonic growth and development. However, in principle, this could result in aberrant expression of genes which may encourage neoplastic transformation. Hypomethylation of some oncogenes has been reported in colorectal tumors, including the *KRAS* and *HRAS* oncogenes (Feinberg and Vogelstein 1983a, b) and of a noncoding element in *MYC* genes (Sharrard et al. 1992), that shows increasing frequency with progression.

DNA hypomethylation is frequent in colonic tumors and is thought to lower the threshold for the establishment of neoplasia. It is the likely initiator of spatial abnormalities in chromatin which result in faulty segregation at mitosis and ultimately determine cancer cell ploidy. Overall cytosine DNA methylation content is reported to decrease with increasing tumor advancement. A well-recognized result of DNA hypomethylation is the induction of global genetic instability and, though the evidence is mostly indirect, this has been observed in cases of extreme global hypomethylation in murine models (Gaudet et al. 2003; Karpf et al. 2005) and in multiple studies of human cancers particularly those of the colorectum (Eden et al. 2003; Matsuzaki et al. 2005; Rodriguez et al. 2006). DNA hypomethylation-related instability is primarily of a chromosomal nature, and may contribute to the high levels of loss of heterozygosity seen in colorectal tumors. Indirect consequences of demethylation-induced instability may stem from the transposition of a normally silent gene into proximity with an active promoter.

Studies of colon cancer cell lines that used a selectable retroviral reporter were among the first to demonstrate, at a fundamental molecular level, that the mechanisms that drive carcinogenesis in the colon are heterogeneous, and that one of these mechanisms governed chromosomal instability (Lengauer et al. 1997a, b). In these experiments, where all cell lines were selectable with G418, the retroviral 5'-LTR was methylated in approximately one-half of the lines, rendering the beta-galactosidase reporter gene undetectable. An interesting observation from this work was that cell lines with a mismatch-repair defect were able to methylate ectopic DNA, whereas cell lines that were wild type for MMR were vulnerable to genome-wide chromosomal aberrations and aneuploidy associated with decreased methylation ability. However, subsequent work did not confirm the link between functional mismatch repair and a reduced methylating capability (Pao et al. 2000).

Loss of Imprinting in Colorectal Cancer

Genomic imprinting is a type of non-Mendelian inheritance, confined to mammals, in which only one allele of a pair is expressed. Imprinted genes are silenced in a parent-of-origin-specific manner, and are classified as maternally or paternally imprinted according to the silenced allele. There are approximately 50–100 imprinted genes in the human genome, and these genes acquire epigenetic marks during gametogenesis. It is known that DNA methylation participates in the maintenance of monoallelic chromatin sites. Genomic imprinting is coordinated within a regulatory region known as the imprinting control region (ICR), which is often an example of a differentially methylated region (DMR). Once imprinted, the allele-specific differential epigenetic state is preserved through cell division, in part, by a maintenance methyl-transferase known as *DNMT1* (Vilkaitis et al. 2005; Jelinic et al. 2007). Interpretation of the imprint is controlled by one of two mechanisms: chromatin barrier formation through the binding protein CTCF which has been described for the imprinted locus *IGF2/H19* (Bell et al. 2000, 2001); or a similar barrier via untranslated RNAs associated with *KCNQ1* (Mancini-DiNardo et al. 2003, 2006), in either case ensuring that only the maternal or paternal allele is expressed.

Imprinted genes tend to be regulators of embryonic growth, placental growth, or adult metabolism that require precise control of their expression for normal development. Congenital overgrowth syndromes, including Beckwith–Wiedemann syndrome, result from the disruption of imprinting controls (Butler 2002). Recent interest has centered on loss of imprinting (LOI) in a large variety of human cancers, including that of the colorectum. In colorectal-cancer patients, LOI can manifest as activation of the normally silent copy of the growth promoting gene *IGF2*. In murine models, the resultant upregulation leads to aberrant proliferative defects which include expanded colonic crypts (Sakatani et al. 2005). Further, murine models where LOI has been introduced have an increased susceptibility to tumor formation (Holm et al. 2005). In humans, LOI at *IGF2* is observed in 54–66% of colorectal cancers and, informatively, in most corresponding normal mucosae (Nakagawa et al. 2001a, b; Ohlsson 2004; Maenaka et al. 2006) and in peripheral blood lymphocytes. Patients with past or present colorectal cancer are more likely to have LOI at *IGF2* as are patients with a family history of colorectal cancer in first-degree relatives (Cui et al. 2003) suggesting epigenetic predisposition. Such findings suggest that LOI may be used as an indicator of risk, and even a point of therapeutic intervention for colorectal neoplasia, as such individuals will have a greatly expanded pool of cells that are vulnerable to neoplasia.

Polycomb Proteins and DNA Methylation in Colorectal Cancer

Several hundred genes have been reported to accumulate de novo DNA methylation within their CpG-island promoters in cancer. A previous report suggested that this process may be associated with a directive mechanism, and that such genes shared common sequence features (Keshet et al. 2006). Recently, one mechanism of CpG-island hypermethylation was suggested from an observed link between the chromatin state in stem cells of genes involved in normal development, on the one hand, and the subsequent acquisition of cancer-associated DNA methylation on the other, thus reinforcing the concept of progenitor or "stem-like" properties in cancer cells (Ohm et al. 2007; Schlesinger et al. 2007; Widschwendter et al. 2007).

Histone modifications characteristic of heterochromatin accompany promoter DNA methylation during normal postimplantation development. One particular chromatin mark, namely trimethylation of histone H3 at lysine 27 (H3K27), has been associated with promoter DNA methylation in colon cancer cells. An important observation from Schlesinger and colleagues (2007) was that marking of H3K27 was already in place in both embryonic stem cells and other undifferentiated cell types, suggesting that these marks may render certain genes more vulnerable to cancer-associated promoter DNA hypermethylation. However, it is important to note that stem-cell H3K27 trimethylation targets do not acquire promoter DNA methylation in normal differentiated tissues. Thus, DNA methylation of H3K27 targets is an abnormal event that occurs in oncogenesis, and not in normal development. However, not all genes that are methylated in colorectal cancer have this epigenetic mark, i.e., H3K27 is not a general marker of gene silencing (Schlesinger et al. 2007); therefore, the possibility of other mechanisms must be given consideration. Genes other than those that carry this epigenetic mark can undergo de novo DNA methylation in cancer, including *MLH1* and *STK11*. These events may represent stochastic promoter DNA methylation, either as an early event, even detectable in histologically normal colonic mucosa adjacent to the colorectal tumor (Nakagawa et al. 2001a, b), or as a late event in the case of methylation of the second allele following mutation or deletion of the first allele; this may represent an adaptive growth response of the neoplasm (Wong et al. 1999; Esteller et al. 2001).

H3K27 undergoes marking by a polycomb repressor complex, PRC2, which is composed of enhancer of zeste homolog 2 (EZH2) working in concert with its cofactors EED and SUZ12 (Zhang et al. 2004). The PRC2 complex plays an important role in the suppression of differentiation-inducing gene products in embryonic stem cells inasmuch as PRC2 components silence genes that, by inducing differentiation, would otherwise abrogate the ability of a progenitor cell to proliferate (Lee et al. 2006). The findings of several studies have suggested that genes marked by PRC2 components in stem cells are significantly more likely to undergo DNA methylation in a cancer-specific manner in colon, breast, and ovarian cancer cells (Widschwendter et al. 2007), and that they contain potential PRC2 binding elements within their DNA sequence (Keshet et al. 2006; Ohm et al. 2007; Schlesinger et al. 2007; Widschwendter et al. 2007). Therefore, genes targeted by polycomb modification in stem cells contain promoter motifs that act as genetic signals for de novo DNA methylation in cancer (Schlesinger et al. 2007; Tanay et al. 2007).

It is important to note that polycomb repressors appear to play a different role in tumor cells than they do in the stem cells that give rise to the cancer. PRC2 targets in stem cells are largely transcription factors that are master regulators of differentiation and development (Lee et al. 2006), whereas, in cancer cells, PRC2 targets feature genes encoding glycoproteins, receptors, and immunoglobulin-related genes (Squazzo et al. 2006), which are not frequent cancer-specific DNA methylation targets. How and when polycomb stem-cell targets acquire abnormal DNA methylation is currently unclear, but it seems likely that this occurs early, perhaps preneoplastically. Therefore, studies of DNA methyltransferases and of polycomb proteins in advanced cancer cells may be of limited relevance to our understanding of this event. Reports of DNA methyltransferase upregulation in cancer cells have been contradictory (Eads et al. 1999; Robertson et al. 1999; Robertson 2001), and it seems unlikely that mere overexpression of DNA methyltransferases is at the root of abnormal DNA methylation in cancer.

There have been reports of tumor-specific increases in polycomb components (Varambally et al. 2002; Kleer et al. 2003), with *EZH2*, in particular, having been reported to be upregulated in multiple cancers (Fiskus et al. 2006; Beke et al. 2007; Bryant et al. 2007; Lu et al. 2007; Marker 2007; Mattioli et al. 2007; Shi et al. 2007). However, as noted earlier, polycomb repression may play a role in advanced cancer cells other than DNA methylation recruitment. One of the hallmarks of neoplasia is resistance to induction of apoptosis. One of the most important biochemical responses to apoptotic stimuli is the induction of a cascade of proteolysis initiated by a family of enzymes called caspases. Recent links between polycomb components and caspases have been reported: Wong and colleagues (2007a, b) have demonstrated that Ring1B, a component of polycomb protein complexes that modulate chromatin structures, is a direct substrate of active caspase-3 and caspase-9 both in vitro and in vivo. Though not directly applicable to the discussion here, such a link between polycombs and a fundamental process which is disrupted in the establishment of neoplasia is nonetheless of great interest.

The CpG-Island Methylator Phenotype

The concept of a concerted "epigenetic instability" in colorectal cancers driving the progression of tumors via de novo DNA methylation of gene promoters was introduced in 1999. As a result of their own observations, Jean-Pierre Issa and colleagues proposed a dichotomous scheme of molecular classification in colorectal cancers, based not upon genetic changes, such as had been identified in 1993 with the recognition of microsatellite instability (MSI) (Ionov et al. 1993), but upon an analogous system involving epigenetic alterations (Toyota et al. 1999). Colorectal cancer included a subset of tumors in which there was *widespread and concordant* DNA hypermethylation of specific gene promoters, which was named CpG-island methylator phenotype or CIMP. CIMP supports neoplastic progression by inactivation of tumor suppressor genes in a similar fashion to the global Darwinian-style somatic evolution model which was used to explain MSI (Issa 2004).

Like the colorectal-cancer milestones before it: the Vogelstein model in 1988 (Vogelstein et al. 1988), and the recognition of MSI some 4–5 years later (Aaltonen et al. 1993; Ionov et al. 1993; Thibodeau et al. 1993), publication of the novel concept of CIMP generated much interest and research activity, the overwhelming majority of which supported the proposition that CIMP was a recognizable phenotype within colorectal cancer (van Rijnsoever et al. 2002; Samowitz et al. 2005a, b, 2006; Song et al. 2005; Ogino et al. 2006a, b, 2007a, b; Tanaka et al. 2006). CIMP tumors were found to occur in other organs (An et al. 2005; Chang et al. 2006; Marsit et al. 2006) and, in the colon, showed associations with female sex (Hawkins et al. 2001, 2002; An et al. 2005; Chang et al. 2006; Marsit et al. 2006), advanced age at presentation (Jass et al. 1998; Iacopetta et al. 2006), a tendency to occur proximally in the colon (Thibodeau et al. 1993; Young et al. 2001a, b; van Rijnsoever et al. 2002; McGivern et al. 2004; Rashid et al. 2004; Samowitz et al. 2005a, b), increased incidence of somatic *BRAF*- and *KRAS*-activating mutations (Toyota et al. 2000; Kambara et al. 2004; Tanaka et al. 2006; Weisenberger et al. 2006), and distinctive histological features such as clonal heterogeneity, mucinous or poorly differentiated histology, and contiguous serrated precursor lesions (Young et al. 2001a, b). In contrast, mutations in *TP53*, upregulation of beta-catenin, and chromosomal instability were relatively rare, suggesting that CIMP tumors developed via an alternative nonoverlapping pathway to that proposed by Vogelstein and colleagues (1988).

The original marker set used to define CIMP was subsequently improved by the introduction of the MINT (methylated in tumor) clones which had been derived from MCA (methylated CpG-island analysis). In addition, this study made the important step of partitioning methylated gene promoters into type-A genes (genes frequently methylated in tumors, but also in normal tissue as a function of aging) and type-C genes (genes which underwent cancer-specific DNA methylation) (Toyota et al. 2002). At least several hundred gene promoters are affected by DNA hypermethylation in cancer (Shames et al. 2007). Until very recently, the subset of these gene promoters which define CIMP had been a moving target, but had consistently been chosen on the basis of their being type-C genes. Type-C gene promoters define a subgroup of colorectal cancers that are characterized by a level of epigenetic instability that is 3- to 5-fold higher than the remainder of colorectal cancers (Issa 2004). It is important to note here that although some of the genes within CIMP are those premarked by polycomb repressor complexes, the genes that define CIMP also include genes such as *MLH1* which are not premarked. Conversely, many premarked genes are not found within the CIMP cluster; therefore, even though there is some overlap between the gene groups, it is likely that CIMP arises through a later-onset and independent mechanism.

Despite multiple supportive publications, CIMP remained a controversial concept. Several groups failed to find a dichotomous distribution in the analysis of promoter DNA methylation in colorectal cancers (Eads et al. 1999; Yamashita et al. 2003; Anacleto et al. 2005). In 2003, a publication appeared which cast significant doubt on the concept of CIMP, suggesting that it was an arbitrary discontinuity in a continuous distribution of hypermethylated gene promoters (Yamashita et al. 2003). This was followed by a further publication that supported those findings (Anacleto et al. 2005), and this doubt has continued until the present day (Wong et al. 2007a, b).

Several factors may have influenced these reports. Firstly, 70–80% of de novo methylated promoters in humans are age-related and need to be excluded from the analysis. Secondly, the use of a nonquantitative detection method and multiple different panels has also added to the confusion surrounding the classification of CIMP.

In an extensive study in 2006, Weisenberger and colleagues attempted to address the controversy as to whether CIMP was a classification that could be demonstrated in a more global and objective manner, as suggested by Issa (2004). To this end, they assayed a total of 295 colorectal tumors for DNA methylation using a stepwise approach, starting with 195 DNA methylation markers, identifying 92 type-C markers, and through unsupervised hierarchical cluster analysis demonstrated unequivocally that CIMP represented a distinct subset of colorectal tumors (Weisenberger et al. 2006). Further, CIMP could be classified by using a subset of just five markers (*NEUROG1*, *CACNA1G*, *IGF2*, *RUNX3*, and *SOCS1*), and this was validated in an independent set of colorectal cancers. In line with previous findings, there was a slight trend toward female sex, and significant associations with location in the proximal colon, MSI-H status, and *MLH1* promoter DNA methylation. This CIMP cluster was also detected by the original panel, although with significantly decreased specificity. The new marker panel identified tumors accounting for almost all sporadic MSI-H colorectal cancers, but also for as many again of colorectal cancers which were not MSI-H. The strongest of all associations was with activating somatic mutation in the *BRAF* oncogene (odds ratio for association was >200). Activating mutations in *BRAF* were rarely seen outside the CIMP cluster. However, only 70–80% of CIMP colorectal cancers have a *BRAF* mutation. This very tight association between CIMP and *BRAF* mutation, and the mutual exclusivity of *BRAF* and *KRAS* mutations resulted in the unexpected finding that CIMP was inversely associated with *KRAS* mutation, in contrast to some previous publications.

In approaching the question of CIMP as causal, there are several aspects of the argument which should be taken into account. Firstly it is clear that, in sporadic colorectal cancer, CIMP precedes the onset of MSI, as it is the epigenetic inactivation of the DNA mismatch-repair gene *MLH1* that gives rise to the MSI-H phenotype at malignant transformation. Reversal of this process has been achieved in cell-line experiments (Herman et al. 1998), and the absence of CIMP in Lynch syndrome tumors also supports the premise that MSI does not accelerate epigenetic instability (McGivern et al. 2004). Secondly, epigenetic changes such as those seen in CIMP colorectal cancers are present in normal mucosa in patients with CIMP cancers (Young et al. 2001a, b; Wynter et al. 2004; Kawakami et al. 2006), and also in a colorectal-cancer predisposition known as hyperplastic polyposis syndrome (HPS) (Wynter et al. 2004; Minoo et al. 2006) where DNA methylation of the normal colonic tissue is extraordinarily dense. Thirdly, patients with HPS exhibit multiple cancer and polyps in their colorectum which are concordant for CIMP and somatic *BRAF* mutation (Chan et al. 2002a, b; Beach et al. 2005), indicating that dysregulation of epigenetic controls may have resulted from a preceding genetic event. Finally, colorectal-cancer families have been reported where somatic *BRAF* mutation and CIMP feature prominently in the tumor phenotype (Frazier et al. 2003; Young et al. 2005; Vandrovcova et al. 2006).

Mechanistic questions also arise during the study of CIMP. Unlike MSI, where a defined genetic cause has been identified, namely that of an inactivated mismatch-repair gene, there have been no analogous mutations found in the epigenetic machinery of human cells which could account for CIMP. However, consistent with MSI, CIMP is likely to be driven by the relaxation of an important cellular control mechanism, to account for the widespread methylated promoters present in CIMP tumors, rather than a stochastic selection of DNA methylation inactivated genes. The close association of *BRAF* mutation with CIMP suggests that one may give rise to the other. Minoo and colleagues have developed a novel mechanistic model to investigate the effect of forced expression of a *BRAF*-activating mutation in a near-normal colorectal cell line. Findings included increased resistance to apoptosis, maintenance of a transformed phenotype, and the stimulation of promoter DNA methylation in the mismatch-repair gene *MLH1* (Minoo et al. 2007). The close relationship between somatic *BRAF* mutation and CIMP is likely to yield more novel insights in the near future.

The Histological Context of CIMP in the Colorectum

An explanation for the distinctive molecular and histologic, and even epidemiologic features of colorectal cancers characterized by CIMP and somatic *BRAF* mutation lies in their origins within a particular subset of serrated polyps (Jass 2001, 2003; Kambara et al. 2004). However, in parallel with the recognition and acceptance of CIMP as a distinct subset of colorectal cancer, its histologic features remained controversial for at least a decade. (For a full review of the serrated pathway of colorectal-cancer development see Chap. 4.)

Colorectal cancer arises in precursor lesions or epithelial polyps of which there are two common types. For decades, one of these, the adenoma, has been considered to have malignant potential. However, the other type (hyperplastic polyps) has been dismissed as innocuous. During the last decade, a subset of hyperplastic polyps called sessile serrated adenomas (SSA) which are large and atypical has risen to prominence as being the major precursor lesion for CIMP colorectal cancers (Goldstein et al. 2003; Jass 2003, 2004; Goldstein 2005). Located frequently in the proximal colon, SSA are characterized by somatic *BRAF* mutation and CIMP (Kambara et al. 2004), consistent with their precursor status (Yang et al. 2004). Patients with Hyperplastic Polyposis Syndrome (HPS) have numerous serrated polyps and frequently harbor SSA in their proximal colon. HPS patients have played a major part in the elucidation of the serrated pathway to colorectal cancers with CIMP and *BRAF* mutation. It was in patients with HPS that the serrated pathway as we understand it currently, was first recognized at both a histomorphologic (Torlakovic et al. 1996) and molecular level (Jass et al. 2000).

Gene-promoter DNA methylation is found in a variety of precursor lesions including adenomas; however, SSA are characterized by the dense DNA methylation of type-C genes that are associated with CIMP, and this is an important distinction.

The markers used to define CIMP (MINT clones) have been analyzed in MSI-H sporadic colorectal tumors as a comparison with the MSI-H colorectal cancers in Lynch syndrome which have a genetic basis (McGivern et al. 2004). Included in this analysis was a gene of the TGF-beta superfamily, *HPP1*, which is frequently methylated in a wide variety of colonic neoplasms including adenomas, serrated polyps, and cancers with and without CIMP (Young et al. 2001a, b). There was a striking difference in the level of DNA methylation seen in the MINT clones, with almost negligible levels being present in the Lynch-syndrome cancers. *HPP1* undergoes marking by PRC2 in progenitor cells and, in contrast with the MINT clones, showed significant DNA methylation in both groups, suggesting that a group of genes are methylated in the early establishment of neoplasia, possibly in microscopically normal colonic mucosa, and that CIMP is a superimposed alteration upon this early epigenetically unstable field change (Shen et al. 2005). *BRAF* mutation is present very early in neoplasia – at the stage of the aberrant crypt focus – and either may synergize with as-yet-unknown factors to potentiate the development of CIMP (Minoo et al. 2007), or may, as a result of CIMP, be able to exert its proliferative effects (Minoo et al. 2006).

Epidemiology of CIMP in Colorectal Neoplasia

Several studies, including one involving almost 400 patients with colorectal cancer, have investigated the epidemiology of CIMP colorectal tumors and confirmed associations with proximal location, advanced age, and female sex (Hawkins et al. 2002). When microsatellite unstable cases were excluded from the analysis, the association with female sex was eliminated. An even larger study investigating the epidemiology of CIMP colorectal tumors was carried out in a population-based panel of over 800 cases from North America (Samowitz et al. 2005a, b). Here again, CIMP was unequivocally demonstrated within the population (Issa et al. 2005) and shown to be tightly associated with somatic *BRAF* mutation, consistent with the findings of others (Kambara et al. 2004; Weisenberger et al. 2006). Here, family history of colon cancer was statistically significantly associated with *BRAF* mutation positive microsatellite-stable cancers (odds ratio for association was 4.2) suggesting a genetic predisposition toward developing colorectal cancers with somatic *BRAF* mutation. Subsequently, this same population was examined for a previously reported finding of an association between MSI-H and smoking (Slattery et al. 2004). The findings on this occasion were more definitive and showed that smoking was significantly associated with CIMP and with *BRAF* mutation irrespective of microsatellite instability status (Samowitz et al. 2006). In contrast, studies of the effect of diet, particularly of folate in the diet, on the propensity to develop colorectal cancers with CIMP and *BRAF* mutation have produced no consistent findings (Slattery et al. 2007). The findings listed earlier confirm and extend our understanding of colorectal tumors with *BRAF* mutation and CIMP as a distinct subtype of colorectal cancer with both genetic and environmental etiologies.

Rare Events: Germline Epimutation and Colorectal Cancer

Epimutation is a term used for the abnormal silencing of a gene due to epigenetic factors rather than a sequence variant (Holliday 1987). If an epimutation in a particular gene were to be present in the germline of an individual, it would be expected that this individual would have an equivalent phenotype to those individuals with germline-inactivating mutations in the same gene. Lynch syndrome is an autosomal dominant cancer predisposition which manifests as familial clustering of predominantly colorectal and endometrial cancers (Douglas et al. 2005). The underlying genetic cause is an inactivating mutation in one of four DNA mismatch-repair genes, namely *MLH1*, *MSH2*, *MSH6*, or *PMS2*. Tumors developing in Lynch syndrome show high-level microsatellite instability (MSI-H) as a result of DNA mismatch-repair deficiency (Lagerstedt Robinson et al. 2007).

Recently, germline epimutations associated with promoter DNA methylation have been reported in two genes in the DNA mismatch-repair system, with consequences to the patients consistent with a Lynch syndrome phenotype. In 2002, Gazzoli and colleagues (2002) described a case of a young-onset female patient with an MSI-H colorectal cancer, where one allele of the *MLH1* gene was methylated in the germline, and allelic loss in the tumor removed the wild-type allele in accordance with the typical inactivation sequence of a tumor-suppressor gene; this patient showed no evidence of carrying a germline genetic mutation in *MLH1*. Vertical transmission was not investigated, as parental DNA was not available for testing. The authors concluded that this case represented an alternative mechanism for Lynch syndrome, but was a relatively rare occurrence. Subsequently, Miyakura and colleagues (2004) described four cases of apparently nonfamilial but early-onset MSI-H colorectal cancer where germline methylation of *MLH1* was prevalent in leukocyte DNA. In two cases, endometrial cancer was also present. An informative polymorphism in the *MLH1* promoter allowed the investigators to demonstrate that DNA methylation was hemiallelic. This report concluded that germline epimutation was an explanation for a minority of cases of young-onset apparently nonfamilial MSI-H cancer, and this has been recently confirmed by Valle and colleagues (2007).

Suter and coworkers (2004) took a more detailed approach and addressed the concept that epimutation might be transmissible between generations. They analyzed two cases where evidence of a germline epimutation in *MLH1* was present in multiple cell lineages from each patient. Both cases had multiple primary cancers, most of which were part of the spectrum of cancers of Lynch syndrome, and allelic loss could be demonstrated in the majority of the tumors. However, methylation of *MLH1* was not detected in the germline of relatives, but was observed in the spermatozoa of the male case suggesting that transmission to subsequent generations was a real possibility. DNA methylation, though extensive in the somatic normal tissues, was mosaic, typical of epigenetic silencing. The authors concluded that germline epimutation was an alternative mechanism of inactivation of *MLH1* in the germline, but unlike Mendelian genetic traits, inheritance was weak as a result of mosaicism.

In a further report from the same group of investigators (Hitchins et al. 2007), an instance of germline vertical transmission of this trait to an offspring was described. In two cases of germline epimutation of *MLH1*, transmission occurred to only one son despite the affected allele being passed to multiple offspring. Evidence of meiotic erasure of epigenetic marks occurred in the constitutive DNA of recipients of the methylated allele and in the spermatozoa of the son with evidence of inheritance of the germline epimutation.

MLH1 is frequently methylated in sporadic colorectal cancer, marking this gene as vulnerable to epigenetic promoter abnormalities. However, a companion gene in the MMR complex, *MSH2*, is almost never methylated even though it has a large CpG-island promoter. Despite this, a family has been reported where three successive generations exhibit germ-line, allele-specific mosaic hypermethylation of *MSH2* in the absence of any evidence of genetic mutation (Chan et al. 2006). Colorectal and endometrial cancers are MSI-H and show immunohistochemical absence of *MSH2*. The highest level of DNA methylation was seen in the rectal mucosa, and the lowest in the leukocytes. These individuals have multiple primary cancers with an early age of onset, and have tumor features consistent with those seen in Lynch syndrome in that they are MSI-H and show absence of an MMR protein on immunohistochemistry (IHC). However, none of the affected individuals has any deleterious sequence variation that could be invoked as the cause of their cancer predisposition. This is a very rare occurrence, occurring in less than 1% of those cases of Lynch syndrome where no deleterious mutations can be found (Hitchins et al. 2005). The mechanism of induction of a germline epimutation is currently unknown. The possibilities include occurrence as a secondary event to a cis- or trans-activating mutation, or as a primary event following fertilization (Horsthemke 2006). DNA methylation is reversible and mosaic in nature and, therefore, is unlikely to be consistent with Mendelian genetic principles. It is even possible that this event is a rare chance occurrence. The current debate concerning these reports is centered around whether the epimutation per se is inherited or whether the epimutation is caused post fertilization by a cis-acting genetic factor or a genetic modifier (Horsthemke 2006; Chong et al. 2007; Leung et al. 2007; Suter and Martin 2007a, b). It is not possible at this time to state definitively that either gene is directly inherited in an epigenetically altered state. The specificity of the epimutation in the *MSH2* family seems to be more consistent with somatic acquisition of epigenetic silencing, whereas *MLH1* germline epimutation is only weakly heritable and may be dependent upon a specific genetic background for expression.

DNA Methylation in the Diagnosis and Therapy of Cancer

There are several key aspects of DNA methylation that make it useful in the diagnosis and treatment of cancer (Laird 2003; Shames et al. 2007). DNA methylation represents a tumor-specific change in the DNA which can be detected in blood, secretions, and tissue biopsies. DNA is stably methylated and can be readily measured by high-throughput PCR techniques, even in archived paraffin-embedded tissue (Laird 2005). Both hyper- and hypomethylation changes are found at a very early stage in neoplasia, and suggest that screening for these changes may identify persons at increased risk for the subsequent development of malignancy. DNA methylation profiles can be used to type subsets of tumors from the same site as well as from different sites. This feature has its utility in the markedly different survival outcomes and responses to treatment present in subgroups of colorectal cancer. It is important to note that DNA methylation is potentially reversible using pharmacologic means, thus offering opportunities for early intervention.

Early detection of cancers is essential to the survival of the patient. The early onset of DNA methylation changes in the often preneoplastic tissue of the colon (Shen et al. 2005) and other parts of the digestive tract presents a unique opportunity to detect cancer early by the development of robust assay systems. Stool tests for molecular markers are under development and hold the promise of being more specific than FOBT for routine colorectal-cancer screening (Belshaw et al. 2004; Suzuki et al. 2005).

Conclusions

There is much evidence for the central importance of DNA methylation in epigenetic processes. Complex and dynamic interdigitation of protein complexes and DNA serves to regulate the spatial arrangement of chromatin in an intricate and highly interdependent, even cyclical manner, rendering it extremely difficult to identify the initiating event (Laird 2005). However, it is currently thought that epigenetic alterations are brought about by the binding of protein complexes to the DNA, mechanisms that regulate transcription or changes in histones. DNA methylation, one of the best studied of the epigenetic alterations, is likely to require a critical seeding density to be established but, once in place, it exerts a powerful influence on subsequent gene expression. Currently, nonetheless, we do not understand the epigenetic regulatory machinery – and the defects within it – that give rise to cancer of the colon and other organs. Our increasing understanding of DNA methylation is likely to be central in its impact on the prevention, detection, and treatment of cancer in the near future.

References

Aaltonen, L. A., P. Peltomaki, et al. (1993). "Clues to the pathogenesis of familial colorectal cancer." *Science* **260**(5109): 812–6.

Ahuja, N. and J. P. Issa (2000). "Aging, methylation and cancer." *Histol Histopathol* **15**(3): 835–42.

Ahuja, N., Q. Li, et al. (1998). "Aging and DNA methylation in colorectal mucosa and cancer." *Cancer Res* **58**(23): 5489–94.

Akiyama, Y., N. Watkins, et al. (2003). "GATA-4 and GATA-5 transcription factor genes and potential downstream antitumor target genes are epigenetically silenced in colorectal and gastric cancer." *Mol Cell Biol* **23**(23): 8429–39.

An, C., I. S. Choi, et al. (2005). "Prognostic significance of CpG island methylator phenotype and microsatellite instability in gastric carcinoma." *Clin Cancer Res* **11**(2 Pt 1): 656–63.

Anacleto, C., A. M. Leopoldino, et al. (2005). "Colorectal cancer "methylator phenotype": fact or artifact?" *Neoplasia* **7**(4): 331–5.

Bariol, C., C. Suter, et al. (2003). "The relationship between hypomethylation and CpG island methylation in colorectal neoplasia." *Am J Pathol* **162**(4): 1361–71.

Baumann, S., G. Keller, et al. (2006). "The prognostic impact of O6-Methylguanine-DNA Methyltransferase (MGMT) promoter hypermethylation in esophageal adenocarcinoma." *Int J Cancer* **119**(2): 264–8.

Baylin, S. B. and J. E. Ohm (2006). "Epigenetic gene silencing in cancer – a mechanism for early oncogenic pathway addiction?" *Nat Rev Cancer* **6**(2): 107–16.

Baylin, S. B., J. W. Hoppener, et al. (1986). "DNA methylation patterns of the calcitonin gene in human lung cancers and lymphomas." *Cancer Res* **46**(6): 2917–22.

Beach, R., A. O. Chan, et al. (2005). "BRAF mutations in aberrant crypt foci and hyperplastic polyposis." *Am J Pathol* **166**(4): 1069–75.

Beke, L., M. Nuytten, et al. (2007). "The gene encoding the prostatic tumor suppressor PSP94 is a target for repression by the Polycomb group protein EZH2." *Oncogene* **26**: 4590–5.

Bell, A. C. and G. Felsenfeld (2000). "Methylation of a CTCF-dependent boundary controls imprinted expression of the Igf2 gene." *Nature* **405**(6785): 482–5.

Bell, A. C., A. G. West, et al. (2001). "Insulators and boundaries: versatile regulatory elements in the eukaryotic." *Science* **291**(5503): 447–50.

Belshaw, N. J., G. O. Elliott, et al. (2004). "Use of DNA from human stools to detect aberrant CpG island methylation of genes implicated in colorectal cancer." *Cancer Epidemiol Biomarkers Prev* **13**(9): 1495–501.

Bird, A. P. (1986). "CpG-rich islands and the function of DNA methylation." *Nature* **321** (6067): 209–13.

Boyer, L. A., K. Plath, et al. (2006). "Polycomb complexes repress developmental regulators in murine embryonic stem cells." *Nature* **441**(7091): 349–53.

Bryant, R. J., N. A. Cross, et al. (2007). "EZH2 promotes proliferation and invasiveness of prostate cancer cells." *Prostate* **67**(5): 547–56.

Butler, M. G. (2002). "Imprinting disorders: non-Mendelian mechanisms affecting growth." *J Pediatr Endocrinol Metab* **15 Suppl 5**: 1279–88.

Chan, A. O., R. R. Broaddus, et al. (2002a). "CpG island methylation in aberrant crypt foci of the colorectum." *Am J Pathol* **160**(5): 1823–30.

Chan, A. O., J. P. Issa, et al. (2002b). "Concordant CpG island methylation in hyperplastic polyposis." *Am J Pathol* **160**(2): 529–36.

Chan, T. L., S. T. Yuen, et al. (2006). "Heritable germline epimutation of MSH2 in a family with hereditary nonpolyposis colorectal cancer." *Nat Genet* **38**(10): 1178–83.

Chang, M. S., H. Uozaki, et al. (2006). "CpG island methylation status in gastric carcinoma with and without infection of Epstein–Barr virus." *Clin Cancer Res* **12**(10): 2995–3002.

Chong, S., N. A. Youngson, et al. (2007). "Heritable germline epimutation is not the same as transgenerational epigenetic inheritance." *Nat Genet* **39**(5): 574–5; author reply 575–6.

Clement, G., R. Braunschweig, et al. (2006). "Methylation of APC, TIMP3, and TERT: a new predictive marker to distinguish Barrett's oesophagus patients at risk for malignant transformation." *J Pathol* **208**(1): 100–7.

Costa, V. L., R. Henrique, et al. (2007). "Epigenetic markers for molecular detection of prostate cancer." *Dis Markers* **23**(1–2): 31–41.

Cui, H., P. Onyango, et al. (2002). "Loss of imprinting in colorectal cancer linked to hypomethylation of H19 and IGF2." *Cancer Res* **62**(22): 6442–6.

Cui, H., M. Cruz-Correa, et al. (2003). "Loss of IGF2 imprinting: a potential marker of colorectal cancer risk." *Science* **299**(5613): 1753–5.

Das, R., N. Dimitrova, et al. (2006). "Computational prediction of methylation status in human genomic sequences." *Proc Natl Acad Sci USA* **103**(28): 10713–6.
Delaval, K., A. Wagschal, et al. (2006). "Epigenetic deregulation of imprinting in congenital diseases of aberrant growth." *Bioessays* **28**(5): 453–9.
Diala, E. S. and R. M. Hoffman (1982). "Hypomethylation of HeLa cell DNA and the absence of 5-methylcytosine in SV40 and adenovirus (type 2) DNA: analysis by HPLC." *Biochem Biophys Res Commun* **107**: 19–26.
Diala, E. S., M. S. Cheah, et al. (1983). "Extent of DNA methylation in human tumor cells." *J Natl Cancer Inst* **71**: 755–64.
Douglas, J. A., S. B. Gruber, et al. (2005). "History and molecular genetics of Lynch syndrome in family G: a century later." *JAMA* **294**(17): 2195–202.
Ducasse, M. and M. A. Brown (2006). "Epigenetic aberrations and cancer." *Mol Cancer* **5**: 60.
Eads, C. A., K. D. Danenberg, et al. (1999). "CpG island hypermethylation in human colorectal tumors is not associated with DNA methyltransferase overexpression." *Cancer Res* **59**(10): 2302–6.
Eden, A., F. Gaudet, et al. (2003). "Chromosomal instability and tumors promoted by DNA hypomethylation." *Science* **300**(5618): 455.
Englander, E. W., A. P. Wolffe, et al. (1993). "Nucleosome interactions with a human Alu element. Transcriptional repression and effects of template methylation." *J Biol Chem* **268**(26): 19565–73.
Esteller, M., M. F. Fraga, et al. (2001). "DNA methylation patterns in hereditary human cancers mimic sporadic tumorigenesis." *Hum Mol Genet* **10**(26): 3001–7.
Feinberg, A. P. (2007). "Phenotypic plasticity and the epigenetics of human disease." *Nature* **447**(7143): 433–40.
Feinberg, A. P. and B. Vogelstein (1983a). "Hypomethylation distinguishes genes of some human cancers from their normal counterparts." *Nature* **301**(5895): 89–92.
Feinberg, A. P. and B. Vogelstein (1983b). "Hypomethylation of ras oncogenes in primary human cancers." *Biochem Biophys Res Commun* **111**(1): 47–54.
Feng, Q., S. E. Hawes, et al. (2007). "Promoter hypermethylation of tumor suppressor genes in urine from patients with cervical neoplasia." *Cancer Epidemiol Biomarkers Prev* **16**(6): 1178–84.
Fiskus, W., M. Pranpat, et al. (2006). "Histone deacetylase inhibitors deplete enhancer of zeste 2 and associated polycomb repressive complex 2 proteins in human acute leukemia cells." *Mol Cancer Ther* **5**(12): 3096–104.
Flatau, E., E. Bogenmann, et al. (1983). "Variable 5-methylcytosine levels in human tumor cell lines and fresh pediatric tumor explants." *Cancer Res* **43**: 4901–5.
Frazier, M. L., L. Xi, et al. (2003). "Association of the CpG island methylator phenotype with family history of cancer in patients with colorectal cancer." *Cancer Res* **63**(16): 4805–8.
Frigola, J., J. Song, et al. (2006). "Epigenetic remodeling in colorectal cancer results in coordinate gene suppression across an entire chromosome band." *Nat Genet* **38**(5): 540–9.
Gama-Sosa, M. A., V. A. Slagel, et al. (1983). "The 5-methylcytosine content of DNA from human tumors." *Nucleic Acids Res* **11**: 6883–94.
Gaudet, F., J. G. Hodgson, et al. (2003). "Induction of tumors in mice by genomic hypomethylation." *Science* **300**(5618): 489–92.
Gazzoli, I., M. Loda, et al. (2002). "A hereditary nonpolyposis colorectal carcinoma case associated with hypermethylation of the MLH1 gene in normal tissue and loss of heterozygosity of the unmethylated allele in the resulting microsatellite instability-high tumor." *Cancer Res* **62**(14): 3925–8.
Goelz, S. E., B. Vogelstein, et al. (1985). "Hypomethylation of DNA from benign and malignant human colon neoplasms." *Science* **228**(4696): 187–90.
Goldstein, N. S. (2005). "Clinical significance of (sessile) serrated adenomas: another piece of the puzzle." *Am J Clin Pathol* **123**(3): 329–30.
Goldstein, N. S., P. Bhanot, et al. (2003). "Hyperplastic-like colon polyps that preceded microsatellite-unstable adenocarcinomas." *Am J Clin Pathol* **119**(6): 778–96.
Hamilton, J. P., F. Sato, et al. (2006). "Reprimo methylation is a potential biomarker of Barrett's-associated esophageal neoplastic progression." *Clin Cancer Res* **12**(22): 6637–42.

Hawkins, N. J. and R. L. Ward (2001). "Sporadic colorectal cancers with microsatellite instability and their possible origin in hyperplastic polyps and serrated adenomas." *J Natl Cancer Inst* **93**(17): 1307–13.

Hawkins, N. J., C. Bariol, et al. (2002). "The serrated neoplasia pathway." *Pathology* **34**(6): 548–55.

Hayslip, J. and A. Montero (2006). "Tumor suppressor gene methylation in follicular lymphoma: a comprehensive review." *Mol Cancer* **5**: 44.

Herman, J. G., A. Merlo, et al. (1995). "Inactivation of the CDKN2/p16/MTS1 gene is frequently associated with aberrant DNA methylation in all common human cancers." *Cancer Res* **55**(20): 4525–30.

Herman, J. G., A. Umar, et al. (1998). "Incidence and functional consequences of hMLH1 promoter hypermethylation in colorectal carcinoma." *Proc Natl Acad Sci USA* **95**(12): 6870–5.

Hitchins, M., R. Williams, et al. (2005). "MLH1 germline epimutations as a factor in hereditary nonpolyposis colorectal cancer." *Gastroenterology* **129**(5): 1392–9.

Hitchins, M. P., J. J. Wong, et al. (2007). "Inheritance of a cancer-associated MLH1 germ-line epimutation." *N Engl J Med* **356**(7): 697–705.

Hochedlinger, K., R. Blelloch, et al. (2004). "Reprogramming of a melanoma genome by nuclear transplantation." *Genes Dev* **18**(15): 1875–85.

Holliday, R. (1987). "The inheritance of epigenetic defects." *Science* **238**(4824): 163–70.

Holm, T. M., L. Jackson-Grusby, et al. (2005). "Global loss of imprinting leads to widespread tumorigenesis in adult mice." *Cancer Cell* **8**(4): 275–85.

Horsthemke, B. (2006). "Epimutations in human disease." *Curr Top Microbiol Immunol* **310**: 45–59.

Iacopetta, B., W. Q. Li, et al. (2006). "BRAF mutation and gene methylation frequencies of colorectal tumors with microsatellite instability increase markedly with patient age." *Gut* **55**(8): 1213–4.

Ionov, Y., M. A. Peinado, et al. (1993). "Ubiquitous somatic mutations in simple repeated sequences reveal a new mechanism for colonic carcinogenesis." *Nature* **363**(6429): 558–61.

Issa, J. P. (1999). "Aging, DNA methylation and cancer." *Crit Rev Oncol Hematol* **32**(1): 31–43.

Issa, J. P. (2000). "CpG-island methylation in aging and cancer." *Curr Top Microbiol Immunol* **249**: 101–18.

Issa, J. P. (2004). "CpG island methylator phenotype in cancer." *Nat Rev Cancer* **4**(12): 988–93.

Issa, J. P., N. Ahuja, et al. (2001). "Accelerated age-related CpG island methylation in ulcerative colitis." *Cancer Res* **61**(9): 3573–7.

Issa, J. P., L. Shen, et al. (2005). "CIMP, at last." *Gastroenterology* **129**(3): 1121–4.

Jackson-Grusby, L., C. Beard, et al. (2001). "Loss of genomic methylation causes p53-dependent apoptosis and epigenetic deregulation." *Nat Genet* **27**(1): 31–9.

Jaenisch, R. and A. Bird (2003). "Epigenetic regulation of gene expression: how the genome integrates intrinsic and environmental signals." *Nat Genet* **33 Suppl**: 245–54.

Jass, J. R. (2001). "Serrated route to colorectal cancer: back street or super highway?" *J Pathol* **193**(3): 283–5.

Jass, J. R. (2003). "Hyperplastic-like polyps as precursors of microsatellite-unstable colorectal cancer." *Am J Clin Pathol* **119**(6): 773–5.

Jass, J. R. (2004). "Hyperplastic polyps and colorectal cancer: is there a link?" *Clin Gastroenterol Hepatol* **2**(1): 1–8.

Jass, J. R., K. A. Do, et al. (1998). "Morphology of sporadic colorectal cancer with DNA replication errors." *Gut* **42**(5): 673–9.

Jass, J. R., H. Iino, et al. (2000). "Neoplastic progression occurs through mutator pathways in hyperplastic polyposis of the colorectum." *Gut* **47**(1): 43–9.

Jelinic, P. and P. Shaw (2007). "Loss of imprinting and cancer." *J Pathol* **211**(3): 261–8.

Jones, P. A. and P. W. Laird (1999). "Cancer epigenetics comes of age." *Nat Genet* **21**(2): 163–7.

Kambara, T., L. A. Simms, et al. (2004). "BRAF mutation is associated with DNA methylation in serrated polyps and cancers of the colorectum." *Gut* **53**(8): 1137–44.

Kang, S., H. S. Kim, et al. (2007). "Inverse correlation between RASSF1A hypermethylation, KRAS and BRAF mutations in cervical adenocarcinoma." *Gynecol Oncol* **105**(3): 662–6.

Karpf, A. R. and S. Matsui (2005). "Genetic disruption of cytosine DNA methyltransferase enzymes induces chromosomal instability in human cancer cells." *Cancer Res* **65**(19): 8635–9.

Kawakami, K., A. Ruszkiewicz, et al. (2006). "DNA hypermethylation in the normal colonic mucosa of patients with colorectal cancer." *Br J Cancer* **94**(4): 593–8.

Keshet, I., Y. Schlesinger, et al. (2006). "Evidence for an instructive mechanism of de novo methylation in cancer cells." *Nat Genet* **38**(2): 149–53.

Kleer, C. G., Q. Cao, et al. (2003). "EZH2 is a marker of aggressive breast cancer and promotes neoplastic transformation of breast epithelial cells." *Proc Natl Acad Sci USA* **100**(20): 11606–11.

Lagerstedt Robinson, K., T. Liu, et al. (2007). "Lynch syndrome (hereditary nonpolyposis colorectal cancer) diagnostics." *J Natl Cancer Inst* **99**(4): 291–9.

Laird, P. W. (2003). "The power and the promise of DNA methylation markers." *Nat Rev Cancer* **3**(4): 253–66.

Laird, P. W. (2005). "Cancer epigenetics." *Hum Mol Genet* **14 Spec No 1**: R65–76.

Lee, T. I., R. G. Jenner, et al. (2006). "Control of developmental regulators by Polycomb in human embryonic stem cells." *Cell* **125**(2): 301–13.

Lengauer, C., K. W. Kinzler, et al. (1997a). "DNA methylation and genetic instability in colorectal cancer cells." *Proc Natl Acad Sci USA* **94**(6): 2545–50.

Lengauer, C., K. W. Kinzler, et al. (1997b). "Genetic instability in colorectal cancers." *Nature* **386**(6625): 623–7.

Leung, S. Y., T. L. Chan, et al. (2007). "Reply to "Heritable germline epimutation is not the same as transgenerational epigenetic inheritance"." *Nat Genet* **39**(5): 576.

Li, S. C. and L. Burgart (2007). "Histopathology of serrated adenoma, its variants, and differentiation from conventional adenomatous and hyperplastic polyps." *Arch Pathol Lab Med* **131**(3): 440–5.

Lippman, Z. and R. Martienssen (2004). "The role of RNA interference in heterochromatic silencing." *Nature* **431**(7006): 364–70.

Lu, C., T. Bonome, et al. (2007). "Gene alterations identified by expression profiling in tumor-associated endothelial cells from invasive ovarian carcinoma." *Cancer Res* **67**(4): 1757–68.

Maenaka, S., T. Hikichi, et al. (2006). "Loss of imprinting in IGF2 in colorectal carcinoma assessed by microdissection." *Oncol Rep* **15**(4): 791–5.

Mancini-DiNardo, D., S. J. Steele, et al. (2003). "A differentially methylated region within the gene Kcnq1 functions as an imprinted promoter and silencer." *Hum Mol Genet* **12**(3): 283–94.

Mancini-DiNardo, D., S. J. Steele, et al. (2006). "Elongation of the Kcnq1ot1 transcript is required for genomic imprinting of neighboring genes." *Genes Dev* **20**(10): 1268–82.

Marker, P. C. (2007). "Essential role for activation of the Polycomb group (PcG) protein chromatin silencing pathway in metastatic prostate cancer Berezovska OP, Glinskii AB, Yang Z, Li XM, Hoffman RM, Glinsky GV, Translational and Functional Genomics Laboratory, Ordway Cancer Center, Ordway Research Institute, Inc., Center for Medical Sciences, Albany, NY." *Urol Oncol* **25**(3): 278–9.

Marsit, C. J., E. A. Houseman, et al. (2006). "Examination of a CpG island methylator phenotype and implications of methylation profiles in solid tumors." *Cancer Res* **66**(21): 10621–9.

Matsuzaki, K., G. Deng, et al. (2005). "The relationship between global methylation level, loss of heterozygosity, and microsatellite instability in sporadic colorectal cancer." *Clin Cancer Res* **11**(24 Pt 1): 8564–9.

Mattioli, E., P. Vogiatzi, et al. (2007). "Immunohistochemical analysis of pRb2/p130, VEGF, EZH2, p53, p16(INK4A), p27(KIP1), p21(WAF1), Ki-67 expression patterns in gastric cancer." *J Cell Physiol* **210**(1): 183–91.

McGivern, A., C. V. Wynter, et al. (2004). "Promoter hypermethylation frequency and BRAF mutations distinguish hereditary non-polyposis colon cancer from sporadic MSI-H colon cancer." *Fam Cancer* **3**(2): 101–7.

Merlo, A., J. G. Herman, et al. (1995). "5′ CpG island methylation is associated with transcriptional silencing of the tumour suppressor p16/CDKN2/MTS1 in human cancers." *Nat Med* **1**(7): 686–92.

Minoo, P., J. R. Jass (2006). "Senescence and Serration: a new twist to an old tale." J Pathol **210**(2): 137–40.

Minoo, P., K. Baker, et al. (2006). "Extensive DNA methylation in normal colorectal mucosa in hyperplastic polyposis." *Gut* **55**(10): 1467–74.

Minoo, P., M. Moyer, et al. (2007). "Role of BRAF-V600E in the serrated pathway of colorectal tumorigenesis." *J Pathol* **212**(2): 124–33.

Mintz, B. and K. Illmensee (1975). "Normal genetically mosaic mice produced from malignant teratocarcinoma cells." *Proc Natl Acad Sci USA* **72**(9): 3585–9.

Miyakura, Y., K. Sugano, et al. (2004). "Extensive but hemiallelic methylation of the hMLH1 promoter region in early-onset sporadic colon cancers with microsatellite instability." *Clin Gastroenterol Hepatol* **2**(2): 147–56.

Nakagawa, H., R. B. Chadwick, et al. (2001a). "Loss of imprinting of the insulin-like growth factor II gene occurs by biallelic methylation in a core region of H19-associated CTCF-binding sites in colorectal cancer." *Proc Natl Acad Sci USA* **98**(2): 591–6.

Nakagawa, H., G. J. Nuovo, et al. (2001b). "Age-related hypermethylation of the 5′ region of MLH1 in normal colonic mucosa is associated with microsatellite-unstable colorectal cancer development." *Cancer Res* **61**: 6991–5.

Ogino, S., M. Cantor, et al. (2006a). "CpG island methylator phenotype (CIMP) of colorectal cancer is best characterised by quantitative DNA methylation analysis and prospective cohort studies." *Gut* **55**(7): 1000–6.

Ogino, S., R. D. Odze, et al. (2006b). "Correlation of pathologic features with CpG island methylator phenotype (CIMP) by quantitative DNA methylation analysis in colorectal carcinoma." *Am J Surg Pathol* **30**(9): 1175–83.

Ogino, S., T. Kawasaki, et al. (2007a). "Evaluation of markers for CpG Island Methylator Phenotype (CIMP) in colorectal cancer by a large population-based sample." *J Mol Diagn* **9**(3): 305–14.

Ogino, S., T. Kawasaki, et al. (2007b). "18q loss of heterozygosity in microsatellite stable colorectal cancer is correlated with CpG island methylator phenotype-negative (CIMP-0) and inversely with CIMP-low and CIMP-high." *BMC Cancer* **7**: 72.

Ohlsson, R. (2004). "Loss of IGF2 imprinting: mechanisms and consequences." *Novartis Found Symp* **262**: 108–21; discussion 121–4, 265–8.

Ohm, J. E., K. M. McGarvey, et al. (2007). "A stem cell-like chromatin pattern may predispose tumor suppressor genes to DNA hypermethylation and heritable silencing." *Nat Genet* **39**(2): 237–42.

Ostertag, E. M. and H. H. Kazazian (2005). "Genetics: LINEs in mind." *Nature* **435**(7044): 890–1.

Pao, M. M., G. Liang, et al. (2000). "DNA methylator and mismatch repair phenotypes are not mutually exclusive in colorectal cancer cell lines." *Oncogene* **19**: 943–52.

Perri, F., R. Cotugno, et al. (2007). "Aberrant DNA methylation in non-neoplastic gastric mucosa of *H. pylori* infected patients and effect of eradication." *Am J Gastroenterol* **102**(11): 1361–71.

Rashid, A. and J. P. Issa (2004). "CpG island methylation in gastroenterologic neoplasia: a maturing field." *Gastroenterology* **127**(5): 1578–88.

Robertson, K. D. (2001). "DNA methylation, methyltransferases, and cancer." *Oncogene* **20**(24): 3139–55.

Robertson, K. D., E. Uzvolgyi, et al. (1999). "The human DNA methyltransferases (DNMTs) 1, 3a and 3b: coordinate mRNA expression in normal tissues and overexpression in tumors." *Nucleic Acids Res* **27**(11): 2291–8.

Rodriguez, J., J. Frigola, et al. (2006). "Chromosomal instability correlates with genome-wide DNA demethylation in human primary colorectal cancers." *Cancer Res* **66**(17): 8462–9468.

Sakatani, T., A. Kaneda, et al. (2005). "Loss of imprinting of Igf2 alters intestinal maturation and tumorigenesis in mice." *Science* **307**(5717): 1976–8.

Samowitz, W. S., H. Albertsen, et al. (2005a). "Evaluation of a large, population-based sample supports a CpG island methylator phenotype in colon cancer." *Gastroenterology* **129**(3): 837–45.

Samowitz, W. S., C. Sweeney, et al. (2005b). "Poor survival associated with the BRAF V600E mutation in microsatellite-stable colon cancers." *Cancer Res* **65**(14): 6063–9.

Samowitz, W. S., H. Albertsen, et al. (2006). "Association of smoking, CpG island methylator phenotype, and V600E BRAF mutations in colon cancer." *J Natl Cancer Inst* **98**(23): 1731–8.

Santos-Reboucas, C. B. and M. M. Pimentel (2007). "Implication of abnormal epigenetic patterns for human diseases." *Eur J Hum Genet* **15**(1): 10–7.

Sathyanarayana, U. G., A. Y. Moore, et al. (2007). "Sun exposure related methylation in malignant and non-malignant skin lesions." *Cancer Lett* **245**(1–2): 112–20.

Schlesinger, Y., R. Straussman, et al. (2007). "Polycomb-mediated methylation on Lys27 of histone H3 pre-marks genes for de novo methylation in cancer." *Nat Genet* **39**(2): 232–6.

Serman, A., M. Vlahovic, et al. (2006). "DNA methylation as a regulatory mechanism for gene expression in mammals." *Coll Antropol* **30**(3): 665–71.

Shames, D. S., J. D. Minna, et al. (2007). "DNA methylation in health, disease, and cancer." *Curr Mol Med* **7**(1): 85–102.

Sharrard, R. M., J. A. Royds, et al. (1992). "Patterns of methylation of the c-myc gene in human colorectal cancer progression." *Br J Cancer* **65**(5): 667–72.

Shen, L., N. Ahuja, et al. (2002). "DNA methylation and environmental exposures in human hepatocellular carcinoma." *J Natl Cancer Inst* **94**: 755–61.

Shen, L., Y. Kondo, et al. (2005). "MGMT promoter methylation and field defect in sporadic colorectal cancer." *J Natl Cancer Inst* **97**(18): 1330–8.

Shi, B., J. Liang, et al. (2007). "Integration of estrogen and Wnt signaling circuits by the polycomb group protein EZH2 in breast cancer cells." *Mol Cell Biol* **27**(14): 5105–19.

Simms, L. A., T. T. Zou, et al. (1997). "Apparent protection from instability of repeat sequences in cancer-related genes in replication error positive gastrointestinal cancers." *Oncogene* **14**(21): 2613–8.

Slattery, M. L., K. Curtin, et al. (2007). "Diet and lifestyle factor associations with CpG island methylator phenotype and BRAF mutations in colon cancer." *Int J Cancer* **120**(3): 656–63.

Slattery, M. L., W. Samowitz, et al. (2004). "CYP1A1, cigarette smoking, and colon and rectal cancer." *Am J Epidemiol* **160**: 842–52.

Song, G. A., G. Deng, et al. (2005). "Mucinous carcinomas of the colorectum have distinct molecular genetic characteristics." *Int J Oncol* **26**(3): 745–50.

Squazzo, S. L., H. O'Geen, et al. (2006). "Suz12 binds to silenced regions of the genome in a cell-type-specific manner." *Genome Res* **16**(7): 890–900.

Suter, C. M. and D. I. Martin (2007a). "Inherited epimutation or a haplotypic basis for the propensity to silence?" *Nat Genet* **39**(5): 573; author reply 576.

Suter, C. M. and D. I. Martin (2007b). "Reply to "Heritable germline epimutation is not the same as transgenerational epigenetic inheritance"." *Nat Genet* **39**(5): 575–6.

Suter, C. M., D. I. Martin, et al. (2004). "Germline epimutation of MLH1 in individuals with multiple cancers." *Nat Genet* **36**(5): 497–501.

Suzuki, M., H. Shigematsu, et al. (2005). "DNA methylation-associated inactivation of TGFbeta-related genes DRM/Gremlin, RUNX3, and HPP1 in human cancers." *Br J Cancer* **93**(9): 1029–37.

Suzuki, M., H. Shigematsu, et al. (2006). "Exclusive mutation in epidermal growth factor receptor gene, HER-2, and KRAS, and synchronous methylation of nonsmall cell lung cancer." *Cancer* **106**(10): 2200–7.

Tanaka, H., G. Deng, et al. (2006). "BRAF mutation, CpG island methylator phenotype and microsatellite instability occur more frequently and concordantly in mucinous than non-mucinous colorectal cancer." *Int J Cancer* **118**(11): 2765–71.

Tanay, A., A. H. O'Donnell, et al. (2007). "Hyperconserved CpG domains underlie Polycomb-binding sites." *Proc Natl Acad Sci USA* **104**(13): 5521–6.

Thibodeau, S. N., G. Bren, et al. (1993). "Microsatellite instability in cancer of the proximal colon." *Science* **260**(5109): 816–9.

Torlakovic, E. and D. C. Snover (1996). "Serrated adenomatous polyposis in humans." *Gastroenterology* **110**(3): 748–55.

Toyota, M. and J. P. Issa (1999). "CpG island methylator phenotypes in aging and cancer." *Semin Cancer Biol* **9**(5): 349–57.

Toyota, M. and J. P. Issa (2002). "Methylated CpG island amplification for methylation analysis and cloning differentially methylated sequences." *Methods Mol Biol* **200**: 101–10.

Toyota, M., M. Ohe-Toyota, et al. (2000). "Distinct genetic profiles in colorectal tumors with or without the CpG island methylator phenotype." *Proc Natl Acad Sci USA* **97**(2): 710–5.

Valle, L., P. Carbonell, et al. (2007). "MLH1 germline epimutations in selected patients with early-onset non-polyposis colorectal cancer." *Clin Genet* **71**(3): 232–7.

van Rijnsoever, M., F. Grieu, et al. (2002). "Characterisation of colorectal cancers showing hypermethylation at multiple CpG islands." *Gut* **51**(6): 797–802.

Vandrovcova, J., K. Lagerstedt-Robinsson, et al. (2006). "Somatic BRAF-V600E mutations in familial colorectal cancer." *Cancer Epidemiol Biomarkers Prev* **15**(11): 2270–3.

Varambally, S., S. M. Dhanasekaran, et al. (2002). "The polycomb group protein EZH2 is involved in progression of prostate cancer." *Nature* **419**(6907): 624–9.

Vilkaitis, G., I. Suetake, et al. (2005). "Processive methylation of hemimethylated CpG sites by mouse Dnmt1 DNA methyltransferase." *J Biol Chem* **280**(1): 64–72.

Vogelstein, B., E. R. Fearon, et al. (1988). "Genetic alterations during colorectal-tumor development." *N Engl J Med* **319**(9): 525–32.

Walsh, C. P., J. R. Chaillet, et al. (1998). "Transcription of IAP endogenous retroviruses is constrained by cytosine methylation." *Nat Genet* **20**(2): 116–7.

Weisenberger, D. J., K. D. Siegmund, et al. (2006). "CpG island methylator phenotype underlies sporadic microsatellite instability and is tightly associated with BRAF mutation in colorectal cancer." *Nat Genet* **38**(7): 787–93.

Widschwendter, M., H. Fiegl, et al. (2007). "Epigenetic stem cell signature in cancer." *Nat Genet* **39**(2): 157–8.

Wong, D. J., S. A. Foster, et al. (1999). "Progressive region-specific de novo methylation of the p16 CpG island in primary human mammary epithelial cell strains during escape from M(0) growth arrest." *Mol Cell Biol* **19**(8): 5642–51.

Wong, J. J., N. J. Hawkins, et al. (2007a). "Colorectal cancer: a model for epigenetic tumorigenesis." *Gut* **56**(1): 140–8.

Wong, C. K., Z. Chen, et al. (2007b). "Polycomb group protein RING1B is a direct substrate of Caspases-3 and -9." *Biochim Biophys Acta* **1773**(6): 844–52.

Wu, D. L., F. Y. Sui, et al. (2006). "Methylation in esophageal carcinogenesis." *World J Gastroenterol* **12**(43): 6933–40.

Wynter, C. V., M. D. Walsh, et al. (2004). "Methylation patterns define two types of hyperplastic polyp associated with colorectal cancer." *Gut* **53**(4): 573–80.

Yamashita, K., T. Dai, et al. (2003). "Genetics supersedes epigenetics in colon cancer phenotype." *Cancer Cell* **4**(2): 121–31.

Yang, S., F. A. Farraye, et al. (2004). "BRAF and KRAS Mutations in hyperplastic polyps and serrated adenomas of the colorectum: relationship to histology and CpG island methylation status." *Am J Surg Pathol* **28**(11): 1452–9.

Yasui, W., K. Sentani, et al. (2006). "Molecular pathobiology of gastric cancer." *Scand J Surg* **95**(4): 225–31.

Yoder, J. A., C. P. Walsh, et al. (1997). "Cytosine methylation and the ecology of intragenomic parasites." *Trends Genet* **13**(8): 335–40.

Young, J., K. G. Biden, et al. (2001a). "HPP1: a transmembrane protein-encoding gene commonly methylated in colorectal polyps and cancers." *Proc Natl Acad Sci USA* **98**(1): 265–70.

Young, J., L. A. Simms, et al. (2001b). "Features of colorectal cancers with high-level microsatellite instability occurring in familial and sporadic settings: parallel pathways of tumorigenesis." *Am J Pathol* **159**(6): 2107–16.

Young, J., M. A. Barker, et al. (2005). "Evidence for BRAF mutation and variable levels of microsatellite instability in a syndrome of familial colorectal cancer." *Clin Gastroenterol Hepatol* **3**(3): 254–63.

Zhang, Y., R. Cao, et al. (2004). "Mechanism of Polycomb group gene silencing." *Cold Spring Harb Symp Quant Biol* **69**: 309–17.

Chapter 5
Pathways and Pathology

Jeremy R. Jass

Introduction

The aim of this chapter is to explain how the traditional tools of the pathologist led first, to the classification of colorectal polyps and, then, to the concept of the adenoma-carcinoma sequence as the dominant tumourigenic pathway in the colorectum. The subsequent discovery of the genetic steps underlying this pathway gave rise to a single linear model to explain the initiation, progression, and final transformation of the adenoma into a carcinoma. A critical analysis of the 'polyp story' will show that, from the outset, the adenoma-carcinoma paradigm was a deliberate oversimplification designed to facilitate clinical decision making. Rigid adherence to the concept, following the genetic revolution in the1980s, delayed the recognition that colorectal cancer is in fact a multi-pathway disease. Furthermore, the most efficient pathways to colorectal cancer are characterized by the co-occurrence of key elements of the archetypal pathways into 'fusion' pathways.

Adenoma-Carcinoma Sequence

The evidence for the adenoma-carcinoma sequence is incontrovertible. For the pathologist, the most convincing proof is the direct observation of cancer arising within an adenoma (Morson 1966). For the clinician the most direct proof is the prevention of colorectal cancer by colonoscopic polypectomy (Winawer et al. 1993). These two kinds of evidence were elegantly combined in this colonoscopic study, inasmuch as all five colorectal cancers that developed during the follow-up of the study group arose within adenomas. A third, but less direct, kind of evidence is provided by various striking relationships between adenomas and carcinomas. For

J.R. Jass
Department of Cellular Pathology, St Mark's Hospital
e-mail: jeremy.jass@nwlh.nhs.uk

J.D. Potter and N.M Lindor (eds.), *Genetics of Colorectal Cancer*,
DOI: 10.1007/978-0-387-09568-4_6,

example, adenomas were shown to be six times as common in surgical specimens of colorectal cancer than in length-, age- and sex-matched colorectal specimens without cancer (Eide 1986a). Nevertheless, uncritical acceptance of these findings may lead to an overly simplistic understanding of the adenoma-carcinoma sequence.

The modern 'unifying' classification of colorectal adenomas encourages the view that these lesions are essentially homogeneous and differ only according to the evolutionary stage at which they are diagnosed. For example, an early adenoma will be small, mildly dysplastic, and have a tubular architecture, while a late or advanced adenoma will be large, severely dysplastic, and more likely to have a villous architecture. Tubular adenoma, tubulovillous adenoma, and villous adenoma were originally termed adenomatous polyp, papillary adenoma, and villous papilloma, respectively. Prior to the internationally agreed adoption of the unifying term 'adenoma' (Morson and Sobin 1976), these lesions were regarded as a set of different entities and not as a biologic continuum. The early controversy regarding the malignant potential of adenomas centred on the fact that the adenomatous polyp (tubular adenoma) was considered, by some, to be harmless (Spratt et al. 1958) and, by others, to have significant malignant potential (Morson et al. 1983). Although the latter view has prevailed, it is nonetheless clear that most adenomas cannot progress to cancer, amply illustrated by experience in different clinical settings.

Malignant Potential of Adenomas in Different Clinical Scenarios

When an autopsy study conducted in Norway was extrapolated to a fixed Norwegian population, the latter cohort was estimated to include 26,419 adenoma-bearing individuals. During a 10-year period, 656 colorectal cancers developed within this population. Assuming that every colorectal cancer developed with an adenoma-bearing individual, it can be inferred that it required 40 adenoma-bearing subjects to produce one colorectal cancer during a 10-year period (Eide 1986b). Some of these subjects will have had two or more adenomas, and the ratio of adenoma to carcinoma will in fact be greater than 40 to 1. This study also showed that the risk of malignancy was much higher for villous adenomas than tubular adenomas (Eide 1986b).

The low risk of progression of tubular adenomas is even more graphically illustrated in familial adenomatous polyposis (FAP), in which many hundreds, if not thousands, of colorectal adenomas begin to develop at around puberty. Colorectal cancer develops at an average age of 39 years in newly diagnosed subjects and at an average age of 33 years in call-up members of known polyposis families (Day et al. 2003). This means that it may take up to 25 years for one or two out of many thousands of colorectal adenomas to become malignant.

It may be argued that the progression from adenoma to carcinoma is an age-related process that will accelerate in subjects developing sporadic adenomas in middle age. This may be countered by the fact that colorectal cancers develop at an early age in Lynch syndrome or hereditary non-polyposis colorectal cancer. This is despite the fact that subjects with Lynch syndrome develop small numbers of colorectal adenomas. For example, in a well-studied series of 22 adenoma-positive patients with Lynch syndrome, most subjects had only one adenoma and only three patients had the maximum of three adenomas. Morphologically, Lynch syndrome adenomas do not differ greatly from the adenomas in familial adenomatous polyposis. Indeed, they cannot be differentiated except by the use of immunohistochemistry to show loss of expression of a DNA mismatch repair protein (Iino et al. 2000), yet there is a strong likelihood that each Lynch syndrome adenoma will not only progress to colorectal cancer but will do so within a short time frame. This rapid evolution may be appreciated when patients with a negative screening colonoscopy develop an interval colorectal cancer before the next screening examination (Vasen et al. 1995).

Based on these simple clinical examples, common sense does not allow us to view adenomas as homogeneous entities differing only in their stage of progression at the time of diagnosis. Although a patient with FAP will inevitably develop colorectal cancer by middle age, an individual adenoma in this syndrome and an individual adenoma in Lynch syndrome have vastly different natural histories. Furthermore, the Norwegian study shows that a non-familial adenoma has a malignant potential that is intermediate between the extremes observed in the two main forms of hereditary colorectal cancer. Despite these observations, adenomas in FAP have been widely accepted as the exact equivalents of sporadic adenomas. Before considering the facts underlying the differing mechanisms of adenoma initiation in the three clinical scenarios outlined, it is necessary to discuss adenoma multiplicity and an apparent paradox.

Adenoma Multiplicity and Malignant Potential

As pointed out earlier, the adenomas occurring in the context of the extreme multiplicity of FAP are, at the individual level, the least aggressive types of adenoma. By contrast, adenoma multiplicity is well established as a marker of aggression in the context of non-familial colorectal neoplasia (Day et al. 2003). The simplistic explanation is that the patient with two adenomas has twice the chance of developing colorectal cancer as the patient with one adenoma, and so on. However, observational studies show that adenoma multiplicity explains aggression only under particular circumstances. For example, one study examined the fate of patients who had adenomas removed sigmoidoscopically but were not followed up (Atkin et al. 1992). There was no increased risk of dying of colorectal cancer in subjects with non-advanced adenomas, *even if these were multiple*. Subjects with advanced adenomas (large, high grade, or villous) had a threefold increase in standardized incidence ratios for mortality due to colorectal cancer. However, in subjects with both multiple and advanced adenomas, the increase in mortality was sixfold (Atkin et al. 1992).

The most parsimonious explanation for these striking observations is that there is a single genetic explanation underlying the predisposition to both adenoma multiplicity and adenoma aggression. An illustration of this possibility is a longitudinal study in which colorectal adenomas were not removed, but their behaviour was observed over time (Almendingen et al. 2003). Most adenomas did not grow, but, when growth occurred, this was more likely to be detected in patients with a family history of colorectal cancer (Almendingen et al. 2003). The existence of colorectal cancer families with multiple adenomas has been known for many years (Lovett 1976). Veale hypothesized an autosomal recessive mechanism (Morson et al. 1983), and this suggestion was vindicated by the discovery of bi-allelic germline alterations in the DNA repair gene *MUTYH* (*MYH*) in some multiple adenoma families (Al-Tassan et al. 2002). *MYH*-associated polyposis also explains how a genetic predisposition could explain both multiplicity and accelerated progression of adenomas. This is because reduced activity of *MYH* predisposes to G to T transition mutations in *APC*, *KRAS*, and possibly other cancer genes that would act synergistically in driving neoplastic evolution (Kambara et al. 2004b). However, the phenotype of *MYH*-associated polyposis approximates to FAP and does not explain the presence of relatively small numbers of adenomas in a familial setting. One of the most promising genetic loci showing linkage to an adenoma and carcinoma susceptibility locus and which could therefore explain the combination of multiple and aggressive adenomas is on chromosome 9q22.32 (Wiesner et al. 2003; Kemp et al. 2006; Skoglund et al. 2006).

Mechanisms Underlying the Initiation of Colorectal Adenomas

For many years, the concept of a *diffuse* hyper-proliferative field change was conceived as the earliest event in colorectal tumourigenesis (Deschner 1982). Interest waned when tests of hyper-proliferation were found to be of limited clinical value. At the same time animal models of colorectal tumourigenesis introduced the concept of minute *focal* lesions with malignant potential. Following the administration of carcinogens, such 'aberrant crypt foci' were visualized by staining the surface of the colonic epithelium with a vital dye such as methylene blue (Bird 1987; Bird et al. 1989). Under the dissecting microscope, the clusters of aberrant crypt openings were recognized by their increased size and increased staining intensity. Using a similar technique, minute lesions resembling aberrant crypt foci were subsequently identified in human colonic mucosa (Roncucci et al. 1991a, b). However, histological examination showed that these were frequently the minute counterparts of the two commonest types of colorectal polyp: adenoma and hyperplastic polyp. In FAP specimens, virtually all such lesions are micro-adenomas. In the colorectum of non-FAP patients, most of these lesions are either micro-hyperplastic polyps with serrated crypts or comprise clusters of slightly widened crypts with surface tufting but minimal epithelial serration (Roncucci et al. 1991a, b). The serrated hyperplastic aberrant crypt foci usually have *BRAF* mutation, while their minimally serrated counterparts usually have *KRAS* mutation (Rosenberg et al. 2007). Outside FAP, probably no more than 5% of these minute lesions are micro-adenomas (Jen et al.

1994). The term 'aberrant crypt focus', without further qualification, confers little meaning in the context of human tissues.

The condition FAP is rare in itself, but provides a highly accessible model for the study of micro-adenomas in humans. It has been shown that a single molecular event, namely disruption of the *APC* gene, is responsible for both the initiation and the subsequent growth of the adenoma in FAP (Lamlum et al. 2000). Loss of the APC protein prevents the normal degradation of the transcriptional co-activator β-catenin, and this, in turn, sends the Wnt signalling pathway into overdrive (Korinek et al. 1997). Nevertheless, in order for an adenoma to be initiated and then to grow into a recognizable lesion, there must be an optimal level of signalling mediated by β-catenin. This depends on a certain level of residual APC function as opposed to the complete loss of APC protein (Lamlum et al. 1999; Albuquerque et al. 2002; Schneikert and Behrens 2006). The APC protein includes a β-catenin regulating domain that comprises seven 20-amino acid repeats; at the gene level, there are a total of 14 such repeats in each normal cell. Most germline mutations (first hit) that cause classical FAP result in a protein containing only a single repeat and therefore a total of eight repeats within each cell. This is adequate for normal cell function. The second hit is usually the result of a mitotic recombination with loss of the wild-type allele (loss of seven repeats) and duplication of the mutant germline allele, leaving only two repeats in the cell. This appears to be the optimum dose for the initiation and subsequent growth of an adenoma. In the situation where the germline mutation causes complete loss of APC function, then the second or somatic hit is not associated with loss of heterozygosity but is typically a mutation causing loss of five repeats. This will again leave a total of two repeats in the cell (Lamlum et al. 1999; Albuquerque et al. 2002). This has been referred to as the 'just right' signalling model in which specific *APC* alterations are selected on the basis that a particular level of residual APC protein function is required to optimally drive the Wnt-signalling cascade and, in turn, tumourigenesis (Albuquerque et al. 2002). A reason for emphasizing the requirement for truncated APC (Schneikert and Behrens 2006) will be discussed later.

'Bottom-up' and 'top-down' models have been used to explain how loss of Wnt pathway regulation leads to the initiation of micro-adenomas (Shih et al. 2001; Preston et al. 2003). There is some confusion in relation to the terms 'bottom-up' and 'top-down', and this is because these terms have been applied to two different (though related) scenarios. The terms have been applied first to the mechanism of *initiation* of the uni-cryptal adenoma (Shih et al. 2001; Preston et al. 2003) and second to the location of the proliferative zone in *established* adenomas (Jass et al. 2002). The pioneering work of Nakamura and Kino (1984) established the 'bottom-up' mechanism at the point of initiation. Their micro-reconstruction studies in FAP specimens showed that the uni-cryptal adenoma begins as a minute bud or outgrowth close to the base of a normal-appearing crypt. Subsequently, the bud migrates upwards in the company of the normal crypt epithelium and, at the same time, extends into the surrounding lamina propria as a dysplastic or adenomatous tubule. Finally, the opening of the dysplastic tubule is relocated to the surface epithelium from which the uni-cryptal adenoma is suspended. The adenomatous

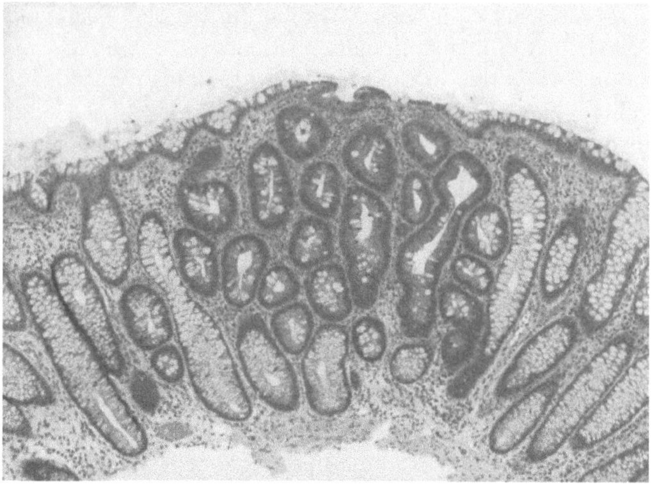

Fig. 5.1 Micro-adenoma in FAP. This lesion arises by 'bottom-up' initiation in which the earliest changes occurring in the lower compartment of a normal-appearing crypt. Ultimately and proliferating tubules come to occupy the superficial compartment giving a 'top-down' distribution of the proliferating adenomatous epithelium. Haematoxylin and Eosin

crypt so-formed is usually considerably shorter than a normal crypt but undergoes more frequent fission to form a micro-adenoma (Fig. 5.1). Through repeated crypt fission or branching, the superficial mucosal compartment is progressively populated by multiple adenomatous crypts. This results in a mass expansion that generates a macroscopically visible small nodule. The adenomatous cells may migrate laterally within the surface epithelium and even down adjacent normal crypts. This downward growth often telescopes or intussuscepts within the normal crypt (snow-plough effect). Therefore, even if the initiation of the neoplastic process is 'bottom-up', 'top-down' growth will occur subsequently (Preston et al. 2003). Additionally, the fact that proliferating adenomatous epithelium occupies the superficial compartment of the polyp while residual normal crypts dominate in the lower mucosal compartment has invited the use of the term 'top-down' to describe the profoundly altered location of the proliferative compartment in an established tubular adenoma (Jass et al. 2002).

Lack of Equivalence of FAP Versus Sporadic Micro-Adenomas

The sporadic model for the evolution of colorectal cancer is loosely based on the premise that carcinomas develop in adenomas and the latter are initiated through inactivation of both copies of the tumour suppressor gene, *APC*. In FAP, the first copy of *APC* is mutated in the germline and the second allele is mutated or lost somatically (see above). In sporadic adenomas, both copies are inactivated somatically. It is likely that that this mechanism does explain the initiation of some sporadic colorectal cancers, but enthusiasm must be tempered by the fact that FAP and sporadic micro-adenomas are not equivalent lesions.

The origin of the monoclonal micro-adenoma in FAP is discussed above. In FAP, further growth appears to depend on the mutually growth-enhancing effects of adjacent monoclonal micro-adenomas, which fuse to form a larger polyclonal mass (Novelli et al. 1996). This mechanism for growth does not apply to the far more isolated crypts of non-FAP micro-adenomas. Despite this early growth advantage for adenomas in FAP, it has been pointed out that the *individual* adenoma in FAP has a much lower potential for malignant transformation than a non-FAP adenoma. These observations raise the possibility that familial and non-familial micro-adenomas could be initiated by different mechanisms.

On the basis of Knudsen's hypothesis, all colorectal cancers should have a mutation of *APC*, and this should also apply to pre-cancerous lesions including the micro-adenoma (or dysplastic aberrant crypt focus). It is often assumed that the 'vast majority' of colorectal cancers have an *APC* mutation, though the mean frequency of *APC* mutation in primary colorectal cancers (as opposed to cell lines) is, in fact, around 60% (Powell et al. 1992; Aaltonen et al. 1993; Huang et al. 1996; Konishi et al. 1996; Olschwang et al. 1997; Salahshor et al. 1999; Shitoh et al. 2000; Kapiteijn et al. 2001; Jass et al. 2003; Samowitz et al. 2007). The frequency of *APC* mutation varies by anatomical location, being highest in the distal, and lowest in the proximal, colorectum (Kapiteijn et al. 2001). In sporadic micro-adenomas, the frequency of an *APC* mutation would be expected to be 100% and, indeed, was 100% in the first published study (Jen et al. 1994). However, there was only one micro-adenoma in that study. Sporadic micro-adenomas are rare and difficult to find. However, among 32 sporadic micro-adenomas identified in four studies, only six (19%) had an *APC* mutation (Jen et al. 1994; Otori et al. 1998; Takayama et al. 2001; Rosenberg et al. 2007). The frequency of *APC* mutation in small sporadic tubular adenomas is intermediate between micro-adenoma and cancer (33%) (Kim et al. 2001; Umetani et al. 2004). The frequency of *APC* mutation approaches that of cancer only in advanced adenomas (De Benedetti et al. 1994; Mulkens et al. 1998).

Taken together, the preceding data are disquieting and difficult to explain, but they cannot be ignored. It is possible that early dysplastic lesions with *APC* mutation are more likely to progress to cancer than those without *APC* mutation. However, the model provided by FAP, a condition in which all adenomas have bi-allelic *APC* alterations but individual adenomas rarely progress, does not support this suggestion. It is also possible that *APC* alterations may occur during progression of sporadic colorectal neoplasia. There is some evidence for this, insofar as in adenomas harbouring a sub-clone with carcinoma-in situ, loss of *APC* was shown to be restricted to this advanced sub-clone (Zauber et al. 1999). This finding links loss of *APC* with progression rather than initiation of sporadic neoplasia.

There is evidence that *APC* inactivation may not occur at any stage in the evolution of certain subsets of colorectal cancer. *APC* mutation occurs with reduced frequency in sporadic colorectal cancers with DNA microsatellite instability (MSI) (Salahshor et al. 1999; Jass et al. 2003; Samowitz et al. 2007), a subset which also shows extensive DNA methylation or the CpG island methylator phenotype (CIMP). *APC* may be inactivated by methylation of its promoter region (Esteller et al. 2000a), and it has been suggested that *APC* methylation could be functionally equivalent to *APC* mutation (Nathke 2004). This suggestion is implausible for three reasons: (1) there is no correlation between CIMP and methylation of *APC* (Esteller et al. 2000a); (2) bi-allelic methylation would result in *complete* silencing of *APC* when some residual functioning APC protein is necessary for initiating tumourigenesis (see earlier); and, (3) the finding of a normal immunohistochemical distribution of β-catenin in colorectal cancers with MSI is indicative of normal Wnt-signalling (Jass et al. 1999; Wong et al. 2002). It is frequently pointed out that mutation of other components of the Wnt-signalling pathway may substitute for *APC* mutation. The main contender for this role is *β-catenin*. However, mutation of *β-catenin* is restricted to colorectal cancers with MSI (Mirabelli-Primdahl et al. 1999) and, within that group, to a subset of Lynch syndrome cancers (Akiyama et al. 2000; Johnson et al. 2004). It is unclear whether or not Lynch syndrome adenomas are initiated by *β-catenin* mutation (Akiyama et al. 2000; Johnson et al. 2004). Nevertheless, it is very clear that *β-catenin* mutation occurs far too infrequently to fill the 40% gap represented by colorectal cancers that lack *APC* mutation.

It may be concluded that *APC* inactivation is not required in all instances of neoplastic initiation in the colorectum and may not be required at all in the evolution of certain subsets of colorectal cancer. Furthermore, although mutation of *KRAS* and mutation and/or aberrant expression of *TP53* are strongly associated with advanced adenomas (Barry et al. 2006; Einspahr et al. 2006), the fact that only around 10% of colorectal cancers are characterized by synchronous mutation of *APC*, *KRAS*, and *TP53* (Smith et al. 2002; Samowitz et al. 2007) leaves a sizeable gap in our knowledge of precancerous pathways.

Precursor Lesions for Sporadic Colorectal Cancer that Are Not Well Represented in FAP

In the preceding sections, it has been shown that the behaviour of colorectal adenomas varies considerably according to the clinical circumstances in which the adenoma presents. The model presented by FAP was used to illustrate the initiation of a micro-adenoma and its subsequent growth into a small tubular adenoma. It was then shown that early adenomas and neoplastic pathways occurring in the non-FAP setting are not necessarily equivalent to those occurring in the setting of FAP. In the following sections, it is argued that there are polypoid precursors to colorectal cancer that cannot be slotted in between the small tubular adenoma and colorectal cancer in order to provide the bridging element within a single linear sequence. These alternative precursor lesions include particular types of adenoma: villous adenoma and serrated adenoma, and particular types of serrated polyp that, until recently, were labelled as hyperplastic polyps with no malignant potential. None of these polyp types is well represented in FAP suggesting that they may be initiated by mechanisms other than inactivation of *APC*.

Villous Adenoma

Villous adenomas are rare, accounting for around 1% of all colorectal polyps. However, they are greatly over-represented within the residual adenomas occurring immediately adjacent to early colorectal cancers. One-third of such adenomas were villous adenomas in the detailed survey that underpinned the adenoma-carcinoma concept (Muto et al. 1975). In placing colorectal adenomas within a morphologic continuum in which the common tubular form is the least aggressive and the villous adenoma has achieved the greatest potential for malignant change, one gains the erroneous impression of the progressive transformation of a tubular adenoma into an adenoma with villi. In the classical villous adenoma, the surface epithelium is greatly increased in area and is folded into broad leaves or folds to produce a complex gyriform or cerebriform pattern. The false concept of finger-like villi results from the two-dimensional appearance of epithelial folds when these are viewed in histological sections. This three-dimensional cerebriform appearance can be appreciated when villous adenomas are studied during colonoscopy. By means of a vital dye and magnifying endoscopy, it is possible to identify minute villous adenomas with a cerebriform surface and no evidence of a pre-existing tubular adenoma (Kudo et al. 1996). This suggests that the villous adenoma develops *de novo* and not from a pre-existing tubular adenoma.

There is increasing evidence in support of the concept that tubular and villous adenomas are fundamentally different types of colorectal neoplastic polyps. The early development of the tubular adenoma, as exemplified in FAP, results in the generation of superficially arranged tubules lined by dysplastic cells. This produces a 'top-down' organization of the proliferative compartment (see above). In classical villous adenomas, the most undifferentiated and highly proliferative cells occur basally (as in normal mucosa), and the cells mature and differentiate as they ascend to their location on the expanded surface epithelium (a 'bottom-up' organization). In tubular adenomas, there is a striking loss of mucin production, but the opposite occurs in villous adenomas, with maturing cells becoming filled with large amounts of secretory mucin. The secretory mucin in villous adenomas is not only intestinal in type (MUC2) but inappropriately features gastric mucin (MUC5AC) (Takata et al. 2003). The transcription factor HATH1 is linked to differentiation of mucin-secreting cell lineages. There is repression of HATH1 in tubular adenomas and non-mucinous adenocarcinomas but strong expression in villous adenomas and mucinous carcinomas of the colorectum (Park et al. 2006).

Despite their importance as precancerous lesions, the rarity of villous adenomas, combined with the practice of lumping them with tubulovillous adenomas, means that little is known of their molecular pathology. *KRAS* mutation is very strongly associated with a villous architecture (Maltzman et al. 2001; Jass et al. 2006; Spring et al. 2006), but this may be occurring in the context of a progressing tubular to tubulovillous adenoma according to the Vogelstein model (Vogelstein et al. 1988). Although it has been argued that pure villous adenomas do not develop with tubular adenomas, there are early descriptions of villous adenomas arising in hyperplastic polyps (Goldman et al. 1970). In retrospect, the published figures appear to show serrated adenomas (see below) with a villous architecture. However, it is interesting that villous adenomas, serrated adenomas, and hyperplastic polyps are all characterized by increased expression of the mucin transcription factor HATH1 (Park et al. 2006) and by increased expression of both intestinal and gastric mucins (Biemer-Hüttmann et al. 1999; Takata et al. 2003).

Serrated Adenoma

Like villous adenomas, serrated adenomas are uncommon and accounted for less than 1% of polyps in the seminal 1990 study that reviewed over 18,000 colorectal polyps (Longacre and Fenoglio-Preiser 1990). This adenoma has a serrated architecture that occurs in hyperplastic polyps, but the epithelial lining has the dysplastic cytology of an adenoma (Fig. 5.2). A relatively high proportion (11%) of serrated adenomas contained foci of intra-mucosal carcinoma, indicating that these lesions do have malignant potential (Longacre and Fenoglio-Preiser 1990). In this seminal report, the authors noted the original diagnoses of the lesions. About one-third were diagnosed as hyperplastic polyps, one-third as adenomas, and one-third as a combination of the two. This illustrates that serrated adenomas do represent a spectrum. Those resembling hyperplastic polyps are more likely to be sessile, to occur in the proximal colon, and to have a tubular architecture. Those resembling adenomas are more likely to occur in the distal colorectum, to be pedunculated, and to have a tubulovillous or villous architecture. In the Japanese literature, these have been classified as 'Type 1' and 'Type 2' serrated adenomas, respectively (Matsumoto et al. 1999; Miwa et al. 2005).

Initially, serrated adenomas were not linked with hyperplastic polyps but were regarded as a type of adenoma that happened to display a superficial likeness to the hyperplastic polyp (Longacre and Fenoglio-Preiser. 1990). A subsequent study of mixed polyps provided a different interpretation and introduced the concept of a 'serrated pathway' to colorectal cancer. Mixed polyps were originally assumed to be simple chance collisions between a hyperplastic polyp and an adenoma (Longacre and Fenoglio-Preiser 1990). However, identical microsatellite mutations were demonstrated in DNA obtained from micro-dissected tissue obtained from the two components of mixed polyps. Furthermore, the adenomatous component frequently comprised serrated adenoma as opposed to the far more common, and therefore expected, tubular adenoma (Iino et al. 1999). Both observations made it highly unlikely that the two components could be chance collisions. The linking of hyperplastic polyps and serrated adenomas within a serrated pathway gained further strong support from the observation that both lesions showed frequent

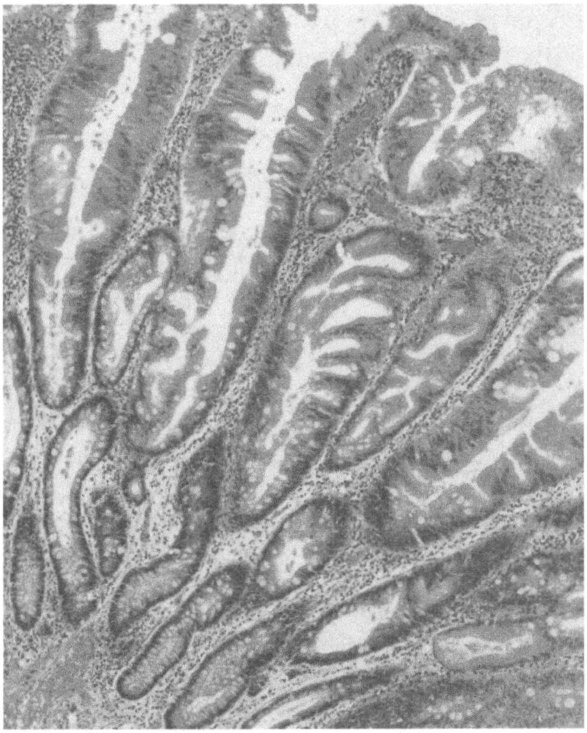

Fig. 5.2 A serrated adenoma in which dysplastic or adenomatous glands show the serrated architectural contour that is characteristic of a hyperplastic polyp. Haematoxylin and Eosin

mutation of the oncogene *BRAF* as well as extensive DNA methylation (Jass 2005). These molecular alterations are believed to be important in the initiation of serrated polyps but are rare events in tubular adenomas (Jass 2005).

There are occasional reports of serrated adenomas occurring in the context of FAP (Matsumoto et al. 2002), suggesting that glandular serration may occur secondarily within an adenoma initiated by inactivation of *APC*. However, most studies of sporadic serrated adenomas emphasize the infrequency of *APC* mutation and loss of the *APC* locus at 5q (Sawyer et al. 2002). In serrated adenomas without mutation of *BRAF*, one frequently finds mutation of *KRAS*. Only a minor subset will lack mutation in either of these oncogenes (Chan et al. 2003; Yang et al. 2004; Jass et al. 2006).

Seeking to assess their contribution to the burden of malignancy, one study found residual serrated adenoma adjacent to 5.8% of colorectal cancers (Mäkinen et al. 2001). Given the frequent destruction of precursor lesions by colorectal cancer, this is inevitably an underestimate of the actual proportion of colorectal cancers that develop in serrated adenomas. As noted earlier, about one-third of residual adenomas were described as villous adenomas at a time when serrated adenomas were not recognized (Muto et al. 1975). Despite the fact that villous adenoma and serrated adenoma may be overlapping categories, it is evident that the importance of both lesions as precursors of colorectal cancer is disproportionate to their rarity.

Hyperplastic Polyps and Allied Lesions with Malignant Potential

In contrast with villous and serrated adenomas, hyperplastic polyps are common lesions that have traditionally been regarded as entirely innocuous. Yet it is now apparent that these lesions may progress to serrated adenoma or to other forms of epithelial dysplasia with malignant potential. Like serrated adenomas, most hyperplastic polyps have mutation of either *KRAS* or *BRAF* (Yang et al. 2004; Jass et al. 2006; Spring et al. 2006). As noted, most micro-hyperplastic polyps (or hyperplastic aberrant crypt foci) also have mutation of *KRAS* or *BRAF* (Beach et al. 2005; Rosenberg et al. 2007). Epithelial polyps that are monoclonal proliferations initiated by mutation of oncogenes may be defined as benign neoplasms (Williams 1997). Nevertheless, hyperplastic polyps show a spectrum of cellular abnormalities that can be described as diametrically opposite to those of conventional adenomas. For example, the cells of hyperplastic polyps show features of both hyper-maturation and even senescence at the ultrastructural level (Kaye et al. 1973; Hayashi et al. 1974), yet cell kinetic studies show reduced rates of proliferation and migration along the crypt column (Hayashi et al. 1974).

It is now clear that mutation of oncogenes in isolation does not result in immediate tumourigenesis but leads to a state of inhibited cell proliferation and senescence (Serrano et al. 1997). The senescent phenotype occurs through the induction of cell regulatory proteins such as $p16^{INK4A}$, $p14^{ARF}$, $p19^{ARF}$, and Rb (Collado and Serrano 2006). In order to switch from a state of senescence to the state of progressive growth that characterizes neoplasia, one or more of the tumour suppressor genes encoding the preceding growth regulators must be inactivated (Michaloglou et al. 2005). Mutational activation of *BRAF* and *KRAS* not only induces a senescent phenotype but, depending on associated factors, is either anti-apoptotic or pro-apoptotic (Cox and Der 2003). There is evidence that a prevailing anti-apoptotic state is fundamental to the pathogenesis of serrated polyps (Tateyama et al. 2002; Komori et al. 2003). Pro-apoptotic molecules downstream of *KRAS* include RASSF1, RASSF2, RASSF5 (NORE1), and MST1. It is noteworthy that most of the genes encoding the preceding cell-cycle and pro-apoptotic proteins are prone to methylation of their promoter regions. Furthermore, methylation of these genes has been demonstrated in hyperplastic polyps (Minoo et al. 2006).

It is therefore likely that the early evolution and progressive growth of hyperplastic polyps depends on mutation of either *KRAS* or *BRAF* and synergies provided by the stepwise silencing of tumour suppressor genes implicated in the control of both cell proliferation and apoptosis. The requisite synergies may differ for *BRAF* and *KRAS*. Oncogenic *KRAS* may be viewed as providing *resistance* to apoptosis through the phosphorylation of pro-survival Akt. *BRAF*- initiated lesions may be more dependent on apoptotic *inhibition* through the methylation of pro-apoptotic genes. This is attested to by the fact that extensive DNA methylation, or the CpG island methylator phenotype-high (CIMP-high), is more evident in hyperplastic polyps with *BRAF* than with *KRAS* mutation (O'Brien et al. 2004). An association

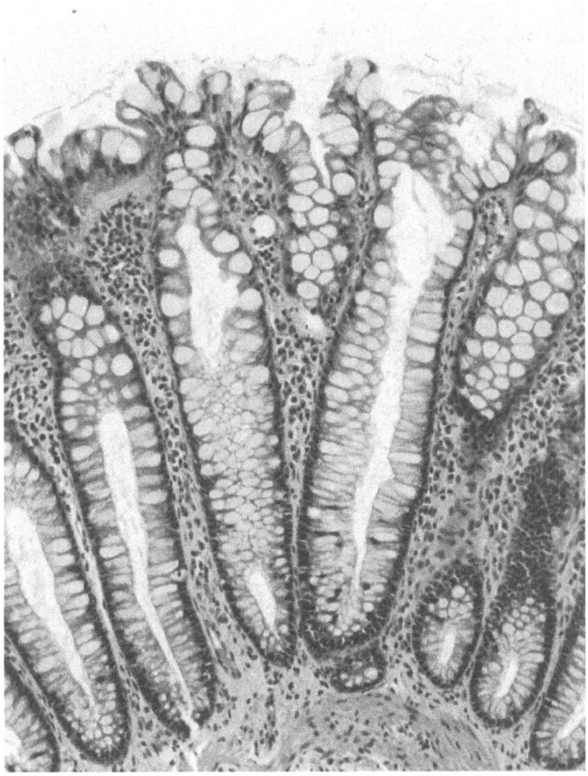

Fig. 5.3 Hyperplastic polyp, goblet cell type (HP-GC). This subtype deviates minimally from the normal and most have *KRAS* mutation. Haematoxylin and Eosin

between *BRAF* but not *KRAS* mutation with CIMP-high is also observed in colorectal cancer (Weisenberger et al. 2006), while *KRAS* mutation in colorectal cancer is associated with CIMP-low (Ogino et al. 2006).

As noted earlier, most hyperplastic polyps and even their minute counterparts (hyperplastic aberrant crypt foci) have either *BRAF* or *KRAS* mutation. Hyperplastic polyps with *KRAS* mutations are usually located in the distal colon and rectum, tend to remain small, and show the least deviation from normal in terms of their histological appearances (Yang et al. 2004; Spring et al. 2006). The retention of conspicuous goblet cells (as in normal colorectal mucosa) accounts for the term 'hyperplastic polyp/goblet cell type' (HP-GC) (Torlakovic et al. 2003) (Fig. 5.3). Hyperplastic polyps with *BRAF* mutation are located both proximally and distally, are relatively large, and show the greatest histologic deviation from the normal (Yang et al. 2004; Spring et al. 2006). The main histologic abnormalities are marked glandular serration and the presence of a prominent population of columnar cells containing mucin-filled microvesicles. This accounts for the term 'hyperplastic polyp/microvesicular type' (HP-MV) (Torlakovic et al. 2003) (Fig. 5.4).

Reports of malignant change in hyperplastic polyps have highlighted particular features of high-risk hyperplastic polyps, notably large size (Urbanski et al. 1984),

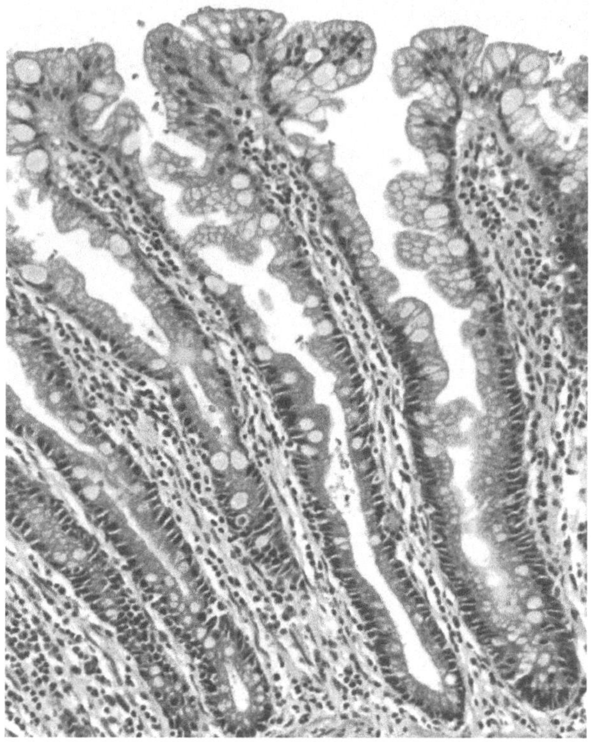

Fig. 5.4 Hyperplastic polyp, microvesicular type (HP-MV). Columnar cells with mucin-filled microvesicles and obvious serration define this subtype, most of which have *BRAF* mutation. Haematoxylin and Eosin

proximal location (Jass et al. 2000a), and multiplicity as seen in the condition, hyperplastic polyposis (Warner et al. 1994). It was through the study of hyperplastic polyposis that Torlakovic and Snover first proposed that some hyperplastic polyps were fundamentally different from the typical benign lesion and should be separated as a type of serrated adenoma (Torlakovic and Snover 1996). The concept of sessile and hyperplastic-like serrated adenomas was already hinted at in the study of Longacre and Torlakovic and subsequently developed in the Japanese literature, but the term serrated adenoma referred to polyps that were unequivocally dysplastic (see earlier). The work of Torlakovic and Snover expanded the concept of sessile serrated adenoma to include lesions that were atypical in architecture and proliferation but lacked the cytologic hallmarks of dysplasia (Fig. 5.5). They showed that 18% of sporadically occurring 'hyperplastic polyps' met the features of 'sessile serrated adenoma' (Torlakovic et al. 2003). In view of the lack of overt dysplasia, others suggested terms such as 'sessile serrated polyp' (Jass 2004) or 'serrated polyp with atypical proliferation' (Jass 2000; O'Brien et al. 2004) for these atypical hyperplastic polyps. The great majority of these atypical serrated polyps have *BRAF* mutation and extensive DNA methylation (Kambara et al. 2004a). As noted earlier, these features also characterize the HP-MV subtype of hyperplastic polyp. Therefore, the sessile serrated adenoma could be viewed as the

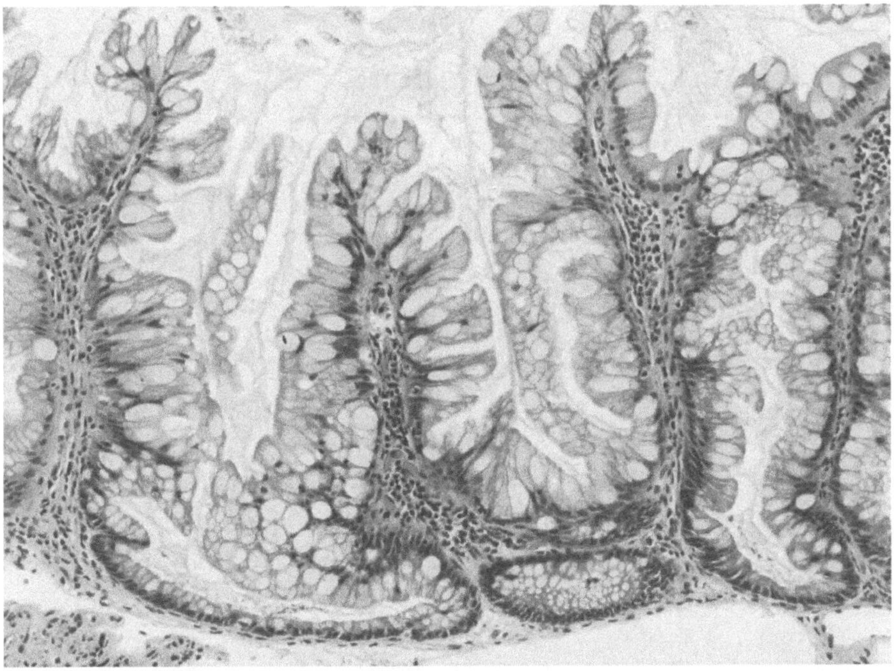

Fig. 5.5 Variant hyperplastic polyp known as sessile serrated adenoma. This differs from hyperplastic polyp/microvesicular type in showing greater architectural disorder and increased mucin production, but there is no overt cytologic atypia. The crypts show dilatation and exaggerated serration, and extend horizontally along the muscularis mucosae. Haematoxylin and Eosin

extreme end of the spectrum of HP-MV. At the other end of the spectrum are the micro-hyperplastic polyps (aberrant crypt foci) with a *BRAF* mutation (Rosenberg et al. 2007).

Colorectal Cancer: A Multi-pathway Disease

There is now a large body of data indicating that sporadic MSI-High (MSI-H) colorectal cancers do not develop through the classical adenoma-carcinoma sequence but within proximally located, atypical hyperplastic polyps or sessile serrated adenomas (Jass 1983; Biemer-Hüttmann et al. 2000; Jass et al. 2000a, b; Hawkins and Ward 2001; Goldstein et al. 2003; Oh et al. 2005). Although it was originally assumed that sporadic adenomas could show MSI-H (Grady et al. 1998), it was subsequently found that this applied almost exclusively to adenomas presenting in Lynch syndrome (Loukola et al. 1999). Conversely, loss of expression of MLH1 and MSI-H was demonstrated within dysplastic subclones in serrated polyps (Jass et al. 2000a; Oh et al. 2005) (Fig. 5.6a, b). Although it was clear that

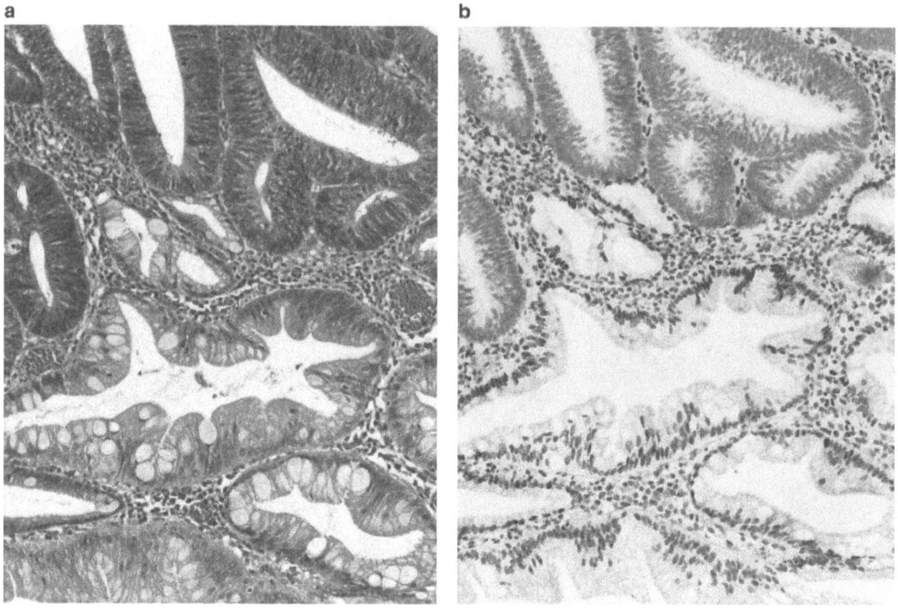

Fig. 5.6 Mixed polyp with 'hyperplastic' and dysplastic components (**a**). The dysplastic component shows loss of expression of MLH1 (**b**). (**a**) Haematoxylin and Eosin; (**b**) Immunohistochemical demonstration of nuclear MLH1 protein product

the 'classic' mutational spectrum implicating *APC*, *KRAS*, and *TP53* was rarely observed in sporadic MSI-H colorectal cancers (Olschwang et al. 1997; Salahshor et al. 1999), it was the demonstration that serrated polyps and sporadic MSI-H colorectal cancers were unique in sharing a genotype encompassing mutation of *BRAF* and extensive DNA methylation (CIMP-high) that cemented the concept of an independent serrated pathway (Chan et al. 2002, 2003; Kambara et al. 2004a, b; O'Brien et al. 2004, 2006; Yang et al. 2004).

Clinico-pathologic and molecular features of the classical and serrated pathways are shown in Table 5.1. Two important points emerge from this pathway subdivision. First, there is little or no overlap between the two pathways. This suggests that the two tumourigenic pathways are completely independent with different underlying causes, epidemiologic associations, natural histories, and clinical correlations. Second, there are clearly many colorectal cancers that do not fit into one or other of these two pathways. Only about 15% of non-familial colorectal cancers are MSI-H, and only about 30% are characterized by mutation of *APC* and *TP53* and chromosomal instability (CIN) (Smith et al. 2002). These groups correspond with Type 1 and 4 cancers in Fig. 5.7 (Jass 2007). Some 55% of colorectal cancers must be characterized by combinations of the molecular features that define the two largely independent pathways. These may be conceived as evolving through 'fusion' pathways (Jass et al. 2006).

Table 5.1 Clinical, molecular and pathologic features of prototype classical and serrated (alternative) pathways to colorectal cancer

Feature	Classical	Serrated
Precursor lesion	Adenoma	Serrated polyp
Gender predilection	Males	Females
Site predilection	Distal colorectum	Proximal colon
Genetic instability	Chromosomal	Microsatellite
DNA methylation	+	+ + +
Loss of heterozygosity	+ + +	−
APC mutation	+ + +	+
TP53 mutation	+ + +	−
BRAF mutation	−	+ + +
Mucin production	+	+ + +
Poor differentiation	+	+ + +
Lymphocytic infiltration	+	+++
Tumour budding[a]	+ + +	+
Dirty necrosis[b]	+ + +	+

[a]De-differentiation at invasive margin
[b]Tumour necrosis with abundant nuclear debris

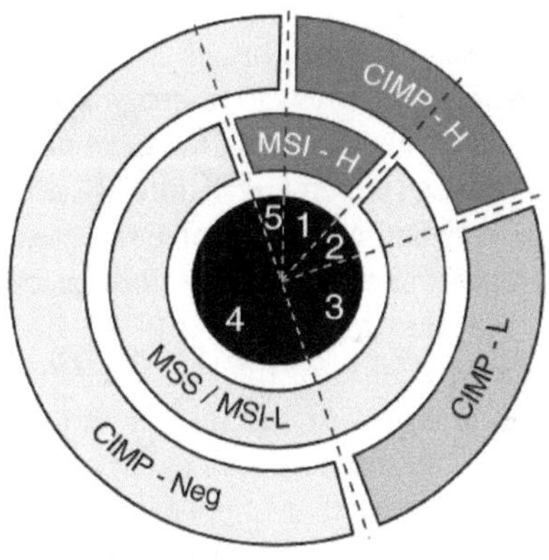

Fig. 5.7 Classification of colorectal cancer based on DNA methylation and DNA microsatellite instability. See text for explanation of Types 1–5

'Fusion Polyps' and 'Fusion Pathways' for the Accelerated Evolution of Colorectal Cancer

In the preceding sections, it was shown that adenomas are lesions characterized by loss of control of epithelial proliferation (for example through inactivation of *APC*) while resistance to apoptosis (achieved, for example, through pro-survival signalling by oncogenic *KRAS* or *BRAF*) is fundamental to the initiation of many serrated polyps. Increased proliferation and inhibition of apoptosis are key components of the malignant phenotype. It is therefore not surprising that advanced colorectal adenomas and a subset of carcinomas (corresponding to Type 3 and some Type

4 cancers in Fig. 5.7) should have alterations of both *APC* and *KRAS* (Vogelstein et al. 1988). This may be viewed as a 'fusion' of the key mutational steps involved in the initiation of different types of epithelial polyp (Jass et al. 2006). Colorectal neoplasms in Lynch syndrome (Type 5 in Fig. 5.7) may be viewed as another type of fusion. *APC* alterations are often found in adenomas in Lynch syndrome (Konishi et al. 1996). However, these neoplasms do not develop CIN as in the conventional adenoma-carcinoma sequence but share the MSI-H phenotype with non-familial cancers that evolve through the serrated pathway.

Although extensive DNA methylation or CIMP-high is present in virtually all non-familial MSI-H colorectal cancers, it is not limited to this subset. Extensive DNA methylation is found in a subset of colorectal cancers with *BRAF* mutation but lacking in MSI-H as well as CIN (Type 2 in Fig. 5.7) (Weisenberger et al. 2006; Goel et al. 2007). Less extensive DNA methylation (CIMP-low) is found in a further subset of colorectal cancers and is associated with *KRAS* mutation (Type 3 in Fig. 5.7) (Issa 2005; Weisenberger et al. 2006; Ogino et al. 2007). Little is known of the mechanisms underlying CIMP. However, it is clear that CIMP-high may be fully developed within hyperplastic polyps and even in the normal colonic mucosa of subjects with hyperplastic polyps (Minoo et al. 2006). On the other hand, CIMP-high is absent in the tubular adenomas of subjects with FAP (Wynter et al. 2006). DNA methylation becomes more extensive in colorectal adenomas in association with progression (Rashid et al. 2000; Kim et al. 2005; O'Brien et al. 2006). Mechanisms underlying the development of CIMP within a progressing adenoma are likely to be radically different from the mechanisms applying in normal mucosa and in hyperplastic polyps (which are manifestly not advanced lesions). Studies that have considered the distribution of methylation across all colorectal cancers, irrespective of pathways and pathogenesis, are therefore lacking in biologic meaning (Yamashita et al. 2003).

The development of MSI-H in sporadic colorectal cancer is largely explained by methylation of the DNA mismatch repair gene *MLH1* (Kane et al. 1997). Loss of expression of MLH1, as shown by immunohistochemistry, occurs in serrated polyps, though as a rare event (Jass et al. 2000a, b). However, once this critical step is established there appears to be a rapid transformation into a malignant phenotype. This is evident from the fact that sub-clones in serrated polyps with loss of expression of MLH1 are usually dysplastic if not frankly malignant (Oh et al. 2005; Goldstein 2006; Sheridan et al. 2006).

O-6-Methylguanine DNA methyltransferase (*MGMT*) is a type of direct DNA repair gene that, like *MLH1*, is inactivated by DNA methylation (Esteller et al. 1999). Unlike MLH1, however, loss of expression of MGMT can be observed relatively frequently in the non-dysplastic crypts of hyperplastic polyps (Whitehall et al. 2001), and *MGMT* methylation even occurs in hyperplastic aberrant crypt foci (Greenspan et al. 2007). Methylation and loss of expression of MGMT also occur within conventional adenomas and particularly in villous adenomas (Rashid et al. 2001). Deficiency of the direct DNA repair protein, MGMT, generates highly mutagenic methylG:T mismatches which may predispose not only to mutation but to CIN through futile cycles of attempted repair (Karran and Bignami 1994). The reason why loss of MGMT may occur relatively frequently in different types of polyp without rapidly driving further tumourigenesis (as is the case for MLH1 deficiency) is because either successful repair of DNA damage is achieved or persisting DNA or chromosomal damage will trigger pro-apoptotic signalling by MLH1 and p53, respectively (Fishel 1999). Further neoplastic evolution may therefore depend on the inactivation of either *MLH1* or *TP53* or a co-functioning alternative to *TP53* such as $p14^{ARF}$ (Esteller et al. 2000b). The cellular response to the presence of DNA mismatches may sometimes lie between the extremes of efficient repair versus programmed cell death. In the presence of an over-extended mismatch repair system, persisting DNA mismatches may give rise to low-level MSI (MSI-L). This would explain the association between *MGMT* methylation and MSI-L (Whitehall et al. 2001; Greenspan et al. 2007; Ogino et al. 2007).

Methylation of *MGMT* is common to both serrated polyps and conventional adenomas and could therefore serve as a unifying mechanism to explain different types of 'fusion pathway'. MethylG:T mismatches that develop as a consequence of *MGMT* inactivation predispose to G:C to A:T mutations. This restricted mutational signature occurs in association with *MGMT* methylation in both *KRAS* (Esteller 2000) and *TP53* (Esteller et al. 2001). In the case of serrated polyps, mutation of *TP53* may underlie the evolution of CIMP-high colorectal cancer without the advent of MSI-H (some Type 2 cancers in Fig. 5.7) (Jass et al. 2006). Conversely, the presence of *MGMT* methylation in villous adenomas (Rashid et al. 2001) explains the frequent finding of *KRAS* mutation in this subset (Maltzman et al. 2001). *KRAS* mutation is generally perceived as either initiating minor and non-progressing serrated lesions (see earlier) or serving as a key step in the progression of established adenomas (O'Brien et al. 2006). However, it is possible that serrated lesions initiated by *KRAS* mutation may occasionally progress, thus explaining the finding of serrated adenomas with *KRAS* mutation (Jass et al. 2006). Indeed there is evidence from both animal (Janssen et al. 2002) and human studies (Takayama et al. 2001) that *KRAS* mutation can initiate the development of dysplastic lesions in the colorectum. Because *KRAS* mutation is associated with CIMP-low, it is conceivable that bi-allelic methylation of *APC* (Esteller et al. 2000) could drive neoplastic progression in lesions previously initiated by *KRAS* mutation and provide a further example of a 'fusion pathway' (Jass et al. 2006).

Conclusion

It is still widely accepted that the vast majority of colorectal cancers develop in adenomas that are in turn initiated by bi-allelic inactivation of the *APC* gene. On this basis, adenomas occurring in the autosomal dominant condition FAP should be the exact counterparts of the benign precursors of most sporadic colorectal cancers. Three lines of inquiry indicate that FAP provides a limited model for explaining the early evolution of sporadic colorectal cancer. These lines of inquiry highlight: (1) the strikingly different malignant potential of adenomas occurring in different clinical presentations; (2) the lack of *APC* mutation in early and advanced sporadic colorectal neoplasms; and (3) the contribution to colorectal malignancy by types of polyp (villous adenoma, serrated adenoma, and variants of hyperplastic polyp) that are biologically distinct and, essentially, do not arise from, the tubular adenoma (the hallmark lesion initiated by *APC* inactivation).

Colorectal cancer is a multi-pathway disease. The existence of two largely non-overlapping pathways to colorectal cancer implies a separation of underlying causes, epidemiologic associations, and natural histories. However, the classical adenoma-carcinoma sequence and the serrated pathway are only prototypes. Molecular features unique to one or another prototype co-occur in various forms and combinations within 'fusion pathways'. Effective cancer control by either chemo-prevention or the endoscopic removal of precursor lesions depends upon a fundamental understanding of the differing early routes to colorectal cancer.

References

Aaltonen, L. A., P. S. Peltomaki, et al. (1993). "Clues to the pathogenesis of familial colorectal cancer." *Science* **260**: 812–816.

Akiyama, Y., H. Nagasaki, et al. (2000). "Beta-catenin and adenomatous polyposis coli (APC) mutations in adenomas from hereditary non-polyposis colorectal cancer patients." *Cancer Lett* **157**: 185–191.

Al-Tassan, N., N. H. Chmiel, et al. (2002). "Inherited variants of MYH associated with somatic G:C – T:A mutations in colorectal tumors." *Nat Genet* **30**: 227–232.

Albuquerque, C., C. Breukel, et al. (2002). "The 'just-right' signaling model: APC somatic mutations are selected based on a specific level of activation of the beta-catenin signaling cascade." *Hum Mol Genet* **11**: 1549–1560.

Almendingen, K., B. Hofstad, et al. (2003). "Does a family history of cancer increase the risk of occurrence, growth, and recurrence of colorectal cancer." *Gut* **52**: 747–751.

Atkin, W. S., B. C. Morson, et al. (1992). "Long-term risk of colorectal cancer after excision of rectosigmoid adenomas." *N Engl J Med* **326**(10): 658–662.

Barry, E. L. R., J. A. Baron, et al. (2006). "K-ras mutations in incident sporadic colorectal adenomas." *Cancer* **106**: 1036–1040.

Beach, R., A. O.-O. Chan, et al. (2005). "BRAF mutations in aberrant crypt foci and hyperplastic polyposis." *Am J Pathol* **166**: 1069–1075.

Biemer-Hüttmann, A.-E., M. D. Walsh, et al. (1999). "Immunohistochemical staining patterns of MUC1, MUC2, MUC4, and MUC5AC mucins in hyperplastic polyps, serrated adenomas, and traditional adenomas of the colorectum." *J Histochem Cytochem* **47**: 1039–1047.

Biemer-Hüttmann, A.-E., M. D. Walsh, et al. (2000). "Mucin core protein expression in colorectal cancers with high levels of microsatellite instability indicates a novel pathway of morphogenesis." *Clin Cancer Res* **6**: 1909–1916.

Bird, R. P. (1987). "Observation and quantification of aberrant crypts in the murine colon treated with a colon carcinogen: preliminary findings." *Cancer Lett* **37**(2): 147–151.

Bird, R. P., E. A. McLellan, et al. (1989). "Aberrant crypts, putative precancerous lesions, in the study of the role of diet in the aetiology of colon cancer." *Cancer Surv* **8**(1): 189–200.

Chan, A. O., J.-P. J. Issa, et al. (2002). "Concordant CpG island methylation in hyperplastic polyposis." *Am J Pathol* **160**(2): 529–536.

Chan, T. L., W. Zhao, et al. (2003). "BRAF and KRAS mutations in colorectal hyperplastic polyps and serrated adenomas." *Cancer Res* **63**: 4878–4881.

Collado, M. and M. Serrano (2006). "The power and promise of oncogene-induced senescence markers." *Nat Rev Cancer* **6**: 472–476.

Cox, A. D. and C. J. Der (2003). "The dark side of Ras: regulation of apoptosis." *Oncogene* **22**: 8999–9006.

Day, D. W., J. R. Jass, et al. (2003). *Morson and Dawson's Gastrointestinal Pathology*. Oxford: Blackwell.

De Benedetti, L., S. Sciallero, et al. (1994). "Association of APC gene mutations and histological characteristics of colorectal adenomas." *Cancer Res* **54**: 3553–3556.

Deschner, E. E. (1982). "Early proliferative changes in gastrointestinal neoplasms." *Am J Gastroenterol* **77**: 207–211.

Eide, T. J. (1986a). "Prevalence and morphological features of adenomas of the large intestine in individuals with and without colorectal carcinoma." *Histopathology* **10**: 111.

Eide, T. J. (1986b). "Risk of colorectal cancer in adenoma-bearing individuals within a defined population." *Int J Cancer* **38**: 173–176.

Einspahr, J. G., M. E. Martinez, et al. (2006). "Associations of Ki-ras proto-oncogene mutation and p53 gene overexpression in sporadic colorectal adenomas with demographic and clinicopathologic characteristics." *Cancer Epidemiol Biomarkers Prev* **15**: 1443–1450.

Esteller, M. (2000). "Epigenetic lesions causing genetic lesions in human cancer: promoter hypermethylation of DNA repair genes." *Eur J Cancer* **36**: 2294–2300.

Esteller, M., S. R. Hamilton, et al. (1999). "Inactivation of the DNA repair gene *O^6-methylguanine-DNA methyltransferase* by promoter hypermethylation is a common event in primary human neoplasia." *Cancer Res* **59**(4): 793–797.

Esteller, M., R. A. Risques, et al. (2001). "Promoter hypermethylation of the DNA repair gene *O^6-methylguanine-DNA methyltransferase* is associated with the presence of G:C to A:T transition mutations in *p53* in human colorectal tumorigenesis." *Cancer Res* **61**(12): 4689–4692.

Esteller, M., A. Sparks, et al. (2000a). "Analysis of adenomatous polyposis coli promoter hypermethylation in human cancer." *Cancer Res* **60**(16): 4366–4371.

Esteller, M., S. Tortola, et al. (2000b). "Hypermethylation-associated inactivation of $p14^{ARF}$ is independent of $p16^{INK4A}$ methylation and *p53* mutational status." *Cancer Res* **60**: 129–133.

Fishel, R. (1999). "Signaling mismatch repair in cancer." *Nat Med* **5**(11): 1239–1241.

Goel, A., T. Nagasaka, et al. (2007). "The CpG island methylator phenotype and chromosomal instability are inversely correlated in sporadic colorectal cancer." *Gastroenterology* **132**: 127–138.

Goldman, H., S. Ming, et al. (1970). "Nature and significance of hyperplastic polyps of the human colon." *Arch Pathol* **89**: 349–354.

Goldstein, N. S. (2006). "Small colonic microsatellite unstable adenocarcinomas and high-grade epithelial dysplasias in sessile serrated adenoma polypectomy specimens. A study of eight cases." *Am J Clin Pathol* **125**: 132–145.

Goldstein, N. S., P. Bhanot, et al. (2003). "Hyperplastic-like colon polyps that preceded microsatellite unstable adenocarcinomas." *Am J Clin Pathol* **119**: 778–796.

Grady, W. M., A. Rajput, et al. (1998). "Mutation of the type II transforming growth factor-β receptor is coincident with the transformation of human colon adenomas to malignant carcinomas." *Cancer Res* **58**: 3101–3104.

Greenspan, E. J., J. L. Cyr, et al. (2007). "Microsatellite instability in aberrant crypt foci from patients without concurrent cancer." *Carcinogenesis* **28**: 769–776.

Hawkins, N. J. and R. L. Ward (2001). "Sporadic colorectal cancers with microsatellite instability and their possible origin in hyperplastic polyps and serrated adenomas." *J Natl Cancer Inst* **93**(17): 1307–1313.

Hayashi, T., R. Yatani, et al. (1974). "Pathogenesis of hyperplastic polyps of the colon: a hypothesis based on ultrastructure and in vitro cell kinetics." *Gastroenterology* **66**: 347–356.

Huang, J., N. Papadopoulos, et al. (1996). "APC mutations in colorectal tumors with mismatch repair deficiency." *Proc Natl Acad Sci U S A* **93**: 9049–9054.

Iino, H., J. R. Jass, et al. (1999). "DNA microsatellite instability in hyperplastic polyps, serrated adenomas, and mixed polyps: a mild mutator pathway for colorectal cancer?" *J Clin Pathol* **52**: 5–9.

Iino, H., L. A. Simms, et al. (2000). "DNA microsatellite instability and mismatch repair protein loss in adenomas presenting in hereditary non-polyposis colorectal cancer." *Gut* **47**: 37–42.

Issa, J.-P. (2005). "CIMP, at last." *Gastroenterology* **129**: 1121–1124.

Janssen, K.-P., F. El Marjou, et al. (2002). "Targetted expression of oncogenic K-ras in intestinal epithelium causes spontaneous tumorigenesis in mice." *Gastroenterology* **123**: 492–504.
Jass, J. R. (1983). "Relation between metaplastic polyp and carcinoma of the colorectum." *Lancet* **i**: 28–30.
Jass, J. R. (2000). "Replication error phenotype in colorectal cancer (letter)." *Gut* **47**: 597–598.
Jass, J. R. (2004). "Hyperplastic polyps and colorectal cancer: is there a link?" *Clin Gastroenterol Hepatol* **2**: 1–8.
Jass, J. R. (2005). "Serrated adenoma of the colorectum and the DNA-methylator phenotype." *Nat Clin Pract Oncol* **2**: 398–405.
Jass, J. R. (2007). "Classification of colorectal cancer based on correlation of clinical, morphological and molecular features." *Histopathology* **50**: 113–130.
Jass, J. R., K. G. Biden, et al. (1999). "Characterisation of a subtype of colorectal cancer combining features of the suppressor and mild mutator pathways." *J Clin Pathol* **52**: 455–460.
Jass, J. R., H. Iino, et al. (2000a). "Neoplastic progression occurs through mutator pathways in hyperplastic polyposis of the colorectum." *Gut* **47**: 43–49.
Jass, J. R., J. Young, et al. (2000b). "Hyperplastic polyps and DNA microsatellite unstable cancers of the colorectum." *Histopathology* **37**: 295–301.
Jass, J. R., V. L. J. Whitehall, et al. (2002). "Emerging concepts in colorectal neoplasia." *Gastroenterology* **123**: 862–876.
Jass, J. R., M. Barker, et al. (2003). "APC mutation and tumour budding in colorectal cancer." *J Clin Pathol* **56**: 69–73.
Jass, J. R., K. Baker, et al. (2006). "Advanced colorectal polyps with the molecular and morphological features of serrated polyps and adenomas: concept of a 'fusion' pathway to colorectal cancer." *Histopathology* **49**: 121–131.
Jen, J., S. M. Powell, et al. (1994). "Molecular determinants of dysplasia in colorectal lesions." *Cancer Res* **54**: 5523–5526.
Johnson, V., E. Volikos, et al. (2004). "Exon 3 beta-catenin mutations are specifically associated with colorectal carcinomas in the hereditary non-polyposis colorectal cancer syndrome." *Gut* **53**: 264–267.
Kambara, T., L. A. Simms, et al. (2004a). "BRAF mutation and CpG island methylation: an alternative pathway to colorectal cancer." *Gut* **53**: 1137–1144.
Kambara, T., V. L. J. Whitehall, et al. (2004b). "Role of inherited defects of MYH in the development of sporadic colorectal cancer." *Genes Chromosomes Cancer* **40**: 1–9.
Kane, M. F., M. Loda, et al. (1997). "Methylation of the hMLH1 promoter correlates with lack of expression of hMLH1 in sporadic colon tumors and mismatch repair-defective human tumor cell lines." *Cancer Res* **57**: 808–811.
Kapiteijn, E., G. J. Liefers, et al. (2001). "Mechanisms of oncogenesis in colon versus rectal cancer." *J Pathol* **195**: 171–178.
Karran, P. and M. Bignami (1994). "DNA damage tolerance, mismatch repair and genome instability." *Bioessays* **16**(11): 833–839.
Kaye, G. E., C. M. Fenoglio, et al. (1973). "Comparative electron microscopic features of normal, hyperplastic and adenomatous human colonic epithelium." *Gastroenterology* **64**: 926–945.
Kemp, Z. E., L. G. Carvajal-Carmona, et al. (2006). "Evidence of linkage to chromosome 9q22.32 in colorectal cancer kindreds in UK." *Cancer Res* **66**: 5003–5006.
Kim, J. C., K. H. Koo, et al. (2001). "Mutations at the APC exon 15 in the colorectal neoplastic tissues of serial array." *Int J Colorectal Dis* **16**(2): 102–107.
Kim, H. C., S. A. Roh, et al. (2005). "CpG island methylation as an early event during adenoma progression in carcinogenesis of sporadic colorectal cancer." *J Gastroenterol Hepatol* **20**: 1920–1926.
Komori, K., Y. Ajioka, et al. (2003). "Proliferation kinetics and apoptosis of serrated adenoma of the colorectum." *Pathol Int* **53**: 277–283.
Konishi, M., R. Kikuchi-Yanoshita, et al. (1996). "Molecular nature of colon tumors in hereditary nonpolyposis colon cancer, familial polyposis, and sporadic colon cancer." *Gastroenterology* **111**: 307–317.

Korinek, V., N. Barker, et al. (1997). "Constitutive transcriptional activation by a β-catenin-Tcf complex in $APC^{-/-}$ colon carcinoma." *Science* **275**(5307): 1784–1787.

Kudo, S., S. Tamura, et al. (1996). "Diagnosis of colorectal tumorous lesions by magnifying endoscopy." *Gastrointest Endosc* **44**(1): 8–14.

Lamlum, H., M. Ilyas, et al. (1999). "The type of somatic mutation of APC in familial adenomatous polyposis is determined by the site of the germline mutation: a new facet to Knudson's 'two-hit' hypothesis." *Nat Med* **5**: 1071–1075.

Lamlum, H., A. Papadopoulou, et al. (2000). "APC mutations are sufficient for the growth of early colorectal adenomas." *Proc Natl Acad Sci U S A* **97**: 2225–2228.

Longacre, T. A. and C. M. Fenoglio-Preiser (1990). "Mixed hyperplastic adenomatous polyps/serrated adenomas. A distinct form of colorectal neoplasia." *Am J Surg Pathol* **14**: 524–537.

Loukola, A., R. Salovaara, et al. (1999). "Microsatellite instability in adenomas as a marker for hereditary nonpolyposis colorectal cancer." *Am J Pathol* **155**: 1849–1853.

Lovett, E. (1976). "Family studies in cancer of the colon and rectum." *Br J Surg* **63**: 13–18.

Mäkinen, M. J., S. M. C. George, et al. (2001). "Colorectal carcinoma associated with serrated adenoma – prevalence, histological features, and prognosis." *J Pathol* **193**: 286–294.

Maltzman, T., K. Knoll, et al. (2001). "Ki-ras proto-oncogene mutations in sporadic colorectal adenomas: relationship to histologic and clinical characteristics." *Gastroenterology* **121**: 302–309.

Matsumoto, T., M. Mizuno, et al. (1999). "Serrated adenoma of the colorectum: colonoscopic and histologic features." *Gastrointest Endosc* **49**: 736–742.

Matsumoto, T., M. Iida, et al. (2002). "Serrated adenoma in familial adenomatous polyposis: relation to germline APC gene mutation." *Gut* **50**: 402–404.

Michaloglou, C., C. W. Vredeveld, et al. (2005). "BRAFE600-associated senescence-like cell cycle arrest of human naevi." *Nature* **436**: 720–724.

Minoo, P., K. Baker, et al. (2006). "Extensive DNA methylation in normal colorectal mucosa in hyperplastic polyposis." *Gut* **55**: 1467–1474.

Mirabelli-Primdahl, L., R. Gryfe, et al. (1999). "Beta-catenin mutations are specific for colorectal carcinomas with microsatellite instability but occur in endometrial carcinomas irrespective of mutator pathway." *Cancer Res* **59**: 3346–3351.

Miwa, S., H. Mitomi, et al. (2005). "Clinicopathologic differences among subtypes of serrated adenomas of the colorectum." *Hepatogastroenterol* **52**: 437–440.

Morson, B. C. (1966). "Factors influencing the prognosis of early cancer of the rectum." *Proc R Soc Med* **59**(7): 607–608.

Morson, B. C. and L. H. Sobin (1976). *Histological Typing of Intestinal Tumours*. Berlin: Springer.

Morson, B. C., H. J. R. Bussey, et al. (1983). "Adenomas of the large bowel." *Cancer Surv* **2**: 451–477.

Mulkens, J., J. Poncin, et al. (1998). "APC mutations in human colorectal adenomas: analysis of the mutation cluster region with temperature gradient gel electrophoresis and clinicopathological features." *J Pathol* **185**(4): 360–365.

Muto, T., H. J. Bussey, et al. (1975). "The evolution of cancer of the colon and rectum." *Cancer* **36**(6): 2251–2270.

Nakamura, S. and I. Kino (1984). "Morphogenesis of minute adenomas in familial polyposis coli." *J Natl Cancer Inst* **73**: 41–49.

Nathke, I. S. (2004). "The adenomatous polyposis coli protein: the Achilles heel of the gut epithelium." *Annu Rev Cell Dev Biol* **20**: 337–366.

Novelli, M. R., J. A. Williamson, et al. (1996). "Polyclonal origin of colonic adenomas in an XO/XY patient with FAP." *Science* **272**: 1187–1190.

O'Brien, M. J., S. Yang, et al. (2004). "Hyperplastic (serrated) polyps of the colorectum. Relationship of CpG island methylator phenotype and K-ras mutation to location and histologic subtype." *Am J Surg Pathol* **28**: 423–434.

O'Brien, M. J., S. Yang, et al. (2006). "Comparison of microsatellite instability, CpG island methylation phenotype, BRAF and KRAS status in serrated polyps and traditional adenomas indicates separate pathways to distinct colorectal carcinoma end points." *Am J Surg Pathol* **30**: 1491–1501.

Ogino, S., M. Cantor, et al. (2006). "CpG island methylator phenotype (CIMP) of colorectal cancer is best characterized by quantitative DNA methylation analysis and prospective cohort studies." *Gut* **55**: 1000–1006.

Ogino, S., T. Kawasaki, et al. (2007). "Molecular correlates with MGMT promoter methylation and silencing support CpG island methylator phenotype-low (CIMP-low) in colorectal cancer." *Gut* **56**: 1564–1571.

Oh, K., M. Redston, et al. (2005). "Support for hMLH1 and MGMT silencing as a mechanism of tumorigenesis in the hyperplastic-adenoma-carcinoma (serrated) carcinogenic pathway in the colon." *Hum Pathol* **36**: 101–111.

Olschwang, S., R. Hamelin, et al. (1997). "Alternative genetic pathways in colorectal carcinogenesis." *Proc Natl Acad Sci U S A* **94**: 12122–12127.

Otori, K., M. Konishi, et al. (1998). "Infrequent somatic mutation of the adenomatous polyposis coli gene in aberrant crypt foci of human colon tissue." *Cancer* **83**: 896–900.

Park, E. T., H. K. Oh, et al. (2006). "HATH1 expression in mucinous cancers of the colorectum and related lesions." *Clin Cancer Res* **12**: 5403–5410.

Powell, S. M., N. Zilz, et al. (1992). "APC mutations occur early during colorectal tumorigenesis." *Nature* **359**: 235–237.

Preston, S. L., W.-D. Wong, et al. (2003). "Bottom-up histogenesis of colorectal adenomas: origin in the monocryptal adenoma and initial expansion by crypt fission." *Cancer Res* **63**: 3819–3825.

Rashid, A., S. Houlihan, et al. (2000). "Phenotypic and molecular characteristics of hyperplastic polyposis." *Gastroenterology* **119**: 323–332.

Rashid, A., L. Shen, et al. (2001). "CpG island methylation in colorectal adenomas." *Am J Pathol* **159**: 1129–1135.

Roncucci, L., A. Medline, et al. (1991a). "Classification of aberrant crypt foci and microadenomas in human colon." *Cancer Epidemiol Biomarkers Prev* **1**: 57–60.

Roncucci, L., D. Stamp, et al. (1991b). "Identification and quantification of aberrant crypt foci and microadenomas in the human colon." *Hum Pathol* **22**: 287–294.

Rosenberg, D. W., S. Yang, et al. (2007). "Mutations of BRAF and KRAS differentially distinguish serrated versus non-serrated hyperplastic aberrant crypt foci in humans." *Cancer Res* **67**: 3551–3554.

Salahshor, S., U. Kressner, et al. (1999). "Colorectal cancer with and without microsatellite instability involves different genes." *Genes Chromosomes Cancer* **26**: 247–252.

Samowitz, W. S., M. L. Slattery, et al. (2007). "APC mutations and other genetic and epigenetic changes in colon cancer." *Mol Cancer Res* **5**: 165–170.

Sawyer, E. J., A. Cerar, et al. (2002). "Molecular characteristics of serrated adenomas." *Gut* **51**: 200–206.

Schneikert, J. and J. Behrens (2006). "Truncated APC is required for cell proliferation and DNA replication." *Int J Cancer* **119**: 74–79.

Serrano, M., A. W. Lin, et al. (1997). "Oncogenic ras provokes premature cell senescence associated with accumulation of p53 and p16INK4a." *Cell* **88**: 593–602.

Sheridan, T. B., H. Fenton, et al. (2006). "Sessile serrated adenomas with low- and high-grade dysplasia and early carcinoma: an immunohistochemical study of serrated lesions 'caught in the act'." *Am J Clin Pathol* **126**: 564–571.

Shih, I.-M., T. L. Wang, et al. (2001). "Top-down morphogenesis of colorectal tumors." *Proc Natl Acad Sci U S A* **98**(5): 2640–2645.

Shitoh, K., F. Konishi, et al. (2000). "Pathogenesis of non-familial colorectal carcinomas with high microsatellite instability." *J Clin Pathol* **53**: 841–845.

Skoglund, J., T. Djureinovic, et al. (2006). "Linkage analysis in a large Swedish family supports the presence of a susceptibility locus for adenoma in colorectal cancer on chromosome 9q22.32-31.1." *J Med Genet* **43**: e7.

Smith, G., F. A. Carey, et al. (2002). "Mutations in APC, Kirsten-ras, and p53 – alternative genetic pathways to colorectal cancer." *Proc Natl Acad Sci U S A* **99**: 9433–9438.

Spratt, J. S., L. V. Ackerman, et al. (1958). "Relationships of polyps of the colon to colonic cancer." *Ann Surg* **148**: 682–696.

Spring, K. J., Z. Z. Zhao, et al. (2006). "High prevalence of sessile serrated adenomas with BRAF mutations: a prospective study of patients undergoing colonoscopy." *Gastroenterology* **131**: 1400–1407.

Takata, M., T. Yao, et al. (2003). "Phenotypic alteration in malignant transformation of colonic villous tumours: with special reference to a comparison with tubular adenomas." *Histopathology* **43**: 332–339.

Takayama, T., M. Ohi, et al. (2001). "Analysis of K-ras, APC, and beta-catenin in aberrant crypt foci in sporadic adenoma, cancer, and familial adenomatous polyposis." *Gastroenterology* **121**(3): 599–611.

Tateyama, H., W. Li, et al. (2002). "Apoptosis index and apoptosis-related antigen expression in serrated adenoma of the colorectum: the saw-toothed structure may be related to inhibition of apoptosis." *Am J Surg Pathol* **26**(2): 249–256.

Torlakovic, E. and D. C. Snover (1996). "Serrated adenomatous polyposis in humans." *Gastroenterology* **110**: 748–755.

Torlakovic, E., E. Skovlund, et al. (2003). "Morphologic reappraisal of serrated colorectal polyps." *Am J Surg Pathol* **27**: 65–81.

Umetani, N., A. Fujimoto, et al. (2004). "Allelic imbalance of APAF-1 locus at 12q23 is related to progression of colorectal carcinoma." *Oncogene* **23**: 8292–8300.

Urbanski, S. J., N. Marcon, et al. (1984). "Mixed hyperplastic adenomatous polyps – an underdiagnosed entity. Report of a case of adenocarcinoma arising within a mixed hyperplastic adenomatous polyp." *Am J Surg Pathol* **8**: 551–556.

Vasen, H. F. A., F. M. Nagengast, et al. (1995). "Interval cancers in hereditary non-polyposis colorectal cancer (Lynch syndrome)." *Lancet* **345**: 1183–1184.

Vogelstein, B., E. R. Fearon, et al. (1988). "Genetic alterations during colorectal-tumor development." *N Engl J Med* **319**: 525–532.

Warner, A. S., M. E. Glick, et al. (1994). "Multiple large hyperplastic polyps of the colon coincident with adenocarcinoma." *Am J Gastroenterol* **89**: 123–125.

Weisenberger, D. J., K. D. Siegmund, et al. (2006). "A distinct CpG island methylator phenotype in human colorectal cancer is the underlying cause of sporadic mismatch repair deficiency and is tightly associated with BRAF mutation." *Nat Genet* **38**: 787–793.

Whitehall, V. L. J., M. D. Walsh, et al. (2001). "Methylation of *O*-6-methylguanine DNA methyltransferase characterises a subset of colorectal cancer with low level DNA microsatellite instability." *Cancer Res* **61**: 827–830.

Wiesner, G. L., D. Daley, et al. (2003). "A subset of familial colorectal neoplasia kindreds linked to chromosome 9q22.2-31.2." *Proc Natl Acad Sci U S A* **100**: 12961–12965.

Williams, G. T. (1997). "Metaplastic (hyperplastic) polyps of the large bowel: benign neoplasms after all?" *Gut* **40**: 691–692.

Winawer, S. J., A. G. Zauber, et al. (1993). "Randomized comparison of surveillance intervals after colonoscopic removal of newly diagnosed adenomatous polyps. The National Polyp Study Workgroup." *N Engl J Med* **328**(13): 901–906.

Wong, N. A. C. S., R. G. Morris, et al. (2002). "Cyclin D1 overexpression in colorectal carcinoma *in vivo* is dependent on β-catenin protein dysregulation, but not *k-ras* mutation." *J Pathol* **197**: 128–135.

Wynter, C. V. A., T. Kambara, et al. (2006). "DNA methylation patterns in adenomas from FAP, multiple adenoma and sporadic carcinoma patients." *Int J Cancer* **118**: 907–915.

Yamashita, K., T. Dai, et al. (2003). "Genetics supersedes epigenetics in colon cancer phenotype." *Cancer Cell* **4**: 121–131.

Yang, S., F. A. Farraye, et al. (2004). "BRAF and KRAS mutations in hyperplastic polyps and serrated adenomas of the colorectum: relationship to histology and CpG island methylation status." *Am J Surg Pathol* **28**: 1452–1459.

Zauber, N. P., M. Sabbath-Solitare, et al. (1999). "K-ras mutation and loss of heterozygosity of the adenomatous polyposis coli gene in patients with colorectal adenomas with in situ carcinoma." *Cancer* **86**: 31–36.

Section III
Germline Susceptibility – Mendelian and Other Syndromes

Chapter 6
Familial Adenomatous Polyposis

Julian A. Sanchez, Graham Casey, and James M. Church

Introduction

Familial adenomatous polyposis (FAP) is a highly penetrant, autosomal dominant syndrome resulting from germline mutations of the adenomatous polyposis coli (*APC*) gene (OMIM 175100) (Galiatsatos and Foulkes 2006). FAP was the first hereditary colorectal cancer syndrome to be recognized clinically (Lockhart-Mummery 1925), and the first hereditary colorectal cancer condition for which a causative gene was identified. It has been the stimulus for the formation of hereditary colorectal cancer registries and international collaborative study groups, and has been one model for understanding the adenoma-carcinoma sequence that occurs frequently in sporadic colorectal cancers. Although FAP is the result of an inactivating germline mutation in a single gene, it is clinically heterogenous, both within and between families. Management of affected individuals can be helped by knowledge of the genotype, and the clinical presentation can often be a clue to the site of the mutation.

Clinical Summary of FAP

Epidemiology

FAP has a frequency of one in every 5,000 to 10,000 live births, and distribution is equal between the sexes (Rozen and Macrae 2006). Polyps develop during the second and third decades of life, and patients usually present either following the development of symptoms or as a result of screening because of a known family history. In a review of over 180 families and 922 affected individuals in the inherited colorectal cancer registry at the Cleveland Clinic, the mean age at presentation was 27 years and the mean age at colectomy was 29 years (Rustin et al. 1990). FAP leads to a nearly inevitable progression to colorectal carcinoma and accounts for 1% of

J.A. Sanchez (✉), G. Casey, and J.M. Church
Department of Colorectal Surgery, Cleveland Clinic)

J.D. Potter and N.M Lindor (eds.), *Genetics of Colorectal Cancer*,
DOI: 10.1007/978-0-387-09568-4_7,

all colorectal cancers (Lipton and Tomlinson 2006). FAP has a variable degree of clinical expression, and the disease presentation can range from very severe to more attenuated forms and has both colonic and extracolonic manifestations.

Clinical Presentation

The majority of patients are asymptomatic while the polyps develop. Patients can present with signs and symptoms of bleeding and abdominal discomfort, although others present with more advanced signs including weight loss, anemia, or intestinal obstruction. Compared to those screened as at-risk, but asymptomatic, family members, symptomatic patients without a family history have a markedly higher chance of having a colorectal cancer at presentation (67% vs. 3%, Bulow 2003). Colorectal polyposis, numbering in the hundreds to thousands, is pathognomonic for disease diagnosis, but the number of polyps can vary substantially. Polyps are generally small, usually less than 1 cm, and can occur throughout the colorectum with a predilection for the sigmoid colon and rectum (Lal and Gallinger 2000). Polyps can be sessile or pedunculated with varied histology from tubular adenoma to villous adenoma. Colorectal carcinoma risk is generally proportional to the size, number, and histology of the polyps, and the genetic progression follows the chromosomal instability pathway. Diagnosis is confirmed with careful documentation of family history, endoscopic evaluation, and testing for an *APC* germline mutation.

Multiple extracolonic manifestations of FAP have been described, representing all three embryological layers. These manifestations can be either benign or malignant. Endodermal lesions include gastric and small bowel polyps and carcinomas. Mesodermal abnormalities include desmoid tumors, osteomas, and dental abnormalities. Ectodermal lesions localize to the eye, brain, and skin appendages.

Gastric Polyps

Histologically, gastric polyps are most often fundic-gland polyps but can also be adenomas or hyperplastic polyps. In a review from the Lahey Clinic, 33% of FAP patients developed gastric polyps but progression to carcinoma was rare (Marcello et al. 1996). Sporadic fundic-gland polyps are generally considered hamartomatous and have little malignant potential. However, patients with FAP often have more of these polyps and up to 41% show dysplasia (Bianci et al. 2007). Although *Helicobacter pylori* gastritis has been shown to have a protective effect on the development of fundic-gland polyps, gastroduodenal reflux with associated bile-acid exposure has been associated with increased polyp development (Choudhry et al. 1998; Marcello et al. 1996). The local bile-rich environment, in the setting of an appropriate *APC* mutation, may promote or protect against polyp formation via epigenetic interactions. Conversely, in a study comparing epigenetic methylation

(CpG Island Methylator Phenotype, or CIMP), in sporadic fundic-gland polyps to FAP-associated polyps, researchers showed that promoter methylation was, overall, a relatively rare event (10%) and that, when it did occur, methylation was more common in sporadic fundic-gland polyps than in syndromic polyps (Abraham et al. 2004). FAP-associated fundic-gland polyps may be less prone to epigenetic events including methylation, and their malignant potential may be due to both germline and somatic mutations.

This difference between sporadic and FAP-associated fundic-gland polyps has been further characterized by the demonstration of differential *APC* mutations. A series of studies comparing sporadic and syndromic polyps demonstrated that, although FAP-associated polyps had a higher overall frequency of both somatic and germline *APC* mutations than sporadic fundic-gland polyps, they had a lower frequency of β-catenin mutations (Abraham et al. 2000, 2001). These data underscore the theory that the FAP-associated lesions arise from a different mechanism than their sporadic counterparts.

In contrast to fundic-gland polyps, gastric adenomas are uncommon (10% of patients undergoing EGD) and are generally found in the gastric antrum (Church et al. 1992). Similar to fundic-gland polyps, there seems to be an association with gastric adenoma formation and gastric bile reflux.

Duodenal Polyps

In treating almost 250 FAP patients at the Cleveland Clinic, Church and colleagues (1992) found that 88% develop duodenal polyps and that these were often located near the ampulla and papilla. A survey of the polyposis registry at St. Mark's Hospital found that patients with severe duodenal polyposis often have germline *APC* mutations at a different locus from that in patients with severe colonic polyposis. Patients with germline *APC* mutations occurring after codon 1400 were more likely to have severe duodenal polyposis, and they tended to show allelic loss. Additionally, severe upper gastrointestinal polyposis was associated with somatic mutations in a cluster region between codons 1400 and 1580 (Groves et al. 2002). Duodenal polyps can become dysplastic and progress to carcinoma, and this represents the third most common cause of death in FAP patients after colorectal cancer and desmoid disease. FAP patients have 100 times the risk of duodenal carcinoma as the normal population (Nivatvongs 1999).

Desmoid Tumors

Desmoid tumors are rare, locally invasive soft-tissue tumors which most commonly arise in patients with FAP. Desmoids are monoclonal, neoplastic processes and have been associated with trisomies, somatic mutations, and translocations. Desmoids often stain very strongly for β-catenin which may explain their local aggressiveness (Zippel and Temple 2007). Although they do not usually metastasize, they are often aggressive locally and are the second leading cause of death in FAP patients and the leading cause of death following colectomy (Arvanitis et al. 1990). Approximately 10–25% of FAP patients will develop desmoids in their lifetime (Sturt and Clark 2006), and FAP patients are 800 times more likely to develop desmoids than the general population (Lynch and Fitzgibbons 1996). Eighty percent of desmoids are intra-abdominal and are often found at the site of surgical incisions. Trauma has been implicated as a predisposing factor in their development, as 84% of FAP-associated desmoids developed within 5 years of abdominal surgery (Bertario et al. 2001). Although surgery seems to be a predisposing factor, desmoids also develop in the absence of trauma. Desmoids have been described in the abdominal mesentery and can grow quite large, displacing abdominal organs and causing significant compression, obstruction, pain, and mortality.

As is the case for colonic FAP disease, the location of the germline *APC* mutation can predict the severity of the desmoid burden. Distal germline *APC* mutations, 3′ of codon 1444, have been associated with more aggressive disease. Such mutations confer a 12-fold increased risk of desmoid development (Caspari et al. 1995). The combination of a distal *APC* mutation and abdominal surgery is associated with an even higher risk of desmoid development (Speake et al. 2007). Desmoids can be hormonally sensitive, due to expression of estrogen, progesterone, and androgen receptors.

Osteomas and Dental Abnormalities

Osteomas may occur in any bone, but are often localized to the facial skeleton. Dental abnormalities affect 70% of FAP patients and include supranumerary teeth, missing teeth, fused roots, and dental osteomas (Lal and Gallinger 2000). These benign tumors may cause symptoms based on their location, and a new diagnosis can prompt a medical consultation for FAP evaluation.

Congenital Hypertrophy of Retinal Pigment Epithelium

Congenital hypertrophy of retinal pigment epithelium (CHRPE) is an asymptomatic hamartoma of the retinal epithelium which can occur in 66–92% of FAP patients (Chen et al. 2006). On indirect ophthalmoscopic evaluation, it characteristically presents as round or oval hyper- or hypopigmented lesions that are often bilateral. Using a cut-off of four or more lesions in both eyes excludes the confusion with sporadic CHRPE and is evidence for a possible *APC* mutation (Chen et al. 2006).

Thyroid Carcinoma

Patients with FAP have an increased risk of thyroid carcinoma which may affect as many as 12% of patients (Herraiz et al. 2007). These cancers are often well-differentiated papillary cancers and predominantly affect young women. Mutant β-catenin may be responsible for the unusual cribriform-morular variant of papillary thyroid carcinoma that is associated with FAP (Xu et al. 2003). These papillary thyroid carcinomas are often associated with the somatic *RET* proto-oncogene translocations and generally have a good prognosis. A review of 97 patients with FAP and papillary thyroid carcinoma showed that 89% had activation of RET/PTC; individual isoforms lead to different tumor behavior (Cetta et al. 2001).

Adrenal and Hepatobiliary Tumors

A review of the Cleveland Clinic FAP registry showed that the presence of an adrenal incidentaloma is 7.4% in FAP patients, significantly higher than in the general population. Although rare, associations with cholangiocarcinoma, pancreatic adenocarcinoma, and hepatoblastoma have also been described with the FAP syndrome.

Attenuated FAP

A subset of FAP patients have an attenuated form, defined as less than 100 synchronous colorectal adenomas. There is a predominantly right-sided colonic distribution and rectal sparing, and patients often present in their fifth to seventh decades of life (Knudsen et al. 2003). Patients with attenuated FAP (AFAP) and 5′ *APC* mutations are less likely to have desmoid disease and other extracolonic manifestations. The cancer risk is similar to that of the classic syndrome, although the cancers occur later, and patients are offered prophylactic surgical therapy. The AFAP phenotype may be a product of gene dosing due to alternative splicing, resulting in subnormal levels of circulating APC protein – see also Chap. 4. This decrease in APC protein production leads to inadequate levels of tumor suppressor activity and promotes polyposis. *APC* mutations in AFAP have been described in both the 5′ and the 3′ gene ends as well as in exon 9 (Knudsen et al. 2003). Allelic variation in *APC*, hormonal and growth-factor influences, and nearby or distant gene interactions may explain the phenotypic differences observed in this attenuated disease (Foulkes 1995). Given the relatively few polyps and right-sided predilection, this phenotype can be confused with Lynch Syndrome and can be similar to that of *MYH*-associated polyposis.

Genetics of FAP

APC and FAP

One of the first clues to the location of the gene involved in FAP was the case report by Herrera et al. (1986), describing a developmentally retarded patient who had polyposis and a deletion of the long arm of chromosome 5 (Herrera et al. 1986). The gene itself, *APC*, maps to 5q21 and was cloned in 1991 following linkage analysis and germline mutation screening of candidate genes in families with FAP (Bodmer et al. 1987; Groden et al. 1991; Kinzler et al. 1991; Nishisho et al. 1991). *APC* consists of 15 transcribed exons and encodes a protein of 2,843 amino acids (Fearnhead et al. 2001). It functions as a tumor suppressor gene and has been implicated in a number of cell processes including transcription regulation, cell cycle control, apoptosis, and maintenance of the fidelity of chromosomal segregation (Fearnhead et al. 2001). Perhaps the best-characterized role for *APC* is as part of a scaffolding protein complex that negatively regulates Wnt signaling (Fearnhead et al. 2001; Goss and Groden 2000; Nathke 2004).

APC is inactivated in the majority of colorectal cancers, and loss of APC function is a hallmark-initiating event of the chromosomal instability pathway (CIN) phenotype of colorectal cancer. *APC* and the transcription co-regulator β-catenin play central roles in the Wingless/Wnt signaling pathway. In normal cells, in the absence of Wnt signaling, *APC* along with Axin, glycogen synthase kinase 3 β (GSK3β), and casein kinase recruit β-catenin into a destruction complex where it is phosphorylated by GSK3β, leading to β-catenin degradation by the ubiquitin-mediated proteosome pathway. This cellular process results in the maintenance of low levels of free cytosolic β-catenin. When the Wnt signaling pathway is activated, the *APC*/Axin/GSK3β complex disassociates, allowing stabilization of cytosolic β-catenin. Accumulated β-catenin associates with T cell factor (TCF) and lymphoid-enhancer factor (LEF), and the resulting complex enters the nucleus and activates transcription. On entering the nucleus, the β-catenin/TCF/LEF proteins provide a potent transcriptional transactivation complex leading to transactivation of a number of critical genes including *MYC* and *cyclin D1* (Nathke 2004, 2006; Watson 2001). Loss of control of this pathway through inactivation of APC leads to aberrant accumulation of β-catenin and transcriptionally active β-catenin/TCF/LEF complexes and abnormal activation of target genes (Fearnhead et al. 2001). The Wingless/Wnt pathway can also be activated by mutations in β-catenin in colorectal cancers without *APC* mutations (Samowitz et al. 1999). APC also participates in a number of other cellular processes related to cytoskeletal organization, in particular microtubule stability (Nathke 2006). The genetic evidence of the importance of derangement of the β-catenin signaling pathway in CRC strongly suggests a central role for the Wnt/APC/β-catenin pathway in CRC development. However, APC loss alone is not sufficient for tumor development, as activating mutations of the *KRAS* and *BRAF* oncogenes, inactivation of *SMAD4*, and inactivating mutations of the *p53* tumor-suppressor gene are also implicated in colorectal cancer development – see Chap. 4.

APC Germline Mutations Leading to FAP

Germline mutations have been identified in the majority of patients with FAP (Fearnhead et al. 2001) and, to date, over 800 different *APC* germline mutations have been reported (Nieuwenhuis and Vasen 2007). The vast majority of *APC* mutations associated with FAP are frame-shift or nonsense mutations that lead to an inactive truncated protein product (Fearnhead et al. 2001; Galiatsatos and Foulkes 2006). Although mutations can occur throughout the gene, the majority of *APC* mutations can be found between codons 1250 and 1464 in the 5′ region of exon 15, a region known as the mutation cluster region (MCR) (Nieuwenhuis and Vasen 2007). The "hot spots" for mutations are at codons 1061 and 1309 and account for approximately 11 and 17%, respectively, of all germline *APC* mutations (Nieuwenhuis and Vasen 2007). The majority of the remaining mutations are spread between codons 200 and 1600 with only a few mutations occurring outside this region (Fearnhead et al. 2001; Nieuwenhuis and Vasen 2007) (Figs. 6.1 and 6.2).

Extensive mutation screening can identify *APC* sequence changes in up to 95% of patients presenting with classical FAP, particularly when standard genetic testing is supplemented with conversion analysis (separation of alleles) or multiplex ligation-dependent probe amplification (MLPA) (Galiatsatos and Foulkes 2006;

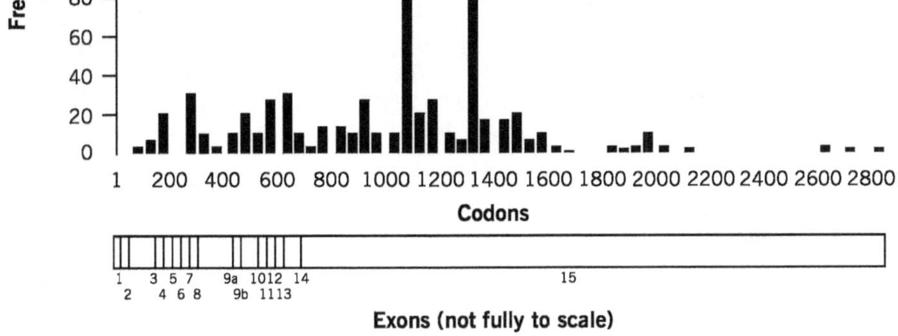

Fig. 6.1 APC germline mutations reported in FAP patients (modified from Nieuwenhuis and Vasen 2007)

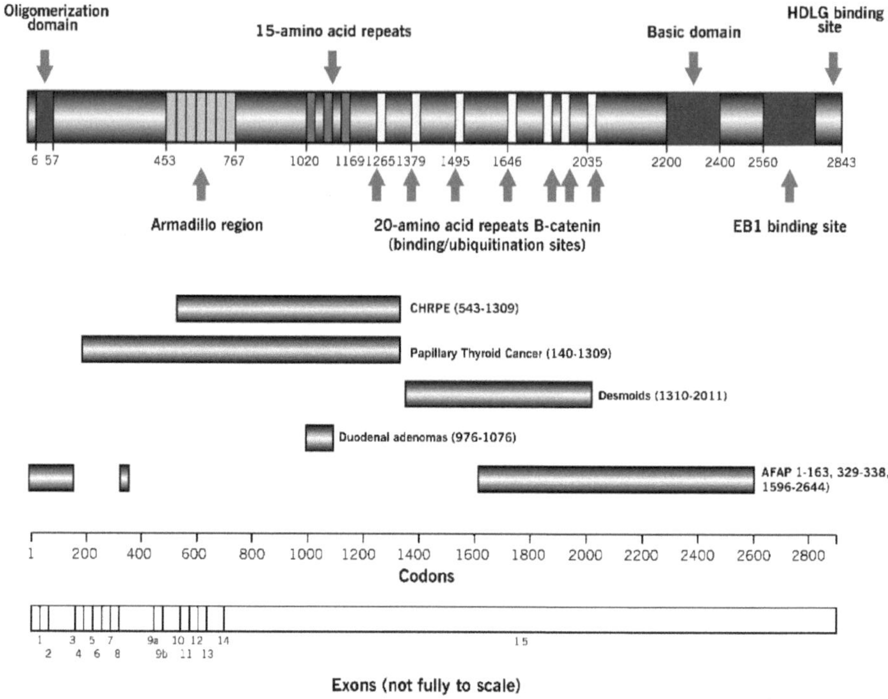

Fig. 6.2 Diagram showing important APC protein motifs, APC exon structure, and FAP phenotype associated with germline mutations (modified from Crabtree et al. 2003; Galiatsatos and Foulkes 2006)

Meuller et al. 2004). Approximately 80% of FAP families have a detectable mutation; however, between 10 and 30% are de novo mutations (Guillem et al. 1999). Although patients with a de novo mutation have no family history and do not get screened, there is some evidence that de novo mutations produce a more severe phenotype, independent of clinical presentation (Bonardi 2006).

Genotype–Phenotype Association: Polyposis Severity

Classical FAP is defined clinically as the presence of >100 synchronous colorectal adenomas, whereas attenuated FAP is defined as <100. Subgroups are also defined, such as profuse FAP (>5,000 synchronous adenomas), and sparse or mild FAP (polyp number between 100s and 1,000s), that have variable ages of onset of colorectal polyposis and age of onset of colorectal cancer (Fearnhead et al. 2001; Nieuwenhuis and Vasen 2007). The severity of disease often correlates with the location of *APC* mutations. For example, patients with mutations in codon 1250 to codon 1464, and particularly at codon 1309, often develop profuse polyposis with symptoms a decade earlier than usual, and colorectal cancer developing at an earlier age (Bertario et al. 2003; Caspari et al. 1994; Enomoto et al. 2000; Ficari et al. 2000; Gayther et al. 1994; Nagase et al. 1992; Nugent et al. 1994). Attenuated FAP, where patients generally develop fewer than 100 colon polyps and cancer onset is delayed, is associated with mutations in the extreme 5′ (exons 1–4) and 3′ (distal to codon 1580) regions of *APC* as well as the alternatively spliced site of exon 9, although exceptions to this have been noted (Brensinger et al. 1998; Friedl et al. 1996; Sieber et al. 2006; Soravia et al. 1998; Walon et al. 1997). An extensive review of relevant literature on this topic has recently been published (Galiatsatos and Foulkes 2006; Nieuwenhuis and Vasen 2007).

The disease phenotype has been showed to vary among populations. A collaborative study of 863 patients from 15 different registries, by researchers from the University of Nebraska, compared the clinical expression and associated *APC* mutations in three ethnic groups: Asians, Europeans, and North Americans (Attard et al. 2007). Investigators found that the risk of gastric cancer in FAP patients was higher in Asian populations than in Europeans and North Americans and that there was a clear difference in the pattern of *APC* gene mutations in North Americans compared to Europeans and Asians. The North American population had a higher frequency of mutations in codons 2 through 811, while the Asian registries reported a greater frequency of *APC* mutations at codons 1099 through 1694. The mutations in the North American population imparted a lower incidence of upper gastrointestinal tumors. This genotype–phenotype association may account for the clinical differences in disease presentation among ethnicities (Attard et al. 2007).

Genotype–Phenotype Association: Extracolonic Manifestations of FAP

Although penetrance nears 100%, there is marked variability in the clinical phenotype of FAP. The majority of FAP patients develop extracolonic manifestations. Desmoid tumors are associated with mutations between codons 1310 and 2011 (Bertario et al. 2003), with the highest severity occurring between codons 1444/5 and 1580/1 (Caspari et al. 1995; Davies et al. 1995; Friedl et al. 2001; Gebert et al. 1999). No consistent genotype correlation has been found with duodenal adenomas, although FAP patients with *APC* mutations in codons 976–1067 have been reported to have a 3- to 4-fold increased risk (Bertario et al. 2003). Congenital hypertrophy of the retinal pigment epithelium (CHRPE) generally precedes the development of polyposis (Galiatsatos and Foulkes 2006) and appears to be associated predominantly with *APC* mutations spanning the region between codons 543 and 1309 (Bertario et al. 2003), but the mutations may extend beyond these boundaries (Caspari et al. 1995; Cetta et al. 2000). Papillary thyroid cancer is associated with *APC* mutations between codons 140 and 1309 (Cetta et al. 2000).

APC is a tumor suppressor gene and follows the two-hit model. Studies suggest that the location of the *APC* germline mutation may influence the location of the second hit (Albuquerque et al. 2002; Crabtree et al. 2003; Lamlum et al. 1999). It has been reported that if the *APC* germline mutation occurs between codons 1194 and 1392, there is strong selection for loss of heterozygosity as the second hit. In contrast, if the germline *APC* mutation occurs outside this region, the second hit will more likely be an inactivating mutation in the second *APC* allele (Fearnhead et al. 2001; Fodde et al. 2001; Sieber et al. 2006). It has been proposed that this may be related to the occurrence of *APC* mutations in relation to β-catenin degradation repeats in *APC*. *APC* contains seven 20-bp repeats that are involved in degrading the transcription cofactor β-catenin and therefore play a role in negatively regulating Wnt signaling. If an *APC* mutation occurs between the first and second repeats, it tends to be associated with loss of heterozygosity, whereas *APC* mutations outside this region tend to be associated with somatic mutations as the second hit (Crabtree et al. 2003; Sieber et al. 2006).

APC Polymorphisms

Missense mutations in the *APC* gene have been described in non-FAP patients with multiple adenomas occurring at earlier ages. One particular variant, a missense polymorphism I1307K, results from a T to A transversion leading to an unstable poly-A stretch, is seen in 6% of Ashkenazim and is associated with increased risk of colorectal cancer (Laken et al. 1997). Although the effect of this missense mutation on APC function has yet to be determined, carriers do have an increased risk of colorectal cancer but not polyposis or other extra colonic manifestations of FAP. Another common variant (E1317Q) in the *APC* gene was reported in 4.3% of FAP patients in one study, associated with a relative risk of colorectal cancer of 11.17 ($p < 0.001$) (Lamlum et al. 2000). However, this finding has not been supported by multiple other studies (Evertsson et al. 2001; Fearnhead et al. 2004; Frayling et al. 1998; Gismondi et al. 2002; Hahnloser et al. 2003).

FAP Modifier Genes

Considerable phenotypic variability occurs even among individuals and families with identical genotypic mutations (Giardiello et al. 1994; Soravia et al. 1998). This variation in clinical presentation suggests that modifier genes or environmental factors can also impact expression of the disease (Houlston et al. 2001; Houlston and Tomlinson 2001). For example, the incidence and severity of duodenal adenomas may be affected by specific *APC* mutations but may be also influenced by a modifier gene on 1p35-36 (Dobbie et al. 1997; Tomlinson et al. 1996), although some studies dispute this finding (Plasilova et al. 2004). N-acetyltransferases (*NAT1* and *NAT2*) are involved in phase 2 reactions that metabolize xenobiotic compounds, and variants have been identified in these genes that have been shown to affect

N-acetyltransferase metabolism reactions. *NAT1* and *NAT2* variants have been associated with a twofold increase in severity of FAP phenotype (Crabtree et al. 2004).

Non-APC-Associated Polyposis

Some patients with multiple colorectal adenomas (generally less than 100 polyps) but no identifiable *APC* gene mutation have been shown to harbor compound heterozygous germline mutations in the base excision repair *MYH* gene (Croitoru et al. 2004; Jenkins et al. 2006). *MYH*-associated polyposis is described in more detail in Chap. 7.

References

Abraham SC, Nobukawa B, Giardiello FM, Hamilton SR, Wu TT (2000) Fundic gland polyps in familial adenomatous polyposis. Am J Pathol 157: 747–54

Abraham SC, Nobukawa B, Giardiello FM, Hamilton SR, Wu TT (2001) Sporadic fundic gland polyps: common gastric polyps arising through activating mutations in the β-catenin gene. Am J Pathol 158: 1005–10

Abraham SC, Park SJ, Cruz-Correa M, Houlihan PS, Half EE, Lynch PM, Wu TT (2004) Frequent CpG island methylation in sporadic and syndromic gastric fundic gland polyps. Am J Clin Pathol 122(5): 740

Albuquerque C, Breukel C, van der Luijt R, Fidalgo P, Lage P, Slors FJ, Leitao CN, Fodde R, Smits R (2002) The 'just-right' signaling model: APC somatic mutations are selected based on a specific level of activation of the beta-catenin signaling cascade. Hum Mol Genet 11: 1549–60

Arvanitis ML, Jagelman DG, Fazio VW, Lavery IC, McGannon E (1990) Mortality in patients with familial adenomatous polyposis. Dis Colon Rectum 33: 639–42

Attard TM, Young RJ, Stoner JA, Lynch HT (2007) Population differences in familial adenomatous polyposis may be an expression of geographic differences in *APC* mutation pattern. Cancer Genet Cytogenet 172: 180–2

Bertario L, Russo A, Sala P, Eboli M, Giarola M, D'amico F, Gismondi V, Varesco L, Pierotti MA, Radice P, Hereditary Colorectal Tumours Registry (2001) Genotype and phenotype factors as determinants of desmoids tumours in patients with familial adenomatous polyposis. Int J Cancer 95: 102–7

Bertario L, Russo A, Sala P, Varesco L, Giarola M, Mondini P, Pierotti M, Spinelli P, Radice P (2003) Multiple approach to the exploration of genotype-phenotype correlations in familial adenomatous polyposis. J Clin Oncol 21: 1698–707

Bodmer WF, Bailey CJ, Bodmer J, Bussey HJ, Ellis A, Gorman P, Lucibello FC, Murday VA, Rider SH, Scambler P, et al. (1987) Localization of the gene for familial adenomatous polyposis on chromosome 5. Nature 328: 614–6

Brensinger JD, Laken SJ, Luce MC, Powell SM, Vance GH, Ahnen DJ, Petersen GM, Hamilton SR, Giardiello FM (1998) Variable phenotype of familial adenomatous polyposis in pedigrees with 3′ mutation in the *APC* gene. Gut 43: 548–52

Bulow S (2003) Results of national registration of familial adenomatous polyposis. Gut 52(5): 742–6

Caspari R, Friedl W, Mandl M, Moslein G, Kadmon M, Knapp M, Jacobasch KH, Ecker KW, Kreissler-Haag D, Timmermanns G, et al. (1994) Familial adenomatous polyposis: mutation at codon 1309 and early onset of colon cancer. Lancet 343: 629–32

Caspari R, Olschwang S, Friedl W, Mandl M, Boisson C, Boker T, Augustin A, Kadmon M, Moslein G, Thomas G, et al. (1995) Familial adenomatous polyposis: desmoid tumours and lack of ophthalmic lesions (CHRPE) associated with *APC* mutations beyond codon 1444. Hum Mol Genet 4: 337–40

Cetta F, Montalto G, Gori M, Curia MC, Cama A, Olschwang S (2000) Germline mutations of the *APC* gene in patients with familial adenomatous polyposis-associated thyroid carcinoma: results from a European cooperative study. J Clin Endocrinol Metab 85: 286–92

Cetta F, Gori M, Montalto G, Zuckermann M, Toti P (2001) Different significance of ret/PTC(1) and ret/PTC(3) rearrangements in thyroid carcinogenesis: lesson from two subgroups of patients with papillary thyroid carcinomas showing the highest incidence of ret/PTC activation. J Clin Endocrinol Metab 86(3): 1429

Chen CS, Phillips KD, Grist S, Bennet G, Craig JE, Muecke JS, Suthers GK (2006) Congenital hypertrophy of the retinal pigmented epithelium in familial colorectal cancer. Fam Cancer 5(4): 397–404

Choudhry U, Boyce HW, Coppola D (1998) Proton pump inhibitor-associated gastric polyps: a retrospective study analysis of their frequency, and endoscopic, histologic, and ultrastructural characteristics. Am J Clin Pathol 110: 615–21

Church JM (1995) Desmoid tumors in patients with familial adenomatous polyposis. Semin Colon Rectal Surg 6: 29–32

Church J, Simmang C (2003) Practice parameters for the treatment of patients with dominantly inherited colorectal cancer (familial adenomatous polyposis and hereditary nonpolyposis colorectal cancer). Dis Colon Rectum 46: 1001–12

Church JM, McGannon E, Hull-Boiner S, Sivak MV, Van Stolk R, Jagelman DG, Fazio VW, Oakley JR, Lavery IC, Milsom JW (1992) Gastroduodenal polyps in patients with familial adenomatous polyposis. Dis Colon Rectum 35: 1170–3

Clark SK, Neale KF, Landgrebe JC, Phillips RK (1999) Desmoid tumours complicating familial adenomatous polyposis. Br J Surg 86: 1185–9

Corman ML (2005) Colon and Rectal Surgery (5th ed.). Philadelphia, PA: Lippincott Williams & Wilkins

Crabtree M, Sieber OM, Lipton L, Hodgson SV, Lamlum H, Thomas HJ, Neale K, Phillips RK, Heinimann K, Tomlinson IP (2003) Refining the relation between 'first hits' and 'second hits' at the *APC* locus: the 'loose fit' model and evidence for differences in somatic mutation spectra among patients. Oncogene 22: 4257–65

Crabtree MD, Fletcher C, Churchman M, Hodgson SV, Neale K, Phillips RK, Tomlinson IP (2004) Analysis of candidate modifier loci for the severity of colonic familial adenomatous polyposis, with evidence for the importance of the N-acetyl transferases. Gut 53: 271–6

Croitoru ME, Cleary SP, Di Nicola N, Manno M, Selander T, Aronson M, Redston M, Cotterchio M, Knight J, Gryfe R, Gallinger S (2004) Association between biallelic and monoallelic germline MYH gene mutations and colorectal cancer risk. J Natl Cancer Inst 96: 1631–4

Davies DR, Armstrong JG, Thakker N, Horner K, Guy SP, Clancy T, Sloan P, Blair V, Dodd C, Warnes TW, et al. (1995) Severe Gardner syndrome in families with mutations restricted to a specific region of the *APC* gene. Am J Hum Genet 57: 1151–8

Dobbie Z, Heinimann K, Bishop DT, Muller H, Scott RJ (1997) Identification of a modifier gene locus on chromosome 1p35-36 in familial adenomatous polyposis. Hum Genet 99: 653–7

Enomoto M, Konishi M, Iwama T, Utsunomiya J, Sugihara KI, Miyaki M (2000) The relationship between frequencies of extracolonic manifestations and the position of *APC* germline mutation in patients with familial adenomatous polyposis. Jpn J Clin Oncol 30: 82–8

Evertsson S, Lindblom A, Sun XF (2001) *APC* I1307K and E1317Q variants are rare or do not occur in Swedish colorectal cancer patients. Eur J Cancer 37: 499–502

Fearnhead NS, Britton MP, Bodmer WF (2001) The ABC of *APC*. Hum Mol Genet 10: 721–33

Fearnhead NS, Wilding JL, Winney B, Tonks S, Bartlett S, Bicknell DC, Tomlinson IP, Mortensen NJ, Bodmer WF (2004) Multiple rare variants in different genes account for multifactorial inherited susceptibility to colorectal adenomas. Proc Natl Acad Sci U S A 101: 15992–7

Ficari F, Cama A, Valanzano R, Curia MC, Palmirotta R, Aceto G, Esposito DL, Crognale S, Lombardi A, Messerini L, Mariani-Costantini R, Tonelli F, Battista P (2000) *APC* gene mutations and colorectal adenomatosis in familial adenomatous polyposis. Br J Cancer 82: 348–53

Fodde R, Smits R, Clevers H (2001) APC, signal transduction and genetic instability in colorectal cancer. Nat Rev Cancer 1: 55–67

Foulkes WD (1995) A tale of four syndromes: familial adenomatous polyposis, gardner syndrome, attenuated *APC* and turcot syndrome. QJM 88(12): 853–63

Frayling IM, Beck NE, Ilyas M, Dove-Edwin I, Goodman P, Pack K, Bell JA, Williams CB, Hodgson SV, Thomas HJ, Talbot IC, Bodmer WF, Tomlinson IP (1998) The *APC* variants I1307K and E1317Q are associated with colorectal tumors, but not always with a family history. Proc Natl Acad Sci U S A 95: 10722–7

Friedl W, Meuschel S, Caspari R, Lamberti C, Krieger S, Sengteller M, Propping P (1996) Attenuated familial adenomatous polyposis due to a mutation in the 3′ part of the *APC* gene. A clue for understanding the function of the APC protein. Hum Genet 97: 579–84

Friedl W, Caspari R, Sengteller M, Uhlhaas S, Lamberti C, Jungck M, Kadmon M, Wolf M, Fahnenstich J, Gebert J, Moslein G, Mangold E, Propping P (2001) Can *APC* mutation analysis contribute to therapeutic decisions in familial adenomatous polyposis? Experience from 680 FAP families. Gut 48: 515–21

Galiatsatos P, Foulkes WD (2006) Familial adenomatous polyposis. Am J Gastroenterol 101: 385–98

Gayther SA, Wells D, SenGupta SB, Chapman P, Neale K, Tsioupra K, Delhanty JD (1994) Regionally clustered *APC* mutations are associated with a severe phenotype and occur at a high frequency in new mutation cases of adenomatous polyposis coli. Hum Mol Genet 3: 53–6

Gebert JF, Dupon C, Kadmon M, Hahn M, Herfarth C, von Knebel Doeberitz M, Schackert HK (1999) Combined molecular and clinical approaches for the identification of families with familial adenomatous polyposis coli. Ann Surg 229: 350–61

Giardiello FM, Krush AJ, Petersen GM, Booker SV, Kerr M, Tong LL, Hamilton SR (1994) Phenotypic variability of familial adenomatous polyposis in 11 unrelated families with identical *APC* gene mutation. Gastroenterology 106: 1542–7

Gismondi V, Bonelli L, Sciallero S, Margiocco P, Viel A, Radice P, Mondini P, Sala P, Montera MP, Mareni C, Quaia M, Fornasarig M, Gentile M, Pietro G, Rossini P, Arrigoni A, Meucci GM, Bruzzi P, Varesco L (2002) Prevalence of the E1317Q variant of the *APC* gene in Italian patients with colorectal adenomas. Genet Test 6: 313–7

Goss KH, Groden J (2000) Biology of the adenomatous polyposis coli tumor suppressor. J Clin Oncol 18: 1967–79

Groden J, Thliveris A, Samowitz W, Carlson M, Gelbert L, Albertsen H, Joslyn G, Stevens J, Spirio L, Robertson M, et al. (1991) Identification and characterization of the familial adenomatous polyposis coli gene. Cell 66: 589–600

Groves C, Lamlum H, Crabtree M, Williamson J, Taylor C, Bass S, Cuthbert-Heavens D, Hodgson S, Phillips R, Tomlinson I (2002) Mutation cluster region, association between germline and somatic mutations and genotype-phenotype correlation in upper gastrointestinal familial adenomatous polyposis. Am J Pathol 160: 2055

Guillem JG, Smith AJ, Calle JP, Ruo L (1999) Gastrointestinal polyposis syndromes. Curr Probl Surg 36: 217–323

Hahnloser D, Petersen GM, Rabe K, Snow K, Lindor NM, Boardman L, Koch B, Doescher D, Wang L, Steenblock K, Thibodeau SN (2003) The *APC* E1317Q variant in adenomatous polyps and colorectal cancers. Cancer Epidemiol Biomarkers Prev 12: 1023–8

Herraiz M, Barbesino G, Faquin W, Chan-Smutko G, Patel D, Shannon KM, Daniels GH, Chung DC (2007) Prevalence of thyroid cancer in familial adenomatous polyposis syndrome and the role of screening ultrasound examinations. Clin Gastroenterol Hepatol 5(3): 367–73

Herrera L, Kakati S, Gibas L, Pietrzak E, Sandberg AA (1986) Brief clinical report: Gardner syndrome in a man with an interstitial deletion of 5q. Am J Med Genet 25: 473–6

Houlston RS, Tomlinson IP (2001) Polymorphisms and colorectal tumor risk. Gastroenterology 121: 282–301

Houlston R, Crabtree M, Phillips R, Crabtree M, Tomlinson I (2001) Explaining differences in the severity of familial adenomatous polyposis and the search for modifier genes. Gut 48: 1–5

Jang Y-S, Steinhagen RM, Heimann TM (1997) Colorectal cancer in familial adenomatous polyposis. Dis Colon Rectum 40: 312

Jenkins MA, Croitoru ME, Monga N, Cleary SP, Cotterchio M, Hopper JL, Gallinger S (2006) Risk of colorectal cancer in monoallelic and biallelic carriers of MYH mutations: a population-based case-family study. Cancer Epidemiol Biomarkers Prev 15: 312–4

Kinzler KW, Nilbert MC, Vogelstein B, Bryan TM, Levy DB, Smith KJ, Preisinger AC, Hamilton SR, Hedge P, Markham A, et al. (1991) Identification of a gene located at chromosome 5q21 that is mutated in colorectal cancers. Science 251: 1366–70

Knudsen AL, Bisgaard ML, Bülow S (2003) Attenuated familial adenomatous polyposis (AFAP): a review of the literature. Fam Cancer 2: 43–55

Laken SJ, Petersen GM, Gruber SB, Oddoux C, Ostrer H, Giardiello FM, Hamilton SR, Hampel H, Markowitz A, Klimstra D, Jhanwar S, Winawer S, Offit K, Luce MC, Kinzler KW, Vogelstein B (1997) Familial colorectal cancer in Ashkenazim due to a hypermutable tract in *APC*. Nat Genet 17: 79–83

Lal G, Gallinger S (2000) Familial adenomatous polyposis. Semin Surg Oncol 18: 314–23

Lamlum H, Ilyas M, Rowan A, Clark S, Johnson V, Bell J, Frayling I, Efstathiou J, Pack K, Payne S, Roylance R, Gorman P, Sheer D, Neale K, Phillips R, Talbot I, Bodmer W, Tomlinson I (1999) The type of somatic mutation at *APC* in familial adenomatous polyposis is determined by the site of the germline mutation: a new facet to Knudson's 'two-hit' hypothesis. Nat Med 5: 1071–5

Lamlum H, Al Tassan N, Jaeger E, Frayling I, Sieber O, Reza FB, Eckert M, Rowan A, Barclay E, Atkin W, Williams C, Gilbert J, Cheadle J, Bell J, Houlston R, Bodmer W, Sampson J, Tomlinson I (2000) Germline *APC* variants in patients with multiple colorectal adenomas, with evidence for the particular importance of E1317Q. Hum Mol Genet 9: 2215–21

Lipton L, Tomlinson I (2006) The genetics of FAP and FAP-like syndromes. Fam Cancer 5: 221–6

Lockhart-Mummery P (1925) Cancer and heredity. Lancet 427–9

Lynch HT, Fitzgibbons R, Jr. (1996) Surgery, desmoid tumors, and familial adenomatous polyposis: case report and literature review. Am J Gastroenterol 91: 2598–601

Marcello PW, Asbun HJ, Veidenheimer MC, Rossi RL, Roberts PL, Fine SN, Coller JA, Murray JJ, Schoetz DJ (1996) Gastroduodenal polyps in familial adenomatous polyposis. Surg Endosc 10: 418–21

Marchesa P, Fazio VW, Church JM, McGannon E (1997) Adrenal masses in patients with familial adenomatous polyposis. Dis Colon Rectum 40(9): 1023–8

Meuller J, Kanter-Smoler G, Nygren AO, Errami A, Gronberg H, Holmberg E, Bjork J, Wahlstrom J, Nordling M (2004) Identification of genomic deletions of the *APC* gene in familial adenomatous polyposis by two independent quantitative techniques. Genet Test 8: 248–56

Nagase H, Miyoshi Y, Horii A, Aoki T, Ogawa M, Utsunomiya J, Baba S, Sasazuki T, Nakamura Y (1992) Correlation between the location of germline mutations in the *APC* gene and the number of colorectal polyps in familial adenomatous polyposis patients. Cancer Res 52: 4055–7

Nathke IS (2004) The adenomatous polyposis coli protein: the Achilles heel of the gut epithelium. Annu Rev Cell Dev Biol 20: 337–66

Nathke I (2006) Cytoskeleton out of the cupboard: colon cancer and cytoskeletal changes induced by loss of APC. Nat Rev Cancer 6: 967–74

Nieuwenhuis MH, Vasen HF (2007) Correlations between mutation site in *APC* and phenotype of familial adenomatous polyposis (FAP): a review of the literature. Crit Rev Oncol Hematol 61: 153–61

Nishisho I, Nakamura Y, Miyoshi Y, Miki Y, Ando H, Horii A, Koyama K, Utsunomiya J, Baba S, Hedge P (1991) Mutations of chromosome 5q21 genes in FAP and colorectal cancer patients. Science 253: 665–9

Nivatvongs S (1999) Benign Neoplasms of the Colon and Rectum. In Gordon PH and Nivatvongs S (eds.) Principles and Practice of Surgery of the Colon, Rectum, and Anus (2nd ed.) pp. 554–73. St. Louis, MO: Quality Medical Publishing

Nugent KP, Phillips RK, Hodgson SV, Cottrell S, Smith-Ravin J, Pack K, Bodmer WF (1994) Phenotypic expression in familial adenomatous polyposis: partial prediction by mutation analysis. Gut 35: 1622–3

Olschwang S, Tiret A, Laurent-Puig P, Muleris M, Parc R, Thomas G (1993) Restriction of ocular fundus lesions to a specific subgroup of *APC* mutations in adenomatous polyposis coli patients. Cell 75(5): 959–68

Plasilova M, Russell AM, Wanner A, Wolf A, Dobbie Z, Muller HJ, Heinimann K (2004) Exclusion of an extracolonic disease modifier locus on chromosome 1p33-36 in a large Swiss familial adenomatous polyposis kindred. Eur J Hum Genet 12: 365–71

Rozen P, Macrae F (2006) Familial adenomatous polyposis: the practical applications of clinical and molecular screening. Fam Cancer 5: 227–35

Rustin RB, Jagelman DG, McGannon E, Fazio VW, Lavery IC, Weakley FL (1990) Spontaneous mutation in familial adenomatous polyposis. Dis Colon Rectum 33(1): 52–5

Samowitz WS, Powers MD, Spirio LN, Nollet F, van Roy F, Slattery ML (1999) Beta-catenin mutations are more frequent in small colorectal adenomas than in larger adenomas and invasive carcinomas. Cancer Res 59: 1442–4

Sekine S, Shimoda T, Nimura S, Nakanishi Y, Akasu T, Katai H, Gotoda T, Shibata T, Sakamoto M, Hirohashi S (2004) High-grade dysplasia associated with fundic gland polyposis in a familial adenomatous polyposis patient, with special reference to *APC* mutation profiles. Modern Pathol 17(11): 1421

Sieber OM, Segditsas S, Knudsen AL, Zhang J, Luz J, Rowan AJ, Spain SL, Thirlwell C, Howarth KM, Jaeger EE, Robinson J, Volikos E, Silver A, Kelly G, Aretz S, Frayling I, Hutter P, Dunlop M, Guenther T, Neale K, Phillips R, Heinimann K, Tomlinson IP (2006) Disease severity and genetic pathways in attenuated familial adenomatous polyposis vary greatly but depend on the site of the germline mutation. Gut 55: 1440–8

Soravia C, Berk T, Madlensky L, Mitri A, Cheng H, Gallinger S, Cohen Z, Bapat B (1998) Genotype-phenotype correlations in attenuated adenomatous polyposis coli. Am J Hum Genet 62: 1290–301

Soravia C, Berk T, McLeod RS, Cohen Z (2000) Desmoid disease in patients with familial adenomatous polyposis. Dis Colon Rectum 43: 363–9

Speake D, Evans DG, Laloo F, Scott NA, Hill J (2007) Desmoid tumours in patients with familial adenomatous polyposis and desmoid region adenomatous polyposis coli mutations. Br J Surg 94: 1009–13

Spigelman AD (1995) Familial adenomatous polyposis and the upper gastrointestinal tract. Semin Colon Rectal Surg 6: 26–8

Spigelman AD, Williams CB, Talbot IC, Domizio P, Phillips RK (1989) Upper gastrointestinal cancer in patients with familial adenomatous polyposis. Lancet 2: 783–5

Sturt NJ, Clark SK (2006) Current ideas in desmoids tumours. Fam Cancer 5: 275–85

Sturt NJ, Gallagher MC, Bassett P, Philp CR, Neale KF, Tomlinson IP, Silver AR, Phillips RK (2004) Evidence for genetic predisposition to desmoid tumours in familial adenomatous polyposis independent of the germline *APC* mutation. Gut 53: 1832–6

Tomlinson IP, Neale K, Talbot IC, Spigelman AD, Williams CB, Phillips RK, Bodmer WF (1996) A modifying locus for familial adenomatous polyposis may be present on chromosome 1p35-p36. J Med Genet 33: 268–73

Walon C, Kartheuser A, Michils G, Smaers M, Lannoy N, Ngounou P, Mertens G, Verellen-Dumoulin C (1997) Novel germline mutations in the *APC* gene and their phenotypic spectrum in familial adenomatous polyposis kindreds. Hum Genet 100: 601–5

Watson SA (2001) Oncogenic targets of B-catenin-mediated transcription in molecular pathogenesis of intestinal polyposis. Lancet 357: 572–3

Xu B, Yoshimoto K, Miyauchi A, Kuma S, Mizusawa N, Hirokawa M, Sano T (2003) Cribriform-morular variant of papillary thyroid carcinoma: a pathological and molecular genetic study with evidence of frequent somatic mutations in exon 3 of the beta-catenin gene. J Pathol 199(1): 58–67

Zippel DB, Temple WJ (2007) When is a neoplasm not a neoplasm? When it is a desmoid. J Surg Oncol 95(3): 190–1

Chapter 7
DNA Mismatch Repair and Lynch Syndrome

Brittany C. Thomas, Matthew J. Ferber, and Noralane M. Lindor

Introduction

Postreplicative DNA mismatch repair (MMR) is a highly conserved molecular mechanism that functions to ensure genomic integrity by repairing mismatched base-pairs that are incorporated into the genetic code during cellular replication. Disruption of this essential function leads to the random accumulation of mutations, resulting in a markedly increased potential for malignancy. Lynch Syndrome is a hereditary predisposition to colon and other types of cancer and the most common hereditary colon cancer syndrome currently known. The association of Lynch Syndrome with defective MMR was elucidated by the demonstration of microsatellite instability (MSI) in colon-tumor DNA and subsequent cloning of *hMSH2* and *hMLH1*, the human homologs of two bacterial MMR genes. Evidence of genomic instability, in the form of MSI induced by deficiencies of the DNA MMR pathway, provided the molecular basis by which to redefine the clinically heterogeneous group of hereditary colon cancer syndromes.

Genomic Instability

Several important early findings led to the discovery of defective DNA mismatch repair as the underlying genetic etiology of Lynch Syndrome. Loeb and colleagues first proposed (1991) and later expanded (2006) on the idea of genomic instability as a *mutator phenotype* initiated by random point mutations early in the colorectal adenoma-carcinoma sequence. They observed that the number of alterations in tumor DNA could not be explained by the well-established spontaneous mutation rate in somatic cells, based on their previous observation of very few errors in newly synthesized DNA of normal (non-neoplastic) daughter cells. They hypothesized that a defect in the DNA replication process, which normally functions to

B.C. Thomas (✉)
Department of Laboratory Medicine & Pathology, Mayo Clinic College of Medicine
email: thomas.brittany@mayo.edu

J.D. Potter and N.M Lindor (eds.), *Genetics of Colorectal Cancer*,
DOI: 10.1007/978-0-387-09568-4_8,

ensure base-pairing accuracy, results in an elevated accumulation of errors in the genetic code (Loeb et al. 1974; Loeb 1991; Venkatesan et al. 2006).

The observation that mutations in oncogenes and tumor suppressor genes accumulate throughout colon cancer development provided the model by which an underlying susceptibility to genomic instability could promote tumorigenesis. A defect in the DNA replication process, allowing an increase in the number of persistent mutations, will inevitably result in alterations in genes involved in cell-cycle regulation. This facilitates the proliferation of neoplastic cells with selective cell-growth advantage and, thus, clonal expansion, a common feature of colorectal tumors. This identified a mechanism for progression to malignancy (Vogelstein et al. 1988; reviewed in Fearon and Vogelstein 1990).

The discovery of microsatellite instability (MSI) in DNA isolated from colon tumors was first reported in 1993 by three independent groups, and observed both in familial and nonfamilial tumors (Aaltonen et al. 1993; Ionov et al. 1993; Thibodeau et al. 1993). Discrepancies in the number of $(CA)_n$ and other dinucleotide repeat sequences were observed within polymorphic repeat segments of DNA, termed microsatellites. These discrepancies were specifically noted as shifts in electrophoretic mobility of the repeat fragments isolated from tumor DNA compared to DNA from normal tissue in the same individual. Expansions and contractions of dinucleotide repeats within tumor microsatellites thus became known as "microsatellite instability" (MIN or MSI) or the "replication error" (RER+) phenotype.

Two other observations confirmed the role of MSI as a phenotypic marker for genomic instability: (1) mutations in yeast genes involved in DNA mismatch repair result in instability of repetitive DNA sequences during cellular replication; and (2) tumor cell lines exhibiting MSI also display elevated spontaneous mutation rates at selected genomic loci (Bhattacharyya et al. 1994).

Fifteen to twenty percent of all colorectal carcinomas exhibit defective MMR in the form of a high level of microsatellite instability (MSI-high or MSI-H). Of these, only about 10% (1.5–2% of all CRCs) can be explained by a germline mutation in an MMR gene (Aaltonen et al. 1998; Cunningham et al. 2001). The vast majority of cases demonstrating defective MMR are explained by somatic hypermethylation of the *hMLH1* gene promoter (Cunningham et al. 1998; Gazzoli et al. 2002). This phenomenon complicates the testing algorithm and ultimate diagnosis of Lynch Syndrome in individuals with colon cancer (see the section "Molecular Screening for Lynch Syndrome").

DNA Mismatch-Repair Mechanism

Evidence of genomic instability in the form of tumor MSI and the cloning of several genes encoding mismatch-repair proteins implicated the DNA MMR complex in the etiology of Lynch Syndrome (see the section "Gene Discovery"). The MMR system serves several functions; the most relevant to Lynch Syndrome tumor development involves the repair of mismatched bases that are incorporated into DNA during cellular replication or DNA insult; this limits the accumulation of potentially deleterious mutations in coding regions of the DNA.

Elucidation of the postreplicative DNA mismatch-repair mechanism in human cells was aided by studies involving the MutHLS system in bacterial *E. coli* and similar systems in the budding yeast, *Saccharomyces cerevisiae*. The bacterial MutHLS repair pathway produces several proteins, including MutS and MutL homodimers, to facilitate methylation-dependent, nick-directed mismatch repair. The identification of several MutS homologs (MSH) and MutL homologs (MLH) in eukaryotes demonstrates conservation between prokaryotic and eukaryotic MMR machinery. Both systems incorporate mismatch recognition, excision of the mispaired segment, and resynthesis of the excised strand; however, the process is not as well characterized in eukaryotic cells. Excision of the mismatched bases in *E. coli* cells is facilitated by methylation of the newly synthesized daughter strand, allowing discrimination between the template and replicated DNA strands during mismatch repair. An analogous signal has not yet been detected in eukaryotic MMR, although it is believed to exist.

The MutSα heterodimer consists of the human MSH2 and MSH6 proteins encoded by the *hMSH2* and *hMSH6* genes. The primary function of the MutSα heterodimer is to initiate the repair process by binding to DNA mismatches detected by MSH6. The errors specifically corrected by this system are single mispaired bases or small insertion/deletion loops (IDLs) that arise as a result of slippage of the primer against the template strand. A second heterodimer complex, MutSβ, consisting of MSH2 and MSH3, also initiates the mismatch-repair mechanism. Although MSH6 and MSH3 have been reported to be functionally redundant (thus explaining the relative lack of observed germline mutations in *hMSH3*), the possible involvement of the MutSβ complex in suppression of deletion and duplication errors has been described (Marsischky et al. 1996; Harrington and Kolodner 2007). Furthermore, MSH3 does not appear to compensate for the loss of MSH6, potentially because the MutSβ complex preferentially repairs IDLs involving two to eight bases versus the base-base mismatches and IDLs containing a smaller number of bases that are repaired by MutSα. Other MutS homologs have been identified and may contribute to the MMR pathway; however, to date, only germline mutations within the *hMSH2* and *hMSH6* genes have been associated with MSH-related Lynch Syndrome.

The MutL homologs, MLH1 and PMS2 (postmeiotic segregation polypeptide), comprise the heterodimer MutLα, which interacts with several proteins including MutSα to facilitate mismatch recognition and reparation. Two other MutL homologs, PMS1 and MLH3, have been described; however, their respective roles in postreplicative DNA MMR and Lynch Syndrome are, at this time, not as well established (reviewed in Kolodner 1995; Jiricny and Nystrom-Lahti 2000; Aquilina and Bignami 2001; Peltomaki 2005).

Gene Discovery

The *hMSH2* gene was the first of the eukaryotic MMR genes to be cloned, mapping to human chromosome 2p22-p21. Homology to the previously identified bacterial *mutS* gene sequence facilitated its discovery. Subsequent detection of germline *hMSH2* mutations in putative Lynch Syndrome patients confirmed its association with hereditary disease (Fishel et al. 1993; Leach et al. 1993). *hMLH1*, *PMS2*, and *PMS1* were cloned shortly thereafter, using similar methods that employed sequences within conserved regions of the MutL family of proteins in yeast and bacteria (Bronner et al. 1994; Nicolaides et al. 1994). Germline mutations in *hMLH1*, located on human chromosome 3p21.3, and *PMS2*, located on human chromosome 7p22, were subsequently detected in affected individuals (Papadopoulos et al. 1994). Cloning of the *hMSH6* gene, located near *hMSH2* on human chromosome 2p16, was the last major causative gene to be associated with Lynch Syndrome (Drummond et al. 1995; Palombo et al. 1995; Akiyama et al. 1997; Miyaki et al. 1997). Despite their suggested roles in the mismatch-repair pathway, germline mutations in two other homologs of the bacterial *mutS* and *mutL* genes, namely *hMSH3* and *PMS1*, do not currently appear to contribute to Lynch Syndrome (Peltomaki and Vasen 2004).

Predisposition to cancer, as conferred by mutations in any one of these four MMR genes, is inherited in an autosomal dominant manner. Consistent with Knudson's "two-hit" hypothesis, germline mutations in *hMLH1* coupled with loss of the wild-type allele, either by loss of heterozygosity or hypermethylation of *hMLH1*, have been observed in tumors of Lynch Syndrome patients. LOH and *hMLH1* hypermethylation are the most common causes of gene inactivation in nonfamilial MSI-H colon cancer, accounting for the nearly 20% of all such colon tumors. Rarely, gene conversion as a mechanism of inactivation has also been observed (Hemminki et al. 1994; Tannergard et al. 1997; Zhang et al. 2006; Ollikainen et al. 2007).

Microsatellite Instability

Microsatellite instability (MSI) is now the widely accepted term used to describe the phenotype observed in both nonfamilial and Lynch Syndrome tumor DNA as a result of defective DNA mismatch repair (Boland et al. 1998). By definition, microsatellites are short repeated segments of DNA that are interspersed randomly across the human genome. They are polymorphic, both in repeat size and number. Repeating units vary in size between one (mononucleotide repeat) and six nucleotides, approximately, and contain 10–50 identical repeats per microsatellite locus (Weber 1990). Microsatellites are, by their repetitive nature, susceptible to instability due to slippage of the DNA polymerase complex during the DNA replication process. Instability, in the form of contractions or expansions in repeat length, occurs when the DNA MMR mechanism fails to correct these mutations. PCR-based analysis of isolated neoplastic (tumor) DNA and non-neoplastic (adjacent normal mucosa) DNA from the same individual, via size-based electrophoretic separation, allows a way of detecting relative microsatellite expansions or contractions, thereby establishing the presence or absence of microsatellite instability (reviewed in Baudhuin et al. 2005a).

Standard designations that describe the various levels of microsatellite instability within colon tumors have been formally adopted as follows: MSI-H (high level of microsatellite instability), MSI-L (low level of microsatellite instability), and MSS (microsatellite stable). MSI-H tumors are characterized by instability detected at 30% or greater of the microsatellite markers analyzed. MSI-L describes tumors that demonstrate instability at less than 30% of markers tested, and MSS tumors are characterized by stability of all markers tested (Dietmaier et al. 1997; Boland et al. 1998; Thibodeau et al. 1998). In addition to these designations, the National Cancer Institute workshop in 1997 recommended a set of five markers comprising two mononucleotide microsatellite markers (*BAT25*, *BAT26*) and three dinucleotide microsatellite markers (*D5S346*, *D2S123*, and *D17S250*) in order to establish standards for microsatellite marker selection and minimize inconsistencies among clinical diagnostic laboratories (Boland et al. 1998).

Currently, the clinical relevance of MSI-L tumors is ambiguous as a certain amount of genomic instability is expected in DNA even in MMR proficient tumors (Laiho et al. 2002). MSI-L tumors, like MSS tumors, reflect MMR proficiency in the majority of cases. However, the use of MSI marker panels that include a preponderance of dinucleotide markers may underestimate the instability demonstrated in MSH6 deficient tumors, given their tendency not to show instability at dinucleotide microsatellite loci (Wagner et al. 2001; Ward et al. 2001); this demonstrates the importance of incorporating mononucleotide markers into a standard MSI testing panel.

Molecular Screening for Lynch Syndrome

Prior to clarification of the molecular etiology of Lynch Syndrome, a set of clinical criteria were adopted by The International Collaborative Group on Hereditary NonPolyposis Colorectal Cancer (ICG-HNPCC), called the Amsterdam Criteria (AC), for the purposes of facilitating early gene linkage and natural history studies (Fig. 7.1) (Vasen et al. 1991). Although their intended purpose was to help distinguish hereditary from nonhereditary cases to facilitate clinical and research studies, fulfillment of the AC became the clinical definition of what was most commonly known as Hereditary NonPolyposis Colon Cancer, HNPCC. Following the identification of several causative genes, however, studies have shown that only about half of families that fulfill the original AC actually have *Lynch Syndrome*, renamed by Boland in 2005 as the molecularly characterized hereditary syndrome defined by the presence of a germline mutation in an MMR gene (see the section "Evolution of a Name: HNPCC Versus Lynch Syndrome"). In support of this distinction, Lindor and colleagues reported disparate clinical features among Amsterdam-criteria-positive families whose tumors demonstrated an MSI-H phenotype (defective MMR) and AC-positive families whose tumors displayed MSI at less than 30% or 0 markers analyzed, calling the latter, Familial Colorectal Cancer Type X (Lindor et al. 2005). The AC have since been proven to lack both sensitivity

Amsterdam Criteria (Vasen et al 1991) *- all four criteria must be met to be considered AC 'positive' -*

- *3 or more relatives with histologically verified colon cancer in which one of the relatives is a first degree relative of the other two*
- *2 successive generations affected*
- *1 relative diagnosed with colon cancer under 50 years of age*
- *familial adenomatous polyposis (FAP) excluded*

Revised Bethesda Guidelines (Umar et al 2004) *- any of the following are sufficient for consideration of MSI studies -*

- *colorectal cancer diagnosed under 50 years of age*
- *presence of synchronous or metachronous colorectal cancer or other HNPCC-related tumor* regardless of age*
- *colorectal cancer in an individual less than 60 years of age, exhibiting tumor infiltrating lymphocytes, Crohn's-like lymphocytic reaction, mucinous/signet-ring differentiation, or medullary growth pattern*
- *colorectal cancer diagnosed in on or more first-degree relatives of an individual with an HNPCC-related tumor* in which one of the two relatives is diagnosed under 50 years of age*
- *colorectal cancer diagnosed in two or more first or second-degree relatives with HNPCC-related tumors* at any age*

**HNPCC-related tumors include colorectal, endometrial, stomach, ovarian, pancreas, ureter and renal pelvis, biliary tract, brain tumors (glioblastoma as seen in Turcot syndrome, a rare variant of Lynch syndrome), sebaceous gland adenomas/adenocarcinomas and keratoacanthomas (as seen in Muir-Torre syndrome, a second variant of Lynch syndrome) and carcinoma of the small bowel.*

Fig. 7.1 The *Amsterdam Criteria* (AC) were originally adopted by The International Collaborative Group on Hereditary NonPolyposis Colorectal Cancer to facilitate gene discovery and natural history research. The AC have since been applied clinically to identify families at risk for hereditary colon cancer. About half of all families that meet the AC actually have Lynch syndrome confirmed by molecular testing.

The *Revised Bethesda Guidelines* are the second iteration of the Bethesda Guidelines created to aid providers in choosing which patients to screen via MSI analysis using clinical and histopathological criteria consistent with Lynch syndrome.

and specificity, missing mutation-positive cases as well as capturing families who do not demonstrate germline MMR mutations. However, the AC are historically and inextricably linked to Lynch Syndrome, representing an important basis by which HNPCC and Lynch Syndrome have been defined for years and continue to facilitate clinical recognition of possible new cases.

Several iterations of criteria and guidelines (Amsterdam criteria I and II, Bethesda guidelines, and revised Bethesda guidelines) have evolved in an effort to guide clinicians toward a diagnosis of Lynch Syndrome (Vasen et al. 1991, 1999; Rodriguez-Bigas et al. 1997; Umar et al. 2004). The advent of molecular screening via microsatellite instability testing and immunohistochemical protein analysis (MSI/IHC) within tumors of affected individuals, provided a new way of screening for Lynch Syndrome, based upon the presence or absence of defective MMR. The revised Bethesda guidelines were developed to aid providers in choosing which patients to screen via MSI analysis by first selecting those with clinical or histopathological features consistent with Lynch syndrome (Fig. 7.1) (Umar et al. 2004). Those cases that demonstrate MSI and evidence of defective MMR, may opt for IHC analysis, if not already performed, to identify the culprit gene within the tumor. Absence of protein expression observed in the tumor is indicative of a potential germline mutation within the corresponding gene; genetic counseling followed by selective gene analysis can then be conducted to identify the familial mutation and establish a diagnosis (as recommended by Umar et al. 2004). Loss of expression of MLH1/PMS2, although reflective of defective DNA MMR, may be explained by either acquired or germline defects in the *hMLH1* or *PMS2* genes. However, loss of MSH2/MSH6 expression within the tumor is generally indicative of a germline mutation within one of the two genes, as there is no alternative explanation to loss of expression of MSH2/MSH6 expression in CRC tumors at this time. For this reason, genetic counseling is recommended prior to IHC analysis, given the high likelihood of a germline mutation in certain cases.

Although this strategy has proved to be a feasible and reliable screening method for Lynch Syndrome (Aaltonen et al. 1998), it is often complicated by other factors. Individuals without striking clinical presentations do not necessarily fall within the revised Bethesda guidelines and are therefore likely to be missed. However, these are few in number and, in general, this screening strategy picks up the majority of Lynch Syndrome cases. Furthermore, tumor from an affected individual is not always available for analysis. In cases with a strong suspicion of Lynch Syndrome, sequence analysis, in addition to analysis for large genomic rearrangements, may be conducted for the asymptomatic/presymptomatic individual to identify a possible underlying germline mutation. However, because the state of genetic testing is imperfect, current methods may be unable to identify a germline mutation and, therefore, a hereditary DNA mismatch-repair defect cannot be ruled out. When feasible, it is beneficial to follow up a negative germline test with tumor analysis (MSI testing) to distinguish hereditary tumors demonstrating defective MMR indicative of an undetectable germline mutation from tumors that developed as a result of other non-MMR related processes.

Lastly, the majority of tumors exhibiting defective MMR are explained by acquired promoter hypermethylation of *hMLH1*, further complicating testing algorithms. MSI-high tumors showing loss of expression of *hMLH1* could be attributable to a germline mutation in *hMLH1* or acquired hypermethylation of the gene. To address this situation, both germline (blood) mutation analysis and tumor methylation studies are available clinically. Recent studies have shown a strong correlation between the loss of *hMLH1* expression by immunohistochemistry and advancing age at diagnosis, right-sided tumor location, and female sex (Kakar et al. 2003). This, in addition to family-history information, can help to determine whether a germline mutation or an epigenetic process is more likely in any specific case. Making this distinction is extremely helpful in guiding which line of testing may be most appropriate. Recent reports demonstrate a remarkable correlation between the presence of a specific somatic *BRAF* V600E mutation and *hMLH1* promoter hypermethylation, thus providing a second testing strategy for differentiating between hereditary and nonfamilial cases (for review see Baudhuin et al. 2005a; Thomas et al. 2005).

Mutation Profile

Several hundred mutations in the MMR genes associated with Lynch Syndrome have been reported. Approximately 50% of disease-causing mutations are within the *hMLH1* gene, 40% in the *hMSH2* gene, and 7% in the *hMSH6* gene; the contribution of mutations in *PMS2* is much smaller (Peltomaki 2004). Other genes have been evaluated for their possible involvement in the pathogenesis of Lynch Syndrome; however, mutations in these four genes account for nearly all Lynch Syndrome cases identified to date.

In general, mutations are found along the entire length of each of *hMLH1*, *hMSH2*, and *hMSH6*, with the exception of exons 1 and 10 of *hMSH6*, in which no mutations have been reported. Several exons harbor mutations more frequently than others, including exons 1 and 16 of *hMLH1*, exons 3 and 12 of *hMSH2*, and exon 4 of *hMSH6*. Despite these apparent mutation "hot spots," the large majority (~80%) of the documented mutations in these genes have been reported as private mutations (Peltomaki 2004). Certain founder mutations do occur repeatedly in specific ethnic groups (see the section "Founder Mutations").

Many of the mutations identified are single base-pair substitutions or small insertions and deletions, both of which typically result in termination of the coding sequence or have marked downstream effects on protein production or function. In addition to the ubiquitous pathogenic mutations including nonsense, frameshift, and splice-site mutations, other types of alterations also make important contributions to the types of mutations frequently observed in Lynch Syndrome.

Large Genomic Rearrangements

Large genomic rearrangements probably account for about 20% of total pathogenic MMR mutations; however, estimates vary widely between 7 and 55% (Wijnen et al. 1998; Yan et al. 2000; Gille et al. 2002; Viel et al. 2002; Wang et al. 2002; Taylor et al. 2003; Baudhuin et al. 2005b; Grabowski et al. 2005; Kurzawski et al. 2006). Discrepancies in reported frequency of these rearrangements are probably due to founder effects, ethnic differences, detection methods, selection criteria, and chance.

Most of the large rearrangements reported to date are large genomic deletions. These large deletions involve deletions of single or multiple exons, including the promoter region in some cases (Charbonnier et al. 2002; Gille et al. 2002; Wang et al. 2002; Nakagawa et al. 2003; Taylor et al. 2003; Baudhuin et al. 2005b;

Grabowski et al. 2005; van der Klift et al. 2005; Kurzawski et al. 2006). Less commonly, whole gene deletions of *hMSH2* have been observed (Gille et al. 2002; Wang et al. 2002). Large genomic duplications have been reported in both *hMSH2* and *hMLH1*, albeit to a much lesser extent than deletions (Charbonnier et al. 2000; Di Fiore et al. 2004; Baudhuin et al. 2005b). Of note, van der Klift (2005) described two other types of large rearrangements including an inversion in *hMSH2* and a 2-kb insertion in intron 7 of the *PMS2* gene (van der Klift et al. 2005).

Large rearrangements most commonly occur in *hMSH2* and *hMLH1*, accounting for one-third of all pathogenic mutations observed in *hMSH2* (Wijnen et al. 1998; Wang et al. 2002). Currently, only four large rearrangements in *hMSH6* have been documented, including three deleterious deletions and one suspected deleterious duplication (Plaschke et al. 2003; van der Klift et al. 2005). However, only one study has actually analyzed an *hMSH6*-mutation enriched population by testing patients whose tumors showed isolated loss of expression of MSH6 by immunohistochemical analysis. They found 2 large rearrangements in the 3 remaining individuals (out of a total of 15) who did not have a detectable *hMSH6* germline mutation by direct DNA sequencing, suggesting that large rearrangements may contribute to the overall mutation spectrum in *hMSH6* at a frequency similar to that observed in *hMSH2* and *hMLH1* (Plaschke et al. 2003). In a similar study of patients whose tumors exhibited isolated loss of PMS2 expression, four of seven (57%) were found to have a large genomic rearrangement involving the *PMS2* gene, including two exonic deletions, a complex rearrangement (due to a genomic deletion or inactivation by gene conversion), and an intronic insertion, all reported as probably pathogenic (Hendriks et al. 2006b). A deletion of exons 1–10 in *PMS2* has also been reported (Rahner et al. 2007).

Large genomic rearrangements first became apparent as part of the mutation spectrum following the identification of a 3.5-kb deletion within the *hMLH1* gene, reported as a Finnish founder mutation (Nystrom-Lahti et al. 1995). A second large deletion of exons 13–16 in *hMLH1* was reported shortly thereafter, leading authors to speculate that, given the large number of *Alu* repeats within the gene, large genomic rearrangements may be common in Lynch Syndrome (Mauillon et al. 1996). Subsequent analysis of the *hMSH2* gene by Wijnen and colleagues identified eight genomic deletions, most probably occurring as a result of a common recombination event, rather than a founder effect, as indicated by haplotype analysis performed on individuals with identical deletions (Wijnen et al. 1998). Follow-up analyses also suggested a high frequency of large rearrangements within *hMSH2* confirming *Alu*-mediated homologous recombination as a major mechanism behind mutation recurrence (Charbonnier et al. 2002; van der Klift et al. 2005). Despite *Alu*-rich genomic structures, however, not all large rearrangements in MMR genes appear to be derived this way. Evidence of nonhomologous recombination, involving *Alu* and *L1* repeat elements, suggests a second, less frequent, mechanism for large rearrangements in *hMSH2*, *hMLH1*, and *hMSH6* (Viel et al. 2002; van der Klift 2005).

Several large rearrangement detection methods have been used; however, the two most widely recommended methods are Southern blot analysis and multiplex ligation-dependent probe amplification (MLPA) because of their sensitivity and simplicity, respectively (Nakagawa et al. 2003; van der Klift et al. 2005). MLPA is a relatively simple method used to detect copy-number mutations such as deletions and duplications with high sensitivity and specificity; however, this method has several limitations. MLPA invariably misses noncopy-number mutations such as large insertions and inversions. Polymorphisms located in the primer regions can disrupt the MLPA reaction as well, causing single-exon deletions which compromise the end result, unless sequencing of the primer regions is also performed. Furthermore, the presence of pseudogenes has made copy-number detection by MLPA more difficult in determining the presence or absence of deletions and duplications in *hPMS2*. Southern blot analysis can also characterize the breakpoints associated with large deletions and duplications more accurately than MLPA, which is important for distinguishing founder mutations from novel mutations, as well as providing an appealing alternative PCR-based approach to surveying DNA from family members for a familial large rearrangement. Notably, Nakagawa and colleagues utilized conversion analysis, a technique initially introduced to increase mutation-detection rates in MMR genes by haploid reduction of host genome, to characterize the breakpoints of large rearrangements detected by MLPA (Yan et al. 2000; Nakagawa et al. 2003). Although Southern blot analysis has been the gold standard method for detecting large rearrangements, MLPA is a less time-consuming and less labor-intensive technique, and consumes smaller amounts of DNA. Despite the debate surrounding choice of method, the development of these reliable tools to detect large genomic rearrangements has led to an overall increase in mutation-detection rates for the MMR genes.

Missense Mutations

As is true of many genetic diseases, base-pair disruptions leading to missense alterations and in-frame deletions represent an elusive set of sequence changes, the pathogenicity of which can be difficult to ascertain. Over one-third of mutations in *hMLH1* and *hMSH6* and nearly 20–25% of mutations in *hMSH2* are missense mutations (Peltomaki and Vasen 2004). Missense and silent mutations are defined by single base-pair substitutions resulting in either an alternate amino acid or the same amino acid, respectively, and may or may not have a deleterious effect on protein function. Traditionally, pathogenic missense mutations in MMR genes affect the local structure or conformation of the encoded protein, producing aberrant protein interaction and function within the MMR pathway (Peltomaki and Vasen 2004). This can result, for example, from a change in the polarity of a specific amino acid. Recent literature suggests that missense and silent mutations may also exert an effect on normal mRNA splicing if the mutation occurs near an intron–exon boundary. It has been well established that disruption of the "invariant" donor and acceptor sites of intron–exon boundaries is pathogenic by causing alternative splicing. However, missense and silent mutations may affect splicing via other methods, including disruption of exonic splicing enhancer consensus sequences or activation of cryptic splice sites (Cartegni et al. 2002; Gorlov et al. 2003; Auclair et al. 2006; Pagenstecher et al. 2006).

Many missense alterations have been successfully classified as disease-causing or susceptibility alleles through studies correlating functional, biochemical, and clinical data (including presence or absence of microsatellite instability in tumor DNA) (Raevaara et al. 2005). The A636P missense alteration prevalent in the Ashkenazi Jewish population, for example, has been established as a pathogenic mutation (Foulkes et al. 2002). However, because missense mutations do not create an obviously truncated product, their effect on protein function is difficult to predict, as the protein may harbor some residual activity. This can lead to unusual clinical manifestations compared to typical Lynch Syndrome patients. Thus, missense alterations are difficult to classify because the biochemical and functional data may not be as compelling as those observed with nonsense or frameshift mutations and, similarly, the clinical data may not be as well defined. Therefore, by their very nature, it can be difficult to distinguish between a "mild" susceptibility allele and a benign missense alteration in a substantial portion of Lynch Syndrome cases, making diagnosis and prognosis equally complex.

Without extensive functional and biochemical studies or reliable clinical data to prove co-segregation with disease in a family, few other resources are available for determining the clinical significance of many of these alterations. However, the development of standardized tools for comparative genomic analysis should serve as valuable resources for understanding this complex category of mutation. Ollila and colleagues assessed how well comparative sequence analysis predicts the results of functional assays as a possible tool to assess the significance of missense mutations (excluding in-frame deletions). Using a specific set of sequences including yeast, parasites, and animals, but excluding plants and bacteria, resulted in an overall predictive value of 92% for *hMSH2* and 82% for *hMLH1* (Ollila et al. 2006). A similar study using computational methods involving *hMLH1* and *hMSH2* found that missense mutations occurring at codons in which the respective amino acid is highly conserved have up to a 97% likelihood of being pathogenic, suggesting the overall predictive value of comparative sequence analysis to be high enough to promote its use in clinical practice (Chan et al. 2007). Although sequence homology probably cannot replace functional studies, it may serve a useful role in the clinical world as a screening method for identifying alterations that warrant confirmatory analysis (Ollila et al. 2006).

The Human Variome Project (HVP) was recently created for the purpose of standardizing a process by which clinicians and laboratorians can publicize mutation/alteration information. Although databases currently exist for the purpose of assimilating genotypic and phenotypic information, they remain incomplete due to issues with compliance, accessibility, and timeliness of data entry. The HVP is devoted to developing a process by which information is collected, stored, and accessed such that all information is captured in an efficient way for optimal clinical use (Cotton et al. 2007). With the advent of the HVP and the promising predictive value of sequence homology, the clinical significance of many MMR missense alterations may be easier to discern in the near future.

Founder Mutations

Included within the collection of over 500 reported MMR gene mutations are several well-documented founder mutations. Nystrom-Lahti and Moisio and colleagues described, and confirmed by haplotype analysis, the first two founder mutations in the Finnish population (called *mutation 1* and *mutation 2*) in the *hMLH1* gene (Nystrom-Lahti et al. 1995; Moisio et al. 1996). These two mutations, specifically a mutation at the splice acceptor site of exon 6 and a large genomic deletion involving exon 16 and surrounding introns, make up a large proportion of the disease-causing mutations in the Finnish population. Together, they are estimated to account for between 63 and 68% of families who fulfill the Amsterdam criteria and 50% of families with verified or putative diagnoses of Lynch Syndrome in this population (Moisio et al. 1996; Nystrom-Lahti et al. 1996).

In 2002, Foulkes and colleagues characterized a previously reported and confirmed pathogenic missense mutation in *hMSH2* as a founder mutation within the Ashkenazi Jewish population (Yuan et al. 1999; Marra et al. 2001; Foulkes et al. 2002). The 1906 G > C mutation which results in an alanine-to-proline amino acid change at codon 636 (A636P) is estimated to represent 10–33% of disease-causing mutations in Ashkenazi Jewish families who meet the Amsterdam criteria (Foulkes et al. 2002). Among other cancer predisposition syndromes, including Bloom syndrome (*BLM*), Fanconi anemia type C (*FANCC*), hereditary breast and ovarian cancer (*BRCA1* and *BRCA2*), and familial adenomatous polyposis (*APC*), founder mutations in the Ashkenazi Jewish population are quite common. However, unlike these other mutations, the frequency of the A636P mutation within the Ashkenazim in general is relatively rare. For instance, the three common founder mutations identified in individuals with Hereditary Breast and Ovarian Cancer (185delAG and 5382insC in *BRCA1* and 6174delT in *BRCA2*) are found in approximately 2.5% of Ashkenazi individuals (Struewing et al. 1997), whereas the A636P has been estimated to occur at a frequency of less than .05% in the general Ashkenazi population (Guillem et al. 2003). Despite its relatively rare occurrence in this population overall (which Faulkes and colleagues attribute to a recent origin or perhaps chance), its prevalence among individuals with Lynch Syndrome in this population is notable.

Recently, a mutation detected in approximately 10% of affected North American families was studied and proven to be a founder mutation through a combined genealogical and molecular approach. The exon 1–6 deletion of *hMSH2*, also known as the *American Founder Mutation* (*AFM*), is characterized molecularly by specific breakpoints not common to all exon 1–6 deletions within this gene. Non-*AFM* exon 1–6 deletions occur commonly as well, probably due to the *Alu*-rich sequences flanking these exons. Therefore, as suggested by haplotype analysis, the *AFM* deletion is distinguishable from other exon 1–6 deletions by its unique end points, providing additional molecular evidence of a common ancestor, now thought to have originated in Germany (Wagner et al. 2003; Lynch et al. 2006).

Reports of founder mutations in other ethnic groups have also been documented. Chan and colleagues have reported two founder mutations, a large deletion in *hMLH1* and a small 4-bp deletion in *hMSH2*, in the southern Chinese population, the latter of which was estimated to account for 21% of deleterious MMR gene mutations in this population. Haplotype analysis suggested a common ancestor in both cases (Chan et al. 2001, 2004). Other founder mutations, as confirmed by haplotype analysis or geneologic studies within the *hMLH1* and *hMSH6* genes, have also been reported as originating in Switzerland, Sweden, Denmark, Finland, the Netherlands, Italy, and Korea (Hutter et al. 1996; Jager et al. 1997; Berends et al. 2002; Caluseriu et al. 2004; Shin et al. 2004; Thiffault et al. 2004; Cederquist et al. 2005; Vahteristo et al. 2005).

Mutations observed more than once are generally thought to arise from a common ancestor (founder mutation) or as a result of a recurrent de novo event. Of interest, the IVS5+3 A>T nucleotide substitution in *hMSH2*, which abolishes the exon 5 splice donor site, accounts for 11–20% of all *hMSH2* mutations worldwide. Although a common haplotype was identified in families originating from Newfoundland, haplotype analysis could not confirm a recent common ancestor in other ethnic populations. This suggests that the mutation, although appearing to be the result of a founder effect in one population, is the most frequently recurring de novo mutation in others, having been reported in the United States, England, Japan, and Italy. Authors suspect that its recurrence as a de novo mutation may be facilitated by a 26-bp mononucleotide repeat sequence that increases the likelihood of misalignment during the replication process (Froggatt et al. 1995, 1999; Desai et al. 2000).

Heritable Epimutations?

Acquired hypermethylation of the *hMLH1* promoter is a well-established mechanism of gene inactivation via transcriptional silencing, in tumors of individuals with non-familial colon cancer (Cunningham et al. 1998; Kuismanen et al. 2000; Miyakura et al. 2001; Gazzoli et al. 2002). Like germline mutations in *hMLH1*, acquired promoter hypermethylation facilitates colon cancer development by disabling an MMR gene as a precursor to carcinogenesis. *hMLH1* hypermethylation is therefore the most common cause of MMR deficiency observed in colon tumors, either via biallelic methylation or methylation and loss of heterozygosity (Ollikainen et al. 2007). Recently, several reports have surfaced of inherited epimutations, epigenetic silencing of a gene that is not normally silenced. Evidence of hypermethylation of *hMLH1* and *hMSH2* in normal tissue (e.g., peripheral blood lymphocytes, buccal mucosa, normal colorectal mucosa, etc.) of individuals with multiple primary cancers and/or early-onset cancer suggests the possibility of heritable epimutations of the MMR genes as a new class of mutations responsible for a Lynch-Syndrome-like presentation. Germline hypermethylation of *hMLH1* is followed by a second "hit" to the opposing allele, initiating tumorigenesis. This is supported by data from several groups that have shown that tumor DNA demonstrated loss of heterozygosity of the unmethylated allele.

Of the handful of cases that have emerged in the literature, all but one have illustrated hemiallelic (one allele) germline methylation of the *hMLH1* gene in individuals with MSI-H tumors showing loss of expression of *hMLH1* by IHC and no detectable germline mutation by sequencing or large rearrangement analyses (Gazzoli et al. 2002; Miyakura et al. 2004; Suter et al. 2004; Hitchins et al. 2005; Chen et al. 2007; Hitchins et al. 2007; Valle et al. 2007). These studies also suggest that the germline epimutation generally arises as a de novo event, either in the parental germline or very early in embryogenesis and is rarely transmitted to subsequent generations. This was supported by evidence that parents and siblings did not demonstrate methylation of the same inherited allele (traceable by SNP or haplotype analysis) as the original proband. Furthermore, it appears that the epimutations usually undergo reversal during gametogenesis to re-establish the normal parent-of-origin methylation pattern. Both Suter et al. (2004) and Hitchins et al. (2007) demonstrated disappearance of germline methylation from one generation to the next by testing, for the presence of methylation, the affected parentally derived allele in the DNA of the offspring of the affected individual. Absence of the epimutation in the proband's children suggests that not only did the epimutation arise de novo, but also that it probably does not confer a predisposition to cancer in subsequent generations (Suter et al. 2004; Hitchins et al. 2007).

Both groups, however, also reported evidence supporting the potential for the epimutation to be passed on to offspring. Hitchins and colleagues described a man who exhibited partial (50%) germline *hMLH1* hypermethylation, whose mother demonstrated the epimutation in all of her somatic cells, showing partial retention of the epimutation in the next generation. Although analysis of his constitutional DNA demonstrated that the man was transcribing RNA only from his paternally derived allele, no evidence of *hMLH1* hypermethylation was detected in a sample of his motile spermatozoa, and germline reactivation of the maternally derived allele was confirmed by RNA analysis. Suter (2004) demonstrated hypermethylation in spermatozoa of an individual who was found to have germline *hMLH1* hypermethylation. However, the proportion of colonies that exhibited the epimutation, <1% (5/526), is probably small enough to imply a negligible risk to future children (Suter et al. 2004). Hitchins and colleagues also hypothesized that oogenesis may be more prone to these epigenetic errors because, in general, the methylated homolog is maternally derived and epimutation reversal has been demonstrated during spermatogenesis in men displaying the germline epimutation in constitutional cells. If this is true, then any transmission of the epimutation will probably be reversed during spermatogenesis as observed in the two cases earlier (Suter et al. 2004; Hitchins et al. 2007). This is also supported by studies conducted in mice that suggested transmission of epimutations due to incomplete reversal of the epimutation during gametogenesis (Roemer et al. 1997; Morgan et al. 1999; Rakyan et al. 2003).

A few groups have hypothesized that changes within the coding or promoter regions of the MMR genes may contribute to epigenetic events such as the *hMLH1* hypermethylation that has been observed both in tumor and normal DNA. One study identified a SNP (−93 A) that was statistically significantly associated with *hMLH1* methylation in endometrial tumors (Chen et al. 2007). Although this probably does not explain the recent reports of *hMLH1* germline methylation, it does provide evidence of cis-factors acting somatically to influence epigenetic events. Another study demonstrated germline hypermethylation of the *hMSH2* gene in ten different individuals across three successive generations, which is, to date, the only report of germline hypermethylation observed in an MMR gene other than *hMLH1*. A potential cis-acting factor present in *hMSH2* in this family was evidenced by perfect segregation of *hMSH2* hypermethylation and a disease haplotype, confirming allele-specific methylation (Chan et al. 2006). This study provided the first evidence of a cis-acting factor that could also be influencing methylation at the germline level (Chan et al. 2006; Hitchins et al. 2007).

Because inactivation of one MMR allele (i.e., of *hMLH1*) is probably present at conception, individuals with MMR germline epimutations (promoter hypermethylation) may be susceptible to cancer risks similar to those observed in Lynch Syndrome who harbor germline mutations within the coding regions of the gene. However, unlike germline mutations in the coding sequence, promoter hypermethylation is probably a transient phenomenon which is, only rarely, transmitted to subsequent generations and, therefore, usually does not pose the same risks to future generations as those associated with inherited MMR mutations. Therefore, germline hypermethylation of an MMR gene could be considered in the evaluation of singleton cases of young onset colon cancer and/or multiple primary cancers whose tumors demonstrate evidence of defective DNA mismatch repair (Volle et al. 2007).

Evolution of a Name: HNPCC Versus Lynch Syndrome

The terms "Lynch Syndrome" and "Hereditary Nonpolyposis Colon Cancer" (HNPCC) have been the most commonly used names for a heterogeneous hereditary colorectal cancer syndrome. This syndrome has been described and subdivided by its various clinical presentations, all of which are predominantly associated with a predisposition to certain types of cancers. The observation of a molecular *mutator phenotype* that manifests as microsatellite instability in tumors of some affected individuals led to discovery of mutations in the MMR genes, allowing for a molecular basis by which a diagnosis of the syndrome could be established. Thus, a redefinition of the syndrome was recently proposed, changing the way in which the terms Lynch Syndrome and HNPCC, are used in order to ensure that proper medical management and counseling are based on an accurate diagnosis of the disease (Boland 2005; Jass 2006).

The syndrome was first described by Dr. S. A. Warthin after reviewing pathology specimens and records of individuals belonging to the striking "Family G" affected by uterine, gastric, and other cancers (Warthin 1913). This family was later followed up by Lynch and colleagues in 1971 who described the high frequency of endometrial and colorectal cancers in the later generations and introduced the *cancer family syndrome* (Lynch et al. 1966; Lynch and Krush 1971). The cancer family syndrome

(CFS) was initially defined by early-onset colon cancer, predominantly proximal in location, with a high frequency of metachronous and/or synchronous cancers, including endometrial and other extracolonic malignancies. Further evaluation of such families revealed two distinct tumor spectra: (1) colon cancer in the context of extracolonic malignancies (CFS) and (2) site-specific colon cancer, called *hereditary site-specific nonpolyposis colon cancer* (*HSSCC*) (Lynch et al. 1991). In 1984, CFS and HSSCC were renamed *Lynch Syndrome I* and *Lynch Syndrome II*, respectively (Boland and Troncale 1984). Together they became known as *hereditary nonpolyposis colon cancer* (*HNPCC*), clinically defined by a family history of predominantly right-sided colon cancer in the absence of the extensive polyposis (Lynch et al. 1985a, b) characteristic of FAP (see Chap. 5).

In the early 1990s, microsatellite instability was described and shown to be related to defective DNA mismatch repair due to germline mutations in the MMR genes, defining the syndrome at a molecular level. Recognizing that not all tumors from families that fulfill the Amsterdam criteria display evidence of the molecular MSI phenotype, suggested the presence of more than one underlying molecular etiology for the pedigree-defined HNPCC. The name, HNPCC, therefore is no longer optimal, as it does not distinguish between the probable two or more molecularly distinct phenotypes (Boland 2005; Jass 2006). Accordingly, in the last 2 years, the term *Lynch Syndrome* has been established as the hereditary cancer syndrome associated with mutations in one of several established MMR genes (Boland 2005). For the families that fulfill the Amsterdam Criteria but do not exhibit evidence of defective MMR as the etiology of tumor formation, the name, *Familial Colorectal Cancer Type X*, has been proposed; although probably still heterogeneous, the clinical manifestations are distinct from those of Lynch Syndrome and are addressed in Chap. 8.2.

The evolution of a "syndrome" and its name has been paralleled by the evolution of "Family G" as first described by Warthin nearly a century before. In addition to the documentation of over 900 family members, the identification of a deleterious *hMSH2* mutation (T>G at the splice acceptor site of exon 4) in affected members of "Family G," as well as a shifting phenotype across the generations (Potter 2001) has transformed this historic cancer family syndrome lineage into a present-day Lynch Syndrome family (Douglas et al. 2005).

Incidence

In the absence of a molecular marker by which to measure disease, the incidence of "HNPCC/Lynch Syndrome" among all CRC cases has historically been based on clinical ascertainment and diagnosis. Thus, early estimates of disease incidence varied widely in the literature, depending on the clinical criteria used for ascertainment. Early population-based studies estimated that incident Lynch Syndrome cases account for about 4–6% of all colorectal cancer (Mecklin 1987). This is probably because the absence of a distinct clinical phenotype, unlike FAP, allowed for more flexibility in the choice of cases that were included in the disease spectrum (Lynch et al. 1985a). Later incidence estimates incorporating the more stringent family-history-based Amsterdam criteria (Vasen et al. 1991), suggested that the incidence of Lynch Syndrome is less than 1% (0.3–0.9%) of all CRC (Aaltonen et al. 1994; Mecklin et al. 1995; Evans et al. 1997; Peel et al. 2000).

The first population-based studies to use MSI analysis to screen for defective MMR in Finland estimated the incidence of Lynch Syndrome associated with a germline mutation in one of several known MMR genes, to be about 2.7% of all CRC in that population (Aaltonen et al. 1998; Salovaara et al. 2000). This translates to a carrier frequency of approximately 1/740 (Salovaara et al. 2000). However, the incidence of Lynch Syndrome in Finland may be inflated by founder effects, suggesting that the incidence may, generally, be lower (Aaltonen et al. 1998; Salovaara et al. 2000; Samowitz et al. 2001). More recent studies in other populations have suggested an overall incidence of about 1% (Ravnic-Glavac et al. 2000; Samowitz et al. 2001) to 2% (Cunningham et al. 2001; Hampel et al. 2005). The last two studies included analysis of the *hMSH6* gene in addition to *hMLH1* and *hMSH2* and added detection of large rearrangements. Therefore, these recent studies probably represent a more accurate estimate of the true incidence of Lynch Syndrome. Limitations of current estimates of disease incidence include characterization of prevalent missense alterations detected in the presence of defective MMR, undetected mutations within regions of the MMR genes that are not or cannot always be analyzed (e.g., introns, promoters), and ascertainment bias involving inclusion of probands with colon cancer only.

Because CRC is estimated to account for only about 45% of cancer diagnoses in individuals with Lynch Syndrome, preliminary studies analyzing the incidence of Lynch Syndrome among all endometrial and other extracolonic cancer cases are beginning to emerge. Hampel and colleagues screened tumors of 543 unselected endometrial cancer patients for the presence of defective MMR. MSI and IHC analysis identified 119 individuals whose tumors exhibited evidence of defective MMR. 10 of the 119 individuals (1.8%) had detectable pathogenic germline mutations in a known MMR gene (Hampel et al. 2006). Thus, the incidence of Lynch Syndrome among endometrial cancer cases appears to be similar to that associated with all CRC cases.

Histopathology

Lynch Syndrome, unlike other hereditary colon cancer syndromes (FAP – see Chap. 5; MAP – see Chap. 8.1), is not associated with a polyposis phenotype (Lynch et al. 1993). The adenoma-carcinoma sequence leading to colonic tumor formation has been illustrated by various researchers (Love 1986; Mecklin et al. 1986b; Lanspa et al. 1990); however, the incidence of adenomas in Lynch Syndrome patients is similar to that observed in the general population, leading researchers to conclude that the high rate of primary and metachronous/synchronous CRCs observed in

HNPCC/Lynch Syndrome is due to an acceleration of the adenoma-carcinoma sequence. This is supported by observations of a greater frequency of adenomas with high-grade dysplasia and/or a villous component in patients with Lynch Syndrome compared to non-Lynch Syndrome patients (Love 1986; Mecklin et al. 1986b; Jass and Stewart 1992; Jass et al. 1994). Because of the accelerated transformation, more frequent clinical screening is required in this population (Jarvinen et al. 2000; Lynch and de la Chapelle 2003; Hendriks et al. 2006a, b; Vasen et al. 2007).

Early studies suggested that associations with certain CRC tumor histopathologic characteristics might also exist. Tumors selected on the basis of family history of colon cancer tended to be poorly differentiated twice as often as in a control group with no family history of colorectal cancer (Mecklin et al. 1986b; Lynch et al. 1993; Jass et al. 1994), and displayed a mucinous component more often than was observed in nonfamilial CRCs, especially when metachronous tumors were included in the evaluation (Mecklin et al. 1986b; Jass and Stewart 1992; Jass et al. 1994). With the discovery of MSI within CRC tumor DNA, studies analyzing histopathologic characteristics of MSI-H tumors (including both nonfamilial and Lynch Syndrome tumors) increased both in number and scope.

MSI-H tumors generally show considerable histopathologic heterogeneity (Greenson et al. 2003; Umar et al. 2004). However, correlations between histopathologic features and MSI-H status have been established. Consistent with previous family-history-based studies, MSI-H tumors tend to be poorly differentiated, mucin producing, and exhibit a medullary growth pattern (Kim et al. 1994; Ward et al. 2001; Greenson et al. 2003). Studies examining those with mucinous tumors, specifically, have demonstrated that MSI-H tumors (defined by abnormal IHC staining or MSI at two or more of the five markers recommended by the Bethesda conference or >30% of markers) comprise 30% of all mucinous colorectal cancers. MSI-H tumors, therefore, make up a greater proportion of mucinous tumors than nonmucinous tumors (Messerini et al. 1997; Kakar et al. 2004). A rare type of mucinous tumor, signet-ring cell carcinoma, has also been observed more frequently among MSI-H tumors (Mecklin et al. 1986b; Lynch et al. 1993; Kim et al. 1994).

Despite the over-representation of mucinous tumors among MSI-H CRCs, early authors observed a tendency toward a better prognosis in those belonging to the hereditary group than generally expected of mucinous tumors (Jass and Stewart 1992; Jass et al. 1994). Kakar et al. (2004) also described better than expected survival rates observed in a group of individuals with MSI-H mucinous tumors (Kakar et al. 2004). In general, tumor microsatellite instability has been associated with favorable prognoses (Thibodeau et al. 1993; Halling et al. 1999; Ward et al. 2001), and it has been suggested that this may be related to the other reported predominant histopathologic features in Lynch Syndrome CRCs, including tumor-infiltrating lymphocytes (TILs), a Crohn's-like lymphocytic response, medullary growth pattern, and diploidy, all of which are associated with a favorable prognosis (Lynch et al. 1993; Jass et al. 1994; Kakar et al. 2004).

Several of the noted histopathologic features have proven to be sensitive predictors of MSI-H status. Based on their apparent predictive value, the Bethesda guidelines were revised to include individuals with "MSI-H histology" diagnosed in those less than 60 years as appropriate for further evaluation as possible Lynch Syndrome patients (Fig. 7.1) (Umar et al. 2004). Recently, a scoring system based on the predictive value of these histologic features was proposed to aid in the selection of cases for screening for Lynch Syndrome. Similar to previous analyses, Jenkins et al. (2007) identified several independent histopathologic predictors of MSI-H phenotype, each with its own specific predictive value (presence of TILs, poor differentiation, mucin production, and Crohn's like response) (Kim et al. 1994; Smyrk et al. 2001; Ward et al. 2001; Greenson et al. 2003). These independent predictors, in addition to age of diagnosis and proximal colon-tumor location, have an overall sensitivity of 93% and specificity of 55% for MSI-H tumors, proving better than any of the individual predictive factors taken alone (Jenkins et al. 2007). Consideration of histopathologic characteristics, as facilitated by the proposed scoring system, MsPath, supports the inclusion of histopathology features in the diagnostic evaluation for Lynch Syndrome.

Clinical Features

Tumor Spectrum

Despite the focus on CRC, Lynch Syndrome involves multiple other organs. The tumor spectrum associated with Lynch Syndrome was originally derived clinically through family studies and subsequently defined further through molecular analysis. Overall, CRC continues to be the most prevalent type of cancer associated with Lynch Syndrome, comprising about 45% of these diagnoses. Lynch Syndrome has traditionally been and continues to be characterized by metachronous and/or synchronous colonic and extracolonic tumors as well as a preponderance of right-sided/proximal colonic tumors (Lynch et al. 1985a; Lanspa et al. 1990; Jass and Stewart 1992; Aaltonen et al. 1993; Ionov et al. 1993; Lothe et al. 1993; Thibodeau et al. 1993; Kim et al. 1994; Greenson et al. 2003). The mean age at diagnosis of colon cancer in Lynch Syndrome patients is about 45 years (Mecklin et al. 1986a; Vasen et al. 1990). More recent studies have suggested differences in mean age at diagnosis and disease penetrance, dependent upon the MMR gene involved.

Early clinical studies revealed multiple extracolonic malignancies associated with Lynch Syndrome. After CRC, the second most common cancer observed is endometrial. The mean age at diagnosis of CRC and endometrial carcinoma associated with germline mutations in *hMLH1* and *hMSH2* remains similar to those observed in earlier clinical studies (45 and 50 years, respectively); however, for *hMSH6*, the average age at diagnosis is about 10 years later for both CRC and endometrial tumors (Wijnen et al. 1999; Peltomaki et al. 2001; Wagner et al. 2001; Hendriks et al. 2004). Other studies have shown greater-than-expected frequencies of cancers of the ovary, stomach, biliary tract, renal pelvis and ureter, and small bowel (Mecklin et al. 1986a; Vasen et al. 1990; Watson and Lynch 1994; Aarnio et al. 1999). The presence of defective MMR manifesting as MSI in gastric, ovarian, and other historical Lynch-Syndrome-related tumors has also been documented (reviewed in Peltomaki 2003).

Penetrance

Similar to observations regarding mean age at diagnosis, the cumulative risk for cancer associated with mutations in *hMLH1* and *hMSH2* differs from that associated with *hMSH6* mutations. The lifetime risk of developing colon cancer (by age 70–75) is approximately 80% (range: 53–90%) associated with *hMLH1* and *hMSH2* mutations. The lifetime risk for CRC associated with a germline *hMSH6* mutation, however, is 50%. Unlike CRC, endometrial cancer appears to be more penetrant in *hMSH6* families with a cumulative lifetime risk of 71% versus the 30–40% associated with *hMLH1* and *hMSH2* (Vasen et al. 1996; Hendriks et al. 2004; Peltomaki 2005). In a recent review of literature on Lynch Syndrome, Vasen et al. (2007) reported lifetime risks for cancer in families with identified MMR gene mutations as follows: CRC in men: 28–75%; CRC in women: 24–54%; endometrial cancer: 27–71%; ovarian cancer: 3–13%; gastric cancer 2–13%; urinary tract cancer: 1–12%; brain tumor: 1–4%; bile duct/gallbladder cancer: 2%; and small bowel cancer: 4–7%. The lifetime risks for various skin cancers have not been well studied, but are certainly increased for sebaceous tumors. Heterozygous *PMS2* mutations, although well established as a mechanism of disease, are less penetrant than mutations in the other three common MMR genes (De Rosa et al. 2000; Hendriks et al. 2006b).

Clinical Variants of Lynch Syndrome

As Lynch Syndrome was being defined both clinically and molecularly, several variants of the disease emerged. Muir-Torre syndrome (MTS) was the first to be reported and was formally defined in 1995. It is characterized by the concurrence of classic Lynch Syndrome tumors with sebaceous gland adenomas/adenocarcinomas and/or keratoacanthomas (Lynch et al. 1991; Schwartz and Torre 1995). Defective MMR was established as the underlying molecular etiology of MTS, through demonstration of MSI in both sebaceous and colorectal tumors (Honchel et al. 1994; Bocker et al. 1997; Entius et al. 2000; Machin et al. 2002). MTS is predominantly caused by deficiencies (including large gene rearrangements) in *hMSH2*, but mutations in *hMLH1* and *hMSH6* have also been reported (Barana et al. 2004; Mangold et al. 2004; Arnold et al. 2007; Mangold et al. 2007).

Biallelic mutations have been reported frequently in association with *PMS2* gene involvement: both homozygous and compound heterozygous mutations have been reported. Turcot syndrome, a clinical variant of both Lynch Syndrome and familial adenomatous polyposis, may be caused by biallelic *PMS2* mutations, resulting in an autosomal recessive syndrome. Clinically, Turcot Syndrome is characterized by the occurrence of primary brain tumors, specifically glioblastomas, in association with colorectal cancer or colorectal adenomas usually at early stages; however, other cancer types have also been observed (Turcot et al. 1959; Agostini et al. 2005).

Biallelic mutations in *hMLH1*, *hMSH2*, and *hMSH6* have become more frequent in the literature; these are generally characterized by a unique phenotype consisting of hematologic malignancies, early-onset CRC, and café-au-lait macules that are essentially indistinguishable from those seen in Neurofibromatosis type 1 (Trimbath et al. 2001; Poley et al. 2007). The term "Lynch Syndrome Type III" has been suggested for this rare phenotype.

References

Aaltonen, L. A., Peltomaki, P., Leach, F. S., et al. 1993. Clues to the pathogenesis of familial colorectal cancer. Science. 260:812–816.

Aaltonen, L. A., Sankila, R., Mecklin, J-P., et al. 1994. A novel approach to estimate the proportion of hereditary nonpolyposis colorectal cancer of total colorectal cancer burden. Cancer Detection and Prevention. 18:57–63.

Aaltonen, L. A., Salovaara, R., Kristo, P., et al. 1998. Incidence of hereditary nonpolyposis colorectal cancer and the feasibility of molecular screening for the disease. New England Journal of Medicine. 338:1481–1487.

Aarnio, M., Sankila, R., Pukkala, E., et al. 1999. Cancer risk in mutation carriers of DNA-mismatch-repair genes. International Journal of Cancer. 81:214–218.

Agostini, M., Tibiletti, M. G., Lucci-Cordisco, E., et al. 2005. Two *PMS2* mutations in a Turcot syndrome family with small bowel cancers. The American Journal of Gastroenterology. 100:1886–1891.

Akiyama, Y., Sato, H., Yamada, T., et al. 1997. Germ-line mutation of the *hMSH6/GTBP* gene in an atypical hereditary nonpolyposis colorectal cancer kindred. Cancer Research. 57:3920–3923.

Aquilina, G., and Bignami, M. 2001. Mismatch repair in correction of replication errors and processing of DNA damage. Journal of Cellular Physiology. 187:145–154.

Arnold, A., Payne, S., Fisher, S., et al. 2007. An individual with Muir-Torre syndrome found to have a pathogenic *MSH6* mutation. Familial Cancer. 6:317–321.

Auclair, J., Busine, M. P., Navarro, C., et al. 2006. Systematic mRNA analysis for the effect of *MLH1* and *MSH2* missense and silent mutations on aberrant splicing. Human Mutation. 27:145–154.

Barana, D., van der Klift, H., Wijnen, J., et al. 2004. Spectrum of genetic alterations in Muir-Torre syndrome is the same as in HNPCC. American Journal of Medical Genetics. 125A:318–319.

Baudhuin, L. M., Burgart, L. J., Leontovich, O., and Thibodeau, S. N. 2005a. Use of microsatellite instability and immunohistochemistry testing for the identification of individuals at risk for Lynch syndrome. Familial Cancer. 4:255–265.

Baudhuin, L. M., Ferber, M. J., Winters, J. L., et al. 2005b. Characterization of *hMLH1* and *hMSH2* gene dosage alterations in Lynch syndrome patients. Gastroenterology. 129:846–854.

Berends, M. J., Wu, Y., Sijmons, R. H., et al. 2002. Molecular and clinical characteristics of *MSH6* variants: an analysis of 25 index carriers of a germline variant. The American Journal of Human Genetics. 70:26–37.

Bhattacharyya, N. P., Skandalis, A., Ganesh, A., et al. 1994. Mutator phenotypes in human colorectal carcinoma cell lines. Proceedings of the National Academy of Sciences of the United States of America. 91:6319–6323.

Bocker, T., Diermann, J., Friedl, W., et al. 1997. Microsatellite instability analysis: a multicenter study for reliability and quality control. Cancer Research. 57:4739–4743.

Boland, C. R. 2005. Evolution of the nomenclature for the hereditary colorectal cancer syndromes. Familial Cancer. 4:211–218.

Boland, C. R., and Troncale, F. J. 1984. Familial colonic cancer without antecedent polyposis. Annals of Internal Medicine. 100:700–701.

Boland, C. R., Thibodeau, S. N., Hamilton, S. R., et al. 1998. A National Cancer Institute workshop on microsatellite instability for cancer detection and familial predisposition: development of international criteria for the determination of microsatellite instability in colorectal cancer. Cancer Research. 58:5248–5257.

Bronner, C. E., Baker, S. M., Morrison, P. T., et al. 1994. Mutation in the DNA mismatch repair gene homologue *hMLH1* is associated with hereditary non-polyposis colon cancer. Nature. 368:258–261.

Caluseriu, O., Di Gregorio, C., Lucci-Cordisco, E., et al. 2004. A founder *MLH1* mutation in families from the districts of Modena and Reggio-Emilia in northern Italy with hereditary non-polyposis colorectal cancer associated with protein elongation and instability. Journal of Medical Genetics. 41:e34.

Cartegni, L., Chew, S. L., and Krainer, A. R. 2002. Listening to silence and understanding nonsense: exonic mutations that affect splicing. Nature Reviews Genetics. 3:285–298.

Cederquist, K. Emanuelsson, M., Wiklund, F., et al. 2005. Two Swedish founder *MSH6* mutations, one nonsense and one missense, conferring high cumulative risk of Lynch syndrome. Clinical Genetics. 68:533–541.

Chan, T. L., Yuen, S. T., Ho, J. W., et al. 2001. A novel germline 1.8-kb deletion of *hMLH1* mimicking alternative splicing: a founder mutation in the Chinese population. Oncogene. 20:2976–2981.

Chan, T. L., Chan, Y. W., Ho, J. W., et al. 2004. *MSH2* c.1452–1455delAATG is a founder mutation and an important cause of hereditary nonpolyposis colorectal cancer in the southern Chinese population. The American Journal of Human Genetics. 74:1035–1042.

Chan, T. L., Yuen, S. T., Kong, C. K., et al. 2006. Heritable germline epimutation of *MSH2* in a family with hereditary nonpolyposis colorectal cancer. Nature Genetics. 38:1178–1183.

Chan, P. A., Duraisamy, S., Miller, P. J., et al. 2007. Interpreting missense variants: comparing computational methods in human disease genes *CDKN2A*, *MLH1*, *MSH2*, *MECP2*, and tyrosinase (*TYR*). Human Mutation. 28:683–693.

Charbonnier, F., Raux, G., Wang, Q., et al. 2000. Detection of exon deletions and duplications of the mismatch repair genes in hereditary nonpolyposis colorectal cancer families using multiplex polymerase chain reaction of short fluorescent fragments. Cancer Research. 60:2760–2763.

Charbonnier, F., Olschwang, S., Wang, Q., et al. 2002. *MSH2* in contrast to *MLH1* and *MSH6* is frequently inactivated by exonic and promoter rearrangements in hereditary nonpolyposis colorectal cancer. Cancer Research. 62:848–853.

Chen, H., Taylor, N. P., Sotamaa, K. M., et al. 2007. Evidence for heritable predisposition to epigenetic silencing of *MLH1*. International Journal of Cancer. 120:1684–1688.

Cotton, R. G., Appelbe, W., Auerbach, A. D., et al. 2007. Recommendations of the 2006 Human Variome Project meeting. Nature Genetics. 39:433–436.

Cunningham, J. M., Christensen, E. R., Tester, D. L., et al. 1998. Hypermethylation of the *hMLH1* promoter in colon cancer with microsatellite instability. Cancer Research. 58:3455–3460.

Cunningham, J. M., Kim, C. Y., Christensen, E. R., et al. 2001. The frequency of hereditary defective mismatch repair in a prospective series of unselected colorectal carcinomas. The American Journal of Human Genetics. 69:780–790.

De Rosa, M., Fasano, C., Panariello, L., et al. 2000. Evidence for a recessive inheritance of Turcot's syndrome caused by compound heterozygous mutations within the PMS2 gene. Oncogene. 19:1719–1723.

Desai, D. C., Lockman, J. C., Chadwick, R. B., et al. 2000. Recurrent germline mutation in *MSH2* arises frequently de novo. Journal of Medical Genetics. 37:646–652.
Dietmaier, W., Wallinger, S., Bocker, T., et al. 1997. Diagnostic microsatellite instability: definition and correlation with mismatch repair protein expression. Cancer Research. 57:4749–4756.
Di Fiore, F., Charbonnier, F., Martin, C., et al. 2004. Screening for genomic rearrangements of the *MMR* genes must be included in the routine diagnosis of HNPCC. Journal of Medical Genetics. 41:18–20.
Douglas, J. A., Gruber, S. B., Meister, K. A., et al. 2005. History and molecular genetics of Lynch Syndrome in family G: a century later. The Journal of the American Medical Association. 294:2195–2202.
Drummond, J. T., Li, G-M., Longley, M. J., and Modrich, P. 1995. Isolation of an hMSH2-p160 heterodimer that restores DNA mismatch repair to tumor cells. Science. 268:1909–1912.
Entius, M. M., Keller, J. J., Drillenburg, P., et al. 2000. Microsatellite instability and expression of hMLH-1 and hMSH-2 in sebaceous gland carcinomas as markers for Muir-Torre syndrome. Clinical Cancer Research. 6:1784–1789.
Evans, D. G., Walsh, S., Jeacock, J., et al. 1997. Incidence of hereditary non-polyposis colorectal cancer in a population-based study of 1137 consecutive cases of colorectal cancer. The British Journal of Surgery. 84:1281–1285.
Fearon, E. R., and Vogelstein, B. 1990. A genetic model for colorectal tumorigenesis. Cell. 61: 759–767.
Fishel, R., Lescoe, M. K., Rao, M. R., et al. 1993. The human mutator gene homolog *MSH2* and its association with hereditary nonpolyposis colon cancer. Cell. 75:1027–1038.
Foulkes, W. D., Thiffault, I., Gruber, S. B., et al. 2002. The founder mutation *MSH2*1906G>C* is an important cause of hereditary nonpolyposis colorectal cancer in the Ashkenazi Jewish population. The American Journal of Human Genetics. 71:1395–1412.
Froggatt, N. J., Joyce, J. A., Davies, R., et al. 1995. A frequent *hMSH2* mutation in hereditary non-polyposis colon cancer syndrome. The Lancet. 345:727.
Froggatt, N. J., Green, J., Brassett, C., et al. 1999. A common MSH2 mutation in English and North American HNPCC families: origin, phenotypic expression, and sex specific differences in colorectal cancer. Journal of Medical Genetics. 36:97–102.
Gazzoli, I., Loda, M., Garber, J., et al. 2002. A hereditary nonpolyposis colorectal carcinoma case associated with hypermethylation of the *MLH1* gene in normal tissue and loss of heterozygosity of the unmethylated allele in the resulting microsatellite instability-high tumor. Cancer Research. 62:3925–3928.
Gille, J. J., Hogervorst, F. B., Pals, G., et al. 2002. Genomic deletions of *MSH2* and *MLH1* in colorectal cancer families detected by a novel mutation detection approach. British Journal of Cancer. 87:892–897.
Gorlov, I. P., Gorlova, O. Y., Frazier, M. L., and Amos, C. I. 2003. Missense mutations in *hMLH1* and *hMSH2* are associated with exonic splicing enhancers. The American Journal of Human Genetics. 73:1157–1161.
Grabowski, M., Mueller-Koch, Y., Grasbon-Frodl, E., et al. 2005. Deletions account for 17% of pathogenic germline alterations in *MLH1* and *MSH2* in hereditary nonpolyposis colorectal cancer (HNPCC) families. Genetic Testing. 9:138–146.
Greenson, J. K., Bonner, J. D., Ben-Yzhak, O., et al. 2003. Phenotype of microsatellite unstable colorectal carcinomas: well-differentiated and focally mucinous tumors and the absence of dirty necrosis correlate with microsatellite instability. The American Journal of Surgical Pathology. 27:563–570.
Guillem, J. G., Rapaport, B. S., Kirchhoff, T., et al. 2003. A636P is associated with early-onset colon cancer in Ashkenazi Jews. Journal of the American College of Surgeons. 196:222–225.
Halling, K. C., French, A. J., McDonnell, S. K., et al. 1999. Microsatellite instability and 8p allelic imbalance in stage B2 and C colorectal cancers. Journal of the National Cancer Institute. 91:1295–1303.
Hampel, H., Frankel, W. L., Martin, E., et al. 2005. Screening for the Lynch syndrome (hereditary nonpolyposis colorectal cancer). The New England Journal of Medicine. 352:1851–1860.

Hampel, H., Frankel, W., Panescu, J., et al. 2006. Screening for Lynch syndrome (hereditary nonpolyposis colorectal cancer) among endometrial cancer patients. Cancer Research. 66:7810–7817.

Harrington, J. M., and Kolodner, R. D. 2007. *Saccharomyces cerevisiae* Msh2-Msh3 acts in repair of base-base mispairs. Molecular and Cellular Biology. 27:6546–6554.

Hemminki, A., Peltomaki, P., Mecklin, J-P., et al. 1994. Loss of the wild type *MLH1* gene is a feature of hereditary nonpolyposis colorectal cancer. Nature Genetics. 8:405–410.

Hendriks, Y. M., Wagner, A., Morreau, H., et al. 2004. Cancer risk in hereditary nonpolyposis colorectal cancer due to *MSH6* mutations: impact on counseling and surveillance. Gastroenterology. 127:17–25.

Hendriks, Y. M., de Jong, A. E., Morreau, H., et al. 2006a. Diagnostic approach and management of Lynch syndrome (hereditary nonpolyposis colorectal carcinoma): a guide for clinicians. CA: A Cancer Journal for Clinicians. 56:213–225.

Hendriks, Y. M., Jagmohan-Changur, S., van der Klift, H. M., et al. 2006b. Heterozygous mutations in *PMS2* cause hereditary nonpolyposis colorectal carcinoma (Lynch syndrome). Gastroenterology. 130:312–322.

Hitchins, M., Williams, R., Cheong, K., et al. 2005. *MLH1* germline epimutations as a factor in hereditary nonpolyposis colorectal cancer. Gastroenterology. 129:1392–1399.

Hitchins, M. P., Wong, J. J., Suthers, G., et al. 2007. Inheritance of a cancer-associated *MLH1* germ-line epimutation. The New England Journal of Medicine. 356:697–705.

Honchel, R., Halling, K. C., Schaid, D. J., et al. 1994. Microsatellite instability in Muir-Torre syndrome. Cancer Research. 54:1159–1163.

Hutter, P., Couturier, A., Scott, R. J., et al. 1996. Complex genetic predisposition to cancer in an extended HNPCC family with an ancestral hMLH1 mutation. Journal of Medical Genetics. 33:636–640.

Ionov, Y. Peinado, M. A., Malkhosyan, S., et al. 1993. Ubiquitous somatic mutations in simple repeated sequences reveal a new mechanism for colonic carcinogenesis. Nature. 363:558–561.

Jager, A. C., Bisgaard, M. L., Myrhoj, T., et al. 1997. Reduced frequency of extracolonic cancers in hereditary nonpolyposis colorectal cancer families with monoallelic *hMLH1* expression. The American Journal of Human Genetics. 61:129–138.

Jarvinen, H. J., Aarnio, M., Mustonen, H., et al. 2000. Controlled 15-year trial on screening for colorectal cancer in families with hereditary nonpolyposis colorectal cancer. Gastroenterology. 118:829–834.

Jass, J. R. 2006. Hereditary non-polyposis colorectal cancer: the rise and fall of a confusing term. World Journal of Gastroenterology. 12:4943–4950.

Jass, J. R., and Stewart, S. M. 1992. Evolution of hereditary non-polyposis colorectal cancer. Gut. 33:783–786.

Jass, J. R., Smyrk, T. C., Stewart, S. M., et al. 1994. Pathology of hereditary non-polyposis colorectal cancer. Anticancer Research. 14:1631–1634.

Jenkins, M. A., Hayashi, S., O'Shea, A. M., et al. 2007. Pathology features in Bethesda guidelines predict colorectal cancer microsatellite instability: a population-based study. Gastroenterology. 133:48–56.

Jiricny, J., and Nystrom-Lahti, M. 2000. Mismatch repair defects in cancer. Current Opinion in Genetics and Development. 10:157–161.

Kakar, S., Burgart, L. J., Thibodeau, S. N., et al. 2003. Frequency of loss of *hMLH1* expression in colorectal carcinoma increases with advancing age. Cancer. 97:1421–1427.

Kakar, S., Aksoy, S., Burgart, L. J., and Smyrk, T. C. 2004. Mucinous carcinoma of the colon: correlation of loss of mismatch repair enzymes with clinicopathologic features and survival. Modern Pathology. 17:696–700.

Kim, H., Jen, J., Vogelstein, B., and Hamilton, S. R. 1994. Clinical and pathological characteristics of sporadic colorectal carcinomas with DNA replication errors in microsatellite sequences. The American Journal of Pathology. 145:148–156.

Kolodner, R. D. 1995. Mismatch repair: mechanisms and relationship to cancer susceptibility. Trends in Biochemical Sciences. 20:397–401.

Kuismanen, S. A., Holmberg, M. T., Salovaara, R., et al. 2000. Genetic and epigenetic modification of *MLH1* accounts for a major share of microsatellite-unstable colorectal cancers. The American Journal of Pathology. 156:1773–1779.

Kurzawski, G., Suchy, J., Lener, M., et al. 2006. Germline *MSH2* and *MLH1* mutational spectrum including large rearrangements in HNPCC families from Poland (update study). Clinical Genetics. 69:40–47.

Laiho, P., Launonen, V., Lahermo, P., et al. 2002. Low-level microsatellite instability in most colorectal carcinomas. Cancer Research. 62:1166–1170.

Lanspa, S. J., Lynch, H. T., Smyrk, T. C., et al. 1990. Colorectal adenomas in the Lynch syndromes. Results of a colonoscopy screening program. Gastroenterology. 98:1117–1122.

Leach, F. S., Nicolaides, N. C., Papadopoulos, N., et al. 1993. Mutations of a mutS homolog in hereditary nonpolyposis colorectal cancer. Cell. 75:1215–1225.

Lindor, N. M., Rabe, K., Petersen, G. M., et al. 2005. Lower cancer incidence in Amsterdam-I criteria families without mismatch repair deficiency. Familial colorectal cancer type X. The Journal of the American Medical Association. 293:1979–1985.

Loeb, L. A. 1991. Mutator phenotype may be required for multistage carcinogenesis. Cancer Research. 51:3075–3079.

Loeb, L. A., Springgate, C. F., and Battula, N. 1974. Errors in DNA replication as a basis of malignant changes. Cancer Research. 34:2311–2321.

Lothe, R. A., Peltomaki, P., Meling, G. I., et al. 1993. Genomic instability in colorectal cancer: relationship to clinicopathological variables and family history. Cancer Research. 53: 5849–5852.

Love, R. R. 1986. Adenomas are precursor lesions for malignant growth in nonpolyposis hereditary carcinoma of the colon and rectum. Surgery, Gynecology, and Obstetrics. 162:8–12.

Lynch, H. T., and de la Chapelle, A. 2003. Hereditary colorectal cancer. The New England Journal of Medicine. 348:919–932.

Lynch, H. T., and Krush, A. J. 1971. Cancer family "G" revisited: 1895–1970. Cancer. 27:1505–1511.

Lynch, H. T., Shaw, M. W., Magnuson, C. W., et al. 1966. Hereditary factors in cancer. Study of two large midwestern kindreds. Archives of Internal Medicine. 117:206–212.

Lynch, H. T., Kimberling, W., Albano, W. A., et al. 1985a. Hereditary nonpolyposis colorectal cancer (Lynch syndromes I and II). I. Clinical description of resource. Cancer. 56:934–938.

Lynch, H. T., Schuelke, G. S., Kimberling, W. J., et al. 1985b. Hereditary nonpolyposis colorectal cancer (Lynch syndromes I and II). II. Biomarker studies. Cancer. 56:939–951.

Lynch, H. T., Lanspa, S., Smyrk, T., et al. 1991. Hereditary nonpolyposis colorectal cancer (Lynch syndromes I and II). Genetics, pathology, natural history, and cancer control, Part I. Cancer Genetics and Cytogenetics. 53:143–160.

Lynch, H. T., Smyrk, T. C., Watson, P., et al. 1993. Genetics, natural history, tumor spectrum, and pathology of hereditary nonpolyposis colorectal cancer: an updated review. Gastroenterology. 104:1535–1549.

Lynch, H. T., de la Chapelle, A., Hampel, H., et al. 2006. American founder mutation for Lynch syndrome. Prevalence estimates and implications. Cancer. 106:448–452.

Machin, P., Catasus, L., Pons, C., et al. 2002. Microsatellite instability and immunostaining for MSH-2 and MLH-1 in cutaneous and internal tumors from patients with Muir-Torre syndrome. Journal of Cutaneous Pathology. 29:415–420.

Mangold, E., Pagenstecher, C., Leister, M., et al. 2004. A genotype-phenotype correlation in HNPCC: strong predominance of *msh2* mutations in 41 patients with Muir-Torre syndrome. Journal of Medical Genetics. 41:567–572.

Mangold, E., Rahner, N., Friedrichs, N., et al. 2007. *MSH6* mutation in Muir-Torre syndrome: could this be a rare finding? British Journal of Dermatology. 156:158–162.

Marra, G., D'Atri, S., Yan, H., et al. 2001. Phenotypic analysis of hMSH2 mutations in mouse cells carrying human chromosomes. Cancer Research. 61:7719–7721.

Marsischky, G. T., Filosi, N., Kane, M. F., and Kolodner, R. 1996. Redundancy of *Saccharomyces cerevisiae MSH3* and *MSH6* in *MSH2*-dependent mismatch repair. Genes and Development. 10:407–420.

Mauillon, J. L., Michel, P., Limacher, J. M., et al. 1996. Identification of novel germline *hMLH1* mutations including a 22 kb Alu-mediated deletion in patients with familial colorectal cancer. Cancer Research. 56:5728–5733.

Mecklin, J-P. 1987. Frequency of hereditary colorectal carcinoma. Gastroenterology. 93: 1021–1025.

Mecklin, J-P., Jarvinen, H. J., and Peltokallio, P. 1986a. Cancer family syndrome. Genetic analysis of 22 Finnish kindreds. Gastroenterology. 90:328–333.

Mecklin, J-P., Sipponen, P., and Jarvinen, H. J. 1986b. Histopathology of colorectal carcinomas and adenomas in cancer family syndrome. Diseases of the Colon and Rectum. 29:849–853.

Mecklin, J-P., Jarvinen, H. J., Hakkiluoto, A., et al. 1995. Frequency of hereditary nonpolyposis colorectal cancer. A prospective multicenter study in Finland. Diseases of the Colon and Rectum. 38:588–593.

Messerini, L., Vitelli, F., De Vitis, L. R., et al. 1997. Microsatellite instability in sporadic mucinous colorectal carcinomas: relationship to clinico-pathological variables. The Journal of Pathology. 182:380–384.

Miyaki, M., Konishi, M., Tanaka, K., et al. 1997. Germline mutation of *MSH6* as the cause of hereditary nonpolyposis colorectal cancer. Nature Genetics. 17:271–272.

Miyakura, Y., Sugano, K., Konishi, F., et al. 2001. Extensive methylation of h*MLH*1 promoter region predominates in proximal colon cancer with microsatellite instability. Gastroenterology. 121:1300–1309.

Miyakura, Y., Sugano, K., Akasu, T., et al. 2004. Extensive but hemiallelic methylation of the h*MLH*1 promoter region in early-onset sporadic colon cancers with microsatellite instability. Clinical Gastroenterology and Hepatology. 2:147–156.

Moisio, A-L., Sistonen, P., Weissenbach, J., et al. 1996. Age and origin of two common MLH1 mutations predisposing to hereditary colon cancer. The American Journal of Human Genetics. 59:1243–1251.

Morgan, H. D., Sutherland, H. G., Martin, D. I., and Whitelaw, E. 1999. Epigenetic inheritance at the agouti locus in the mouse. Nature Genetics. 23:314–318.

Nakagawa, H., Hampel, H., and de la Chapelle, A. 2003. Identification and characterization of genomic rearrangements of *MSH2* and *MLH1* in Lynch syndrome (HNPCC) by novel techniques. Human Mutation. 22:258.

Nicolaides, N. C., Papadopoulos, N., Liu, B., et al. 1994. Mutations of two *PMS* homologues in hereditary nonpolyposis colon cancer. Nature. 371:75–80.

Nystrom-Lahti, M., Kristo, P., Nicolaides, N. C., et al. 1995. Founding mutations and Alu-mediated recombination in hereditary colon cancer. Nature Medicine. 1:1203–1206.

Nystrom-Lahti, M., Wu, Y., Moisio, A-L., et al. 1996. DNA mismatch repair gene mutations in 55 kindreds with verified or putative hereditary non-polyposis colorectal cancer. Human Molecular Genetics. 5:763–769.

Ollikainen, M., Hannelius, U., Lindgren, C. M., et al. 2007. Mechanisms of inactivation of *MLH1* in hereditary nonpolyposis colorectal carcinoma: a novel approach. Oncogene. 26:4541–4549.

Ollila, S., Sarantaus, L., Kariola, R., et al. 2006. Pathogenicity of *MSH2* missense mutations is typically associated with impaired repair capability of the mutated protein. Gastroenterology. 131:1408–1417.

Pagenstecher, C., Wehner, M., Friedl, W., et al. 2006. Aberrant splicing in *MLH1* and *MSH2* due to exonic and intronic variants. Human Genetics. 119:9–22.

Palombo, F., Gallinari, P., Iaccarino, I., et al. 1995. GTBP, a 160-kilodalton protein essential for mismatch-binding activity in human cells. Science. 268:1912–1914.

Papadopoulos, N., Nicolaides, N. C., Wei, Y-F., et al. 1994. Mutation of a *mutL* homolog in hereditary colon cancer. Science. 263:1625–1629.

Peel, D. J., Ziogas, A., Fox, E. A., et al. 2000. Characterization of hereditary nonpolyposis colorectal cancer families from a population-based series of cases. Journal of the National Cancer Institute. 92:1517–1522.

Peltomaki, P. 2003. Role of DNA mismatch repair defects in the pathogenesis of human cancer. Journal of Clinical Oncology. 21:1174–1179.

Peltomaki, P. 2005. Lynch syndrome genes. Familial Cancer. 4:227–232.

Peltomaki, P., and Vasen, H. 2004. Mutations associated with HNPCC predisposition – update of ICG-HNPCC/INSiGHT mutation database. Disease Markers. 20:269–276.

Peltomaki, P., Gao, X., and Mecklin, J-P. 2001. Genotype and phenotype in hereditary nonpolyposis colon cancer: a study of families with different vs. shared predisposing mutations. Familial Cancer. 1:9–15.

Plaschke, J., Ruschoff, J., and Schackert, H. K. 2003. Genomic rearrangements of *hMSH6* contribute to the genetic predisposition in suspected hereditary non-polyposis colorectal cancer syndrome. Journal of Medical Genetics. 40:597–600.

Poley, J-W., Wagner, A., Hoogmans, M., et al. 2007. Biallelic germline mutations of mismatch-repair genes: a possible cause for multiple pediatric malignancies. Cancer. 109:2349–2356.

Potter, J. D. 2001. At the interfaces of epidemiology, genetics, and genomics. Nature Reviews Genetics. 2:142–147.

Raevaara, T. E., Korhonen, M. K., Lohi, H., et al. 2005. Functional significance and clinical phenotype of nontruncating mismatch repair variants of *MLH1*. Gastroenterology. 129: 537–549.

Rahner, N., Friedrichs, N., Wehner, M., et al. 2007. Nine novel pathogenic germline mutations in *MLH1*, *MSH2*, *MSH6* and *PMS2* in families with Lynch syndrome. Acta Oncologica. 46:763–769.

Rakyan, V. K., Chong, S., Champ, M. E., et al. 2003. Transgenerational inheritance of epigenetic states at the murine $Axin^{Fu}$ allele occurs after maternal and paternal transmission. Proceedings of the National Academy of Sciences of the United States of America. 100:2538–2543.

Ravnic-Glavac, M., Potocnik, U., and Glavac, D. 2000. Incidence of germline *hMLH1* and *hMSH2* mutations (HNPCC patients) among newly diagnosed colorectal cancers in a Slovenian population. Journal of Medical Genetics. 37:533–536.

Rodriguez-Bigas, M. A., Boland, C. R., Hamilton, S. R., et al. 1997. A National Cancer Institute workshop on hereditary nonpolyposis colorectal cancer syndrome: meeting highlights and Bethesda guidelines. Journal of the National Cancer Institute. 89:1758–1762.

Roemer, I., Reik, W., Dean, W., and Klose, J. 1997. Epigenetic inheritance in the mouse. Current Biology. 7:277–280.

Salovaara, R., Loukola, A., Kristo, P., et al. 2000. Population-based molecular detection of hereditary nonpolyposis colorectal cancer. Journal of Clinical Oncology. 18:2193–2200.

Samowitz, W. S., Curtin, K., Lin, H. H., et al. 2001. The colon cancer burden of genetically defined hereditary nonpolyposis colon cancer. Gastroenterology. 121:830–838.

Schwartz, R. A., and Torre, D. P. 1995. The Muir-Torre syndrome: a 25 years retrospect. Journal of the American Academy of Dermatology. 33:90–104.

Shin, Y-K., Heo, S-C., Shin, J-H., et al. 2004. Germline Mutations in *MLH1*, *MSH2* and *MSH6* in Korean hereditary non-polyposis colorectal cancer families. Human Mutation. 24:351.

Smyrk, T. C., Watson, P., Kaul, K., and Lynch, H. T. 2001. Tumor-infiltrating lymphocytes are a marker for microsatellite instability in colorectal carcinoma. Cancer. 91:2417–2422.

Struewing, J. P., Hartge, P., Wacholder, S., et al. 1997. The risk of cancer associated with specific mutations of *BRCA1* and *BRCA2* among Ashkenazi Jews. The New England Journal of Medicine. 336:1401–1408.

Suter, C. M., Martin, D. I. K., and Ward, R. L. 2004. Germline epimutation of *MLH1* in individuals with multiple cancers. Nature Genetics. 36:497–501.

Tannergard, P., Liu, T., Weger, A., et al. 1997. Tumorigenesis in colorectal tumors from patients with hereditary non-polyposis colorectal cancer. Human Genetics. 101:51–55.

Taylor, C. F., Charlton, R. S., Burn, J., et al. 2003. Genomic deletions in *MSH2* or *MLH1* are a frequent cause of hereditary non-polyposis colorectal cancer: identification of novel and recurrent deletions by MLPA. Human Mutation. 22:428–433.

Thibodeau, S. N., Bren, G., and Schaid, D. 1993. Microsatellite instability in cancer of the proximal colon. Science. 260:816–819.

Thibodeau, S. N., French, A. J., Cunningham, J. M., et al. 1998. Microsatellite instability in colorectal cancer: different mutator phenotypes and the principal involvement of *hMLH1*. Cancer Research. 58:1713–1718.

Thiffault, I., Foulkes, W. D., Marcus, V. A., et al. 2004. Putative common origin of two *MLH1* mutations in Italian-Quebec hereditary non-polyposis colorectal cancer families. Clinical Genetics. 66:137–143.

Thomas, B. C., Thibodeau, S. N., and Lindor, N. M. 2005. Clinical significance of tumor microsatellite instability and immunohistochemistry for mismatch repair deficiency in colorectal cancers. Current Colorectal Cancer Reports. 1:103–109.

Trimbath, J. D., Petersen, G. M., Erdman, S. H., et al. 2001. Cafe-au-lait spots and early onset colorectal neoplasia: a variant of HNPCC? Familial Cancer. 1:101–105.

Turcot, J., Despres, J. P., and St Pierre, F. 1959. Malignant tumors of the central nervous system associated with familial polyposis of the colon: report of two cases. Diseases of the Colon and Rectum. 2:465–468.

Umar, A., Boland, C. R., Terdiman, J. P., et al. 2004. Revised bethesda guidelines for hereditary nonpolyposis colorectal cancer (Lynch syndrome) and microsatellite instability. Journal of the National Cancer Institute. 96:261–268.

Vahteristo, P., Ojala, S., Tamminen, A., et al. 2005. No *MSH6* germline mutations in breast cancer families with colorectal and/or endometrial cancer. Journal of Medical Genetics. 42:e22.

Valle, L., Carbonell, P., Fernandez, V., et al. 2007. *MLH1* germline epimutations in selected patients with early-onset non-polyposis colorectal cancer. Clinical Genetics. 71:232–237.

van der Klift, H., Wijnen, J., Wagner, A., et al. 2005. Molecular characterization of the spectrum of genomic deletions in the mismatch repair genes *MSH2*, *MLH1*, *MSH6*, and *PMS2* responsible for hereditary nonpolyposis colorectal cancer (HNPCC). Genes, Chromosomes & Cancer. 44:123–138.

Vasen, H. F., Offerhaus, G. J., den Hartog Jager, F. C., et al. 1990. The tumour spectrum in hereditary non-polyposis colorectal cancer: a study of 24 kindreds in the Netherlands. International Journal of Cancer. 46:31–34.

Vasen, H. F., Mecklin, J-P., Khan, P. M., and Lynch, H. T. 1991. The International Collaborative Group on Hereditary Non-Polyposis Colorectal Cancer (ICG-HNPCC). Diseases of the Colon and Rectum. 34:424–425.

Vasen, H. F., Wijnen, J. T., Menko, F. H., et al. 1996. Cancer risk in families with hereditary nonpolyposis colorectal cancer diagnosed by mutation analysis. Gastroenterology. 110:1020–1027.

Vasen, H. F., Watson, P., Mecklin, J-P., and Lynch, H. T. 1999. New clinical criteria for hereditary nonpolyposis colorectal cancer (HNPCC, Lynch syndrome) proposed by the International Collaborative group on HNPCC. Gastroenterology. 116:1453–1456.

Vasen, H. F., Moslein, G., Alonso, A., et al. 2007. Guidelines for the clinical management of Lynch syndrome (hereditary non-polyposis cancer). Journal of Medical Genetics. 44:353–362.

Venkatesan, R. N., Bielas, J. H., and Loeb, L. A. 2006. Generation of mutator mutants during carcinogenesis. DNA Repair. 5:294–302.

Viel, A., Petronzelli, F., Della Puppa, L., et al. 2002. Different molecular mechanisms underlie genomic deletions in the *MLH1* gene. Human Mutation. 20:368–374.

Vogelstein, B., Fearon, E. R., Hamilton, S. R., et al. 1988. Genetic alterations during colorectal-tumor development. The New England Journal of Medicine. 319:525–532.

Wagner, A., Hendriks, Y., Meijers-Heijboer, E. J., et al. 2001. Atypical HNPCC owing to *MSH6* germline mutations: analysis of a large Dutch pedigree. Journal of Medical Genetics. 38:318–322.

Wagner, A., Barrows, A., Wijnen, J. T., et al. 2003. Molecular analysis of hereditary nonpolyposis colorectal cancer in the United States: high mutation detection rate among clinically selected families and characterization of an American founder genomic deletion of the *MSH2* gene. The American Journal of Human Genetics. 72:1088–1100.

Wang, Y., Waltraut, F., Sengteller, M., et al. 2002. A modified multiplex PCR assay for detection of large deletions in *MSH2* and *MLH1*. Human Mutation. 19:279–286.

Ward, R., Meagher, A., Tomlinson, I., et al. 2001. Microsatellite instability and the clinicopathological features of sporadic colorectal cancer. Gut. 48:821–829.

Warthin, A. S. 1913. Heredity with reference to carcinoma. Archives of Internal Medicine. 12: 546–555.

Watson, P., and Lynch, H. T. 1994. The tumor spectrum in HNPCC. Anticancer Research. 14: 1635–1639.

Weber, J. L. 1990. Informativeness of human (dC-dA)*n*.(dG-dT)*n* polymorphisms. Genomics. 7: 524–530.

Wijnen, J., van der Klift, H., Vasen, H., et al. 1998. *MSH2* genomic deletions are a frequent cause of HNPCC. Nature Genetics. 20:326–328.

Wijnen, J., de Leeuw, W., Vasen, H., et al. 1999. Familial endometrial cancer in female carriers of *MSH6* germline mutations. Nature Genetics. 23:142–144.

Yan, H., Papadopoulos, N., Marra, G., et al. 2000. Conversion of diploidy to haploidy. Nature. 403:723–724.

Yuan, Z. Q., Wong, N., Foulkes, W. D., et al. 1999. A missense mutation in both *hMSH2* and *APC* in an Ashkenazi Jewish HNPCC kindred: implications for clinical screening. Journal of Medical Genetics. 36:790–793.

Zhang, J., Lindroos, A., Ollila, S., et al. 2006. Gene conversion is a frequent mechanism of inactivation of the wild-type allele in cancers from *MLH1/MSH2* deletion carriers. Cancer Research. 66:659–664.

Chapter 8
Additional Syndromes with Hereditary Predisposition to Colorectal Cancer

Chapter 8.1
MUTYH-Associated Polyposis

Spring Holter and Steven Gallinger

Introduction

Until recently, *APC* was the only known gene in which germline mutations were shown to lead to the development of colorectal adenomatous polyposis. However, in 2002, Al-Tassan et al. reported "Family N," a British kindred consisting of three of seven siblings affected with multiple adenomatous polyps or colorectal adenocarcinoma without a detectable germline *APC* mutation (2002). Somatic *APC* mutation analysis of the adenomas and adenocarcinoma revealed an unusually high proportion of G:C to T:A transversions, which are characteristic of base-excision repair (BER) pathway defects. Germline analysis of the BER genes *MUTYH*, *OGG1*, and *MTH1* identified compound heterozygous germline mutations, Y165C and G382D, within *MUTYH* in all affected individuals. All unaffected family members were either heterozygous or wild type, indicating an autosomal recessive pattern of inheritance. This was the first report of pathogenic germline mutations within a BER gene as well as the first report of an autosomal recessive inheritance pattern associated with adenomatous polyposis and hereditary colorectal cancer, now known as *MUTYH*-associated polyposis (MAP).

Base-Excision Repair and *MUTYH*

The BER pathway protects against DNA damage due to reactive oxygen species (ROS) produced during cellular metabolism or through environmental exposure to ionizing radiation or chemicals. The most stable and highly mutagenic base lesion is 8-oxo-7,8-dihydro-2'-deoxyguanosine (8-oxodG) that is generated in DNA from guanine. Cells with deficient BER are susceptible to DNA damage by ROS. Key components of the human BER pathway that prevent DNA damage due

---

S. Holter and S. Gallinger (✉)
Department of Surgery, Samuel Lunenfeld Research Institute, Mount Sinai Hospital
University of Toronto

J.D. Potter and N.M Lindor (eds.), *Genetics of Colorectal Cancer*,
DOI: 10.1007/978-0-387-09568-4_9,

to ROS include *MTH1*, *OGG1*, and *MUTYH*. Each enzyme has a specific function following oxidative damage to the cell. *MTH1*, an 8-oxo-dGTPase, hydrolyzes 8-oxoGTP to 8-oxoGMP to prevent the incorporation of the oxidized guanine into nascent DNA. *OGG1* is a DNA glycosylase that detects and removes 8-oxodGs that are paired with cytosine. *MUTYH*, another DNA glycosylase, specifically identifies and removes adenine residues that have been incorrectly paired with 8-oxodG. Failure to correct 8-oxoG:A mispairs leads to the characteristic G:C to T:A transversions in the subsequent round of DNA replication (Lu et al. 2001).

At least 80 germline *MUTYH* variants have been reported to date; the majority are missense mutations, but they also include nonsense, small insertion/deletions, and splice site variants (Cheadle and Sampson 2007). Large rearrangements of *MUTYH* have not been reported (Nielsen et al. 2005). By far the most commonly reported mutations are Y165C and G382D which account for approximately 80% of *MUTYH* mutations reported in Caucasian populations (Sampson et al. 2003; Sieber et al. 2003) with a baseline population frequency of 0.3–2% (Al-Tassan et al. 2002; Sieber et al. 2003; Croitoru et al. 2004; Fleischmann et al. 2004). Specific mutations in certain ethnic groups have also been reported, such as E466X, Y90X, and 1395delGGA in Indian, Pakistani, and Italian populations, respectively (Jones et al. 2002; Gismondi et al. 2004).

The amino acids at the positions of the originally reported *MUTYH* missense mutations, Y165C and G382D, are conserved across multiple species. Disruption of these conserved amino acids may affect the ability of *MUTYH* to remove A from an A:8-oxodG pair. To assess the functional consequences of these mutations and others, we have cloned wild-type *MUTYH* and the mutants Y90X, Y165C, R231H, R260Q, P281L, Q377X, E466X, G382D, 1103delC (generated by site-directed mutagenesis). Bacterially expressed recombinant variant MUTYH proteins were purified and studied for their DNA-binding and glycosylase activities on synthetic double-stranded DNA substrates containing A:8-oxodG mismatches. Mutants R260Q and G382D were found to be partially active in substrate binding and adenine removal (64 and 82% less active than wild type), and Y165C, R231H, and P281L were severely defective in both activities. All of the frameshift mutants Y90X, Q377X, E466X, and 1103delC, were also devoid of DNA-binding and glycosylase activities (unpublished). Additional *MUTYH* mutations, R227W, V232F, and A459D, have also been shown to have decreased glycosylase activity (Cheadle and Sampson 2007).

MAP Cancer Development

Chromosomal instability (CIN) and microsatellite instability (MSI) are well characterized pathways through which the majority of colorectal cancers develop and the hereditary colorectal cancer conditions Familial Adenomatous Polyposis (FAP) (see Chap. 5) and Lynch Syndrome/Hereditary NonPolyposis Colorectal Cancer (HNPCC) (see Chap. 6) are prime examples of each. FAP-related and the majority of nonfamilial colorectal cancers develop through the CIN pathway characterized by mutations in *APC*, *p53*, *K-ras* or *SMAD4*, loss of 18q, and an aneuploid karyotype. In contrast, Lynch-related and approximately 10–20% of sporadic colorectal cancers develop due to defects in the mismatch repair pathway leading to widespread MSI and a near-diploid karyotype (Chung 2000). The pathway in which defective *MUTYH* leads to colorectal cancer is still under investigation. Several studies analyzing somatic events in MAP-associated adenomas and carcinomas have confirmed that MAP tumors are characterized by truncating mutations in *APC* caused by G:C to T:A transversions, particularly at GAA sequences, which lead to a stop codon, TAA (Al-Tassan et al. 2002; Lipton et al. 2003). MAP tumors also have high a frequency of mutations in *K-ras*, particularly a G:C to T:A transversion GGT > TGT (G12C) in exon 1 (Lipton et al. 2003; Jones et al. 2004). MAP tumors are generally microsatellite stable and are nearly diploid (Lipton et al. 2003). MAP tumors may develop through a distinct somatic pathway but one that shares characteristics of both the CIN pathway, such as *APC* mutations and microsatellite stable tumors, and the MSI pathway, with a near-diploid karyotype and low levels of loss of heterozygosity. A robust immunohistochemical assay would be helpful in characterizing the adenoma-to-carcinoma sequence in MAP.

The development of colorectal cancer due to *MUTYH* inactivation may be specific to germline mutations. Halford et al. investigated 75 unselected sporadic colorectal cancers and did not identify somatic *MUTYH* mutations. *MUTYH* mRNA and protein expression was found at normal levels in 35 colorectal cancer cell lines suggesting that epigenetic silencing is unlikely (2003). The role of somatic *MUTYH* mutations in the development of nonfamilial colorectal cancer is yet to be clarified.

Myh$^{-/-}$ mice have failed to recapitulate the phenotype seen in patients with MAP. *Myh*$^{-/-}$ mice do not develop tumors and have only an age-dependent accumulation of 8-oxodG in liver. Only when there is concurrent deficiency of both *myh* and *ogg1* (*myh*$^{-/-}$/*ogg1*$^{-/-}$) do mice have an increased accumulation of 8-oxodG in other organs, including lung and small intestine, and tumors begin to develop (Russo et al. 2007). Thus far, no pathogenic human mutations in the other BER genes, *OGG1* or *MTH1*, have been identified to be associated with tumor development.

Clinical Features

MAP is characterized by the development of multiple adenomatous polyps and a clinical phenotype that is often indistinguishable from Attenuated Familial Adenomatous Polyposis (AFAP), and classic FAP. Establishing the correct genetic diagnosis for the patient with adenomatous polyposis is not only important for that individual, but will also direct cancer surveillance for his or her family members. Polyposis and cancer risk exist for each successive generation in families with autosomal dominantly inherited FAP or AFAP; whereas a single generation is at risk for autosomal recessively inherited MAP.

Polyps in MAP tend to be mainly small tubular or tubulovillous adenomas with mild dysplasia and occasional hyperplastic polyps. Also typically present are microadenomas, which were previously thought to be pathognomonic of FAP (Sieber et al. 2003; Lipton et al. 2003). Diagnosis of polyposis is generally at an older age than classic FAP but similar to AFAP, with a mean age ranging from 45 to 56 (Sampson et al. 2003; Sieber et al. 2003; Wang et al. 2004; Nielsen et al. 2005; Croitoru et al. 2007). There have been some reports of very early-onset polyposis with the youngest diagnosis at 13 years (Sampson et al. 2003). At least 50% of individuals are diagnosed with a colorectal cancer at the time of polyposis diagnosis. Cancer develops throughout the colorectum without a site-specific predilection (Sampson et al. 2003; Sieber et al. 2003; Wang et al. 2004; Nielsen et al. 2005; Croitoru et al. 2007). Very early-onset (<30 years) colorectal cancers have also been reported with MAP (Nielsen et al. 2005; Aretz et al. 2006). MAP colorectal cancers have no pathological features to distinguish them from FAP-related or sporadic colorectal cancers.

Polyp number in MAP is variable. Multiple studies have been reported which assess the frequency of biallelic *MUTYH* mutations in patients with classic polyposis, attenuated polyposis, and sporadic colorectal cancers. Clinic-based series of *APC*-negative polyposis patients have estimated that biallelic *MUTYH* mutations account for 11–42% of attenuated polyposis, defined as 10–100 adenomas (Sieber et al. 2003; Isidro et al. 2004; Wang et al. 2004; Nielsen et al. 2005; Croitoru et al. 2007) and 7.5–29% of classic polyposis, defined as >100 adenomas (Sampson et al. 2003; Sieber et al. 2003; Gismondi et al. 2004; Nielsen et al. 2005). Biallelic *MUTYH* mutations do not appear to be a major contributor in patients with fewer than ten polyps (Wang et al. 2004). However, studies of nonpolyposis early-onset colorectal cancer, as well as nonfamilial colorectal cancers, have identified biallelic mutations in approximately 0.5% of patients (Fleischmann et al. 2004; Wang et al. 2004).

The major distinguishing feature of MAP is family history. MAP follows an autosomal recessive inheritance pattern; thus, a typical family may have several individuals in a single generation affected. A MAP patient may also appear to have nonfamilial colorectal cancer (no family history); this may be difficult to differentiate from a case with a de novo *APC* mutation. Some families may appear to have a dominant family history with successive generations diagnosed with colorectal cancer. These cancers may be phenocopies due to the high incidence of colorectal cancer. *MUTYH* mutations are common enough that 1–2% of carriers will have children with a partner who is also a carrier and the family history will follow a pseudodominant inheritance pattern (Sieber et al. 2003; Croitoru et al. 2007; Nielsen et al. 2007).

Although no detailed genotypic studies of penetrance have been performed for MAP, it has been suggested that penetrance is near 100%, because no true controls have been identified with biallelic *MUTYH* mutations. One study found that biallelic carriers have a 53-fold (95% CI: 14–200, $p < 0.0001$) increased risk of CRC compared to the general population with a cumulative risk by age 70 of 80% (35–100%) (Jenkins et al. 2006). The high estimates of penetrance for MAP may be because it is recessively inherited; therefore, most patients do not have a family history and present, often with symptoms, later than those under surveillance.

Extracolonic cancer risks in MAP are still being elucidated. The most consistently reported feature is upper gastrointestinal polyps: approximately 5% of MAP patients have exhibited duodenal adenomas with or without duodenal adenocarcinoma (Nielsen et al. 2005, 2006). Fundic gland polyps have been reported (Jo et al. 2005) as well as a report of a MAP patient diagnosed with colonic polyposis at age 13 and gastric cancer at age 17 (Sampson et al. 2003). Upper gastrointestinal features may be underestimated as not all individuals with an attenuated polyposis phenotype may be referred for upper endoscopy.

Endogenous reactive oxygen species have been implicated in the carcinogenesis of multiple cancer types, including lung, breast, kidney, liver, and prostate (Okamoto et al. 1994; Malins and Haimanot 1991; Jaruga et al. 1994; Wang et al. 2002; DeMarzo et al. 2003). Therefore, it may be hypothesized that germline *MUTYH* mutations would predispose carriers to some of these malignancies due to defects in BER of endogenous oxidative damage. Several series have evaluated patients with lung cancer, prostate cancer, hepatocellular carcinoma, cholangiocarcinoma, acute myeloid leukemia, acute lymphoblastic leukemia, and squamous cell carcinoma of the head and neck for germline mutations in *MUTYH*. No pathogenic biallelic germline mutations have been identified in any of these patient populations (Akyerli et al. 2003; Al-Tassan et al. 2004; Baudhuin et al. 2006; Gorgens 2007; Shin et al. 2007).

Extracolonic features that are typically associated with FAP such as congenital hypertrophy of the retinal pigment epithelium (CHRPE), osteomas, and dental anomalies (Gismondi et al. 2004) have occasionally been reported in MAP patients. These may not be true associations but purely coincidental as, in the past, MAP patients were clinically diagnosed with FAP and were therefore evaluated for these features which may occur sporadically in the general population.

Many of the hereditary colorectal cancer predisposition syndromes have associated dermatologic manifestations. Several MAP families have been reported with associated skin lesions. Baglioni and colleagues (2005) identified two siblings with MAP and pilomatricomas, benign tumors of the hair follicle. Ponti et al. (2005, 2007) have reported four individuals from three unrelated MAP families with sebaceous gland hyperplasia.

Cancer Risk to Heterozygotes

Increased colorectal cancer risk in heterozygous carriers of *MUTYH* is controversial. Most studies have been unable to detect an association between heterozygous *MUTYH* mutations and increased colorectal cancer risk and those that have been able to show an association are borderline statistically significant (Croitoru et al. 2004; Farrington et al. 2005; Jenkins et al. 2006; Webb et al. 2006). A heterozygous *MUTYH* mutation is probably a low-penetrance allele; the studies carried out to date are underpowered to detect any association between heterozygous mutations and colorectal cancer risk. It is possible that *MUTYH* carriers could acquire a somatic mutation or inactivation of the second wild-type allele and develop colorectal cancer. However, somatic *MUTYH* mutations are infrequent in colorectal cancers (Halford et al. 2003). Low levels of heterozygosity at the MUTYH locus on chromosome 1p have been demonstrated in small numbers by different groups (Croitoru et al. 2004; Kambara et al. 2004).

Conclusion

There is still much to be learned about the molecular basis, clinical features, and appropriate management of MAP. Why do biallelic *MUTYH* mutations result in tumors of the gastrointestinal tract and not other organs that are more susceptible to damage by reactive oxygen species? With the low rate of somatic *MUTYH* mutations in sporadic colorectal cancer, do biallelic *MUTYH* mutations simply confer an increased mutation rate in genes known to be associated with colorectal cancer development, e.g., *APC* and *K-ras*? Why do not germline *MTH1* or *OGG1* mutations cause a similar tumor phenotype? Are there modifier genes that affect the clinical phenotype in individuals with MAP, which may, in turn, explain why some individuals have numerous polyps and some have only early-onset colorectal cancer? Are there genotype-phenotype correlations that may predispose to a more severe presentation? Are there histopathological features of MAP tumors that may aid in the identification of individuals at risk? Are there other extracolonic cancer risks? Do heterozygous mutation carriers have an increased risk of colorectal cancer? What is the best management strategy for biallelic, as well as heterozygous, *MUTYH* mutation carriers? These issues are important for the understanding of the pathogenesis of colorectal cancer, the approach to the patient with multiple polyps, and the management of individuals with MAP.

References

Aaltonen, L. A., Peltomaki, P., Leach, F. S., Sistone, P., Pylkkanen, L., Mecklin, J. P., Jarvinen, H., Powell, S. M., Jen, J., Hamilton, S. R. and et al. 2007. Clues to the pathogenesis of familial colorectal cancer. Science 260: 812–16.

Akyerli, C. B., Ozbek, U., Aydin-Sayitoglu, M., Sirma, S., Ozcelik, T. 2003. Analysis of *MYH* Tyr165Cys and Gly382Asp variants in childhood leukemias. Journal of Cancer Research and Clinical Oncology 129: 604–605.

Al-Tassan, N., Chmiel, N. H., Maynard, J., Fleming, N., Livingston, A. L., Williams, G. T., Hodges, A. K., Davies, D. R., David, S. S., Sampson, J. R., Cheadle, J. P. 2002. Inherited variants of MYH associated with somatic G:C → T:A mutations in colorectal tumors. Nature Genetics 30: 227–232.

Al-Tassan, N., Eisen, T., Maynard, J., Bridle, H., Shah, B., Fleischmann, C., Sampson, J. R., Cheadle, J. P., Houlston, R.S. 2004. Inherited variants in *MYH* are unlikely to contribute to the risk of lung carcinoma. Human Genetics 114: 207–210.

Aretz, S., Uhlhaas, S., Goergens, H., Siberg, K., Vogel, M., Pagenstecher, C., Mangold, E., Caspari, R., Propping, P., Friedl, W. 2006. *MUTYH*-associated polyposis: 70 of 71 patients with biallelic mutations present with an attenuated or atypical phenotype. International Journal of Cancer 119: 807–814.

Baglioni, S., Melean, G., Gensini, F., Santucci, M., Scatizzi, M., Papi, L., Genuardi, M. 2005. A kindred with *MYH*-associated polyposis and pilomatricomas. American Journal of Human Genetics 134A: 212–214.

Baudhuin, L. M., Roberts, L. R., Enders, F. T. B., Swanson, R. L., Mettler, T. A., Aderca, I., Stadheim, L. M., Highsmith, W. E. 2006. MYH Y165C and G382D mutations in hepatocellular carcinoma and cholangiocarcinoma patients. Journal of Cancer Research and Clinical Oncology 132: 159–162.

Cheadle, J. P. and Sampson, J. R. 2007. *MUTYH*-associated polyposis – from defect in base excision repair to clinical genetic testing. DNA Repair 6: 274–279.

Chung, D. C. 2000. The genetic basis of colorectal cancer: insights into critical pathways of tumorigenesis. Gastroenterology 119: 854–865.

Croitoru, M. E., Cleary, S. P., Di Nicola, N., Manno, M., Selander, T., Aronson, M., Redston, M., Cotterchio, M., Knight, J., Gryfe, R., Gallinger, S. 2004. Association between biallelic and monoallelic germline MYH gene mutations and colorectal cancer risk. Journal of the National Cancer Institute 96: 1631–1634.

Croitoru, M. E., Cleary, S. P., Berk, T., Di Nicola, N., Kopolovic, I., Bapat, B., Gallinger, S. 2007. Germline MYH mutations in a clinic-based series of Canadian multiple colorectal adenoma patients. Journal of Surgical Oncology 95: 499–506.

DeMarzo, A. M., Nelson, W. G., Isaacs, W. B., Epstein, J. I. 2003. Pathological and molecular aspects of prostate cancer. Lancet 361: 955–964.

Farrington, S. M., Tenesa, A., Barneston, R., Wiltshire, A., Prendergast, J., Porteous, M., Campbell, H., Dunlop, M. G. 2005. Germline susceptibility to colorectal cancer due to base-excision repair gene defects. American Journal of Human Genetics 77: 112–119.

Fleischmann, C., Peto, J., Cheadle, J., Shah, B., Sampson, J., Houlston, R. S. 2004. Comprehensive analysis of the contribution of germline *MYH* variation to early-onset colorectal cancer. International Journal of Cancer 109: 554–558.

Gismondi, V., Meta, M., Bonelli, L., Radice, P., Sala, P., Bertario, L., Viel, A., Fornasarig, M., Arrigoni, A., Gentile, M., Ponz de Leon, M., Anselmi, L., Mareni, C., Bruzzi, P., Varesco, L. 2004. Prevalence of the Y165C, G382D, and 1395delGGA germline mutations of the *MYH* gene in Italian patients with adenomatous polyposis coli and colorectal adenomas. International Journal of Cancer 109: 680–684.

Gorgens, H., Muller, A., Kruger, S., Kuhlisch, E., Konig, I. R., Ziegler, A., Schackert, H. K., Eckelt, U. 2007. Analysis of the base excision repair genes MTH1, OGG1 and MUTYH in patients with squamous oral carcinomas. Oral Oncology 43: 791–795.

Halford, S. E. R., Rowan, A. J., Lipton, L., Sieber, O. M., Pack, K., Thomas, H. J. W., Hodgson, S. V., Bodmer, W. F., Tomlinson, I. P. M. 2003. Germline mutations but not somatic changes at the MYH locus contribute to the pathogenesis of unselected colorectal cancers. American Journal of Pathology 162: 1545–1548.

Isidro, G., Laranjeira, F., Pires, A., Leite, J., Regateiro, F., e Sousa, F. C., Soares, J., Castro, C., Giria, J., Brito, M. J., Medeira, A., Teixeira, R., Morna, H., Gaspar, I., Marinho, C., Jorge, R., Brehm, A., Ramos, J. S., Boavida, M. G. 2004. Germline MUTYH (MYH) mutations in Portuguese individuals with multiple colorectal adenomas. Human Mutation 24: 353–534.

Jaruga, P., Zastawny, T. H., Skokowski, J., Dizdarglu, M., Olinski, R. 1994. Oxidative DNA base damage and antioxidant enzyme activities in human lung cancer. FEBS Letters 341: 59–64.

Jenkins, M. A., Croitoru, M. E., Monga, N., Cleary, S. P., Cotterchio, M., Hopper, J. L., Gallinger, S. 2006. Risk of colorectal cancer in monoallelic and biallelic carriers of MYH mutations: a population-based case-family study. Cancer Epidemiology Biomarkers and Prevention 15: 312–314.

Jo, W. S., Bandipalliam, P., Shannon, K. M., Niendorf, K. B., Chan-Smutko, G., Hur, C., Syngal, S., Chung, D. C. 2005. Correlation of polyp number and family history of colon cancer with germline MYH mutations. Clinical Gastroenterology and Hepatology 3: 1022–1028.

Jones, S., Emmerson, P., Maynard, J., Best, J. M., Jordan, S., Williams, G. T., Sampson, J. R., Cheadle, J. P. 2002. Biallelic germline mutations in MYH predispose to multiple colorectal adenoma and somatic G:C → T:A mutations. Human Molecular Genetics 11: 2961–2967.

Jones, S., Lambert, S., Williams, G. T., Best, J. M., Sampson, J. R., Cheadle, J. P. 2004. Increased frequency of the k-ras G12C mutation in MYH polyposis colorectal adenomas. British Journal of Cancer 90: 1591–1593.

Kambara, T., Whitehall, V.L.J., Spring, K.J., Barker, M.A., Arnold, S., Wynter, C.V.A., Matsubara, N., Tanaka, N., Young, J.P., Leggett, B.A., Jass, J.R. 2004. Role of inherited defects of MYH in the development of sporadic colorectal cancer. Genes, Chromosomes and Cancer 40: 1–9.

Lindor, N. M., Rabe, K., Petersen, G. M., Haile, R., Casey, G., Baron, J., Gallinger, S., Bapat, B., Aronson, M., Hopper, J., Jass, J., LeMarchand, L., Grove, J., Potter, J., Newcomb, P., Terdiman, J. P., Conrad, P., Moslein, G., Goldberg, R., Ziogas, A., Anton-Culver, H., de Andrade, M., Siegmund, K., Thibodeau, S. N., Boardman, L. A., Seminara, D. 2005. Lower cancer incidence in Amsterdam-I criteria families without mismatch repair deficiency: familial colorectal cancer type X. JAMA 293(16): 2028–30.

Lipton, L., Halford, S. E., Johnson, V., Novelli, M. R., Jones, A., Cummings, C., Barclay, E., Sieber, O., Sadat, A., Bisgaard, M. L., Hodgson, S. V., Aaltonen, L. A., Thomas, H. J. W., Tomlinson, I. P. M. 2003. Carcinogenesis in MYH-associated polyposis follows a distinct genetic pathway. Cancer Research 63: 7595–7599.

Lu, A. L., Li, X., Gu, Y., Wright, P. M., Chang, D. Y. 2001. Repair of oxidative DNA damage: mechanisms and functions. Cell Biochemistry and Biophysics 35: 141–170.

Malins, D. C. and Haimanot R. 1991. Major alterations in the nucleotide structure of DNA in cancer of the female breast. Cancer Research 51: 5430–5432.

Nielsen, M., Franken, P. F., Reinards, T. H. C. M., Weiss, M. M., Wagner, A., van der Klift, H., Kloosterman, S., Houwing-Duistermaat, J. J., Aalfs, C. M., Ausems, M. G. E. M., Brocker-Vriends, A. H. J. T., Gomez-Garcia, E. B., Hoogerbrugge, N., Menko, F. H., Sijmons, R. H., Verhoef, S., Kuipers, E. J., Morreau, H., Breuning, M. H., Tops, C. M. J., Wijnen, J. T., Vasen, H. F. A., Fodde, R., Hes, F. J. 2005. Multiplicity in polyp count and extracolonic manifestations in 40 Dutch patients with *MYH* associated polyposis coli (MAP). Journal of Medical Genetics 42: e54.

Nielsen, M., Poley, J. W., Verhoef, S., van Puijenbroek, M., Weiss, M. M., Burger, G. T., Dommering, C. J., Vasen, H. F. A., Kuipers, E. J., Wagner, A., Morreau, H., Hes, F. J. 2006. Duodenal carcinoma in MUTYH-associated polyposis. Journal of Clinical Pathology 59: 1212–1215.

Nielsen, M., Hes, F. J., Nagengast, F. M., Weiss, M. M., Mathus-Vliegen, E. M., Morreau, H., Breuning, M. H., Wijnen, J. T., Tops, C. M. J., Vasen, H. F. A. 2007. Germline mutations in APC and MUTYH are responsible for the majority of families with attenuated familial adenomatous polyposis. Clinical Genetics 71: 427–433.

Okamoto, K., Toyokuni, S., Uchida, K., Ogawa, O., Takenewa, J., Kakehi, Y., Kinoshita, H., Hattori-Nakakuki, Y., Hiai, H., Yoshida, O. 1994. Formation of 8-hydroxy-2'-deosyguanosine and 4-hydroxy-2-nonenal-modified proteins in human renal-cell carcinoma. International Journal of Cancer 58: 825–829.

Ponti, G., Ponz de Leon, M., Maffei, S., Pedroni, M., Losi, L., Di Gregorio, C., Gismondi, V., Scarselli, A., Benatti, P., Roncari, B., Seidenari, S., Pellacani, G., Varotti, C., Prete, E., Varesco, L., Roncucci, L. 2005. Attenuated familial adenomatous polyposis and Muir-Torre syndrome linked to compound biallelic constitutional MYH gene mutations. Clinical Genetics 68: 442–447.

Ponti, G., Venesio, T., Losi, L., Pellacani, G., Bertario, L., Sala, P., Pedroni, M., Petti, C., Maffei, S., Varesco, L., Lerch, E., Baggio, A., Bassoli, S., Longo, C., Seidenari, S. 2007. BRAF mutations in multiple sebaceous hyperplasias of patients belonging to MYH-associated polyposis pedigrees. Journal of Investigative Dermatology 127: 1387–1391.

Russo, M. T., De Luca, G., Degan, P., Bignami, M. 2007. Different DNA repair strategies to combat the threat from 8-oxoguanine. Mutation Research 614: 69–76.

Sampson, J. R., Dolwani, S., Jones, S., Eccles, D., Ellis, A., Evans, D. G., Frayling, I., Jordan, S., Maher, E. R., Mak, T., Maynard, J., Pigatto, F., Shaw, J., Cheadle, J. P. 2003. Autosomal recessive colorectal adenomatous polyposis due to inherited mutations of MYH. Lancet 362: 39–41.

Shin, E. J., Chappell, E., Pethe, V., Hersey, K., van der Kwast, T., Fleshner, N., Bapat, B. 2007. *MYH* mutations are rare in prostate cancer. Journal of Cancer Research and Clinical Oncology 133: 373–378.

Sieber, O. M., Lipton, L., Crabtree, M., Heinimann, K., Fidalgo, P., Phillips, R. K. S., Bisgaard, M. L., Orntoft, T. F., Aaltonen, L. A., Hodgson, S. V., Thomas, H. J. W., Tomlinson, I. P. M. 2003. Multiple colorectal adenomas, classic adenomatous polyposis, and germ-line mutations in MYH. New England Journal of Medicine 348: 791–799.

Wang, X. W., Hussain, S. P., Huo, T. I., Wu, C. G., Forgues, M., Hofseth, L. J., Brechot, C., Harris, C. C. 2002. Molecular pathogenesis of human hepatocellular carcinoma. Toxicology 181–182: 43–47.

Wang, L., Baudhuin, L. M., Boardman, L. A., Steenblock, K. J., Petersen, G. M., Halling, K. C., French, A. J., Johnson, R. A., Burgart, L. J., Rabe, K., Lindor, N. M., Thibodeau, S. N. 2004. *MYH* mutations in patients with attenuated and classic polyposis and with young-onset colorectal cancer without polyps. Gastroenterology 127: 9–16.

Webb, E. L., Rudd, M. F., Houlston, R. S. 2006. Colorectal cancer risk in monoallelic carriers of MYH variants. American Journal of Human Genetics 79: 768–771.

Chapter 8.2
Familial Colorectal Cancer Type X

Noralane M. Lindor

In 1991, the International Collaborative Group on Hereditary NonPolyposis Colon Cancer published what became known as the Amsterdam I criteria (AC-I) for the definition of HNPCC (Vasen et al. 1991). The AC-I are fulfilled if all four of the following conditions are met: (1) three cases of colorectal cancer (CRC) in which two of the affected individuals are first-degree relatives of the third; (2) CRCs occurring in two generations; (3) one CRC diagnosed before the age of 50 years; and (4) familial adenomatous polyposis not diagnosed in the family. Prior to discovery of the molecular basis of Lynch Syndrome or hereditary nonpolyposis colon cancer (HNPCC), Dr. Henry Lynch and others, had already defined this as a syndrome with autosomal dominant inheritance, characterized by greatly increased risks for colorectal carcinoma (CRC) that tended to occur two decades younger and was more likely to be located in the right colon than nonfamilial CRC. In addition, it was recognized that risks for the following carcinomas were also increased: endometrium, stomach, small intestine, hepatobiliary tract, kidney, ureter, and ovary. Application of the stringent AC-I to colorectal cancer patients and families facilitated the identification of the genetic lesions underlying HNPCC: a germline mutation in one of several DNA mismatch-repair genes (see Chap. 6 for details).

Following this major discovery, the term "HNPCC" began to be used inconsistently in the medical literature: multiple papers continued to use HNPCC to mean probands who had pedigrees that fulfilled AC-I, regardless of molecular findings; others began to use HNPCC to refer only to families with hereditary DNA mismatch-repair deficiency; lastly, there began to be papers containing the implicit assumption that if a pedigree fulfilled the AC-I, then a DNA mismatch-repair defect must be present in that pedigree, and this too, was called HNPCC. This confusion existed despite that fact that research had already showed that fulfillment of AC-I was *not* equivalent to having a hereditary DNA mismatch-repair gene mutation (Bisgaard et al. 2002; Wijnen et al. 1998 and others). For example, Wijnen et al. (1998), reported on 184 probands of which 92 had family histories that met AC-I. Mutations in *MLH1* or *MSH2* were found in only 45% of those meeting the criteria. Note that

N.M. Lindor
Department of Medical Genetics, Mayo Clinic College of Medicine
e-mail: nlindor@mayo.edu

J.D. Potter and N.M Lindor (eds.), *Genetics of Colorectal Cancer*,
DOI: 10.1007/978-0-387-09568-4_10,

both younger age at diagnosis of CRC and the presence of endometrial cancer in the family increased the likelihood that a mismatch-repair mutation would be found.

For those families that met AC-I but did not have DNA mismatch-repair gene defect, the magnitude of CRC risk and the tumor spectrum were unknown. As a consequence it was unknown whether the extremely rigorous cancer-screening recommendations for "HNPCC" (Burke et al. 1997) were appropriate for those who did not have DNA mismatch-repair defects.

To explore this subject, cancer risks were studied among relatives of 161 probands from AC-I-positive families from the USA, Canada, Australia, and Germany (Lindor et al. 2005). All were characterized according to whether their tumors showed evidence of DNA mismatch-repair deficiency using tumor microsatellite instability testing. Ninety families had DNA mismatch-repair deficiency and 71 had normal DNA mismatch repair on tumor microsatellite-instability testing. The Amsterdam-defining "triad" of three affected individuals, which always included the proband, was removed from the analysis so as to be maximally conservative. The remaining 3,422 relatives were either first- or second-degree relatives of a triad member. The incidence of cancer in these relatives was calculated as the ratio of observed to expected cases to the number of at-risk person-years (standardized incidence ratio; SIR). In the 90 families with DNA mismatch repair-deficient tumors, the risk for cancers was statistically significantly elevated in complete accordance with expectation of the syndrome described by Dr. Henry Lynch and colleagues: risks were increased for colorectal, endometrial, gastric, small intestine, and kidney/ureter cancers. However, in the 71 families without DNA mismatch-repair deficiency, there was a modestly increased risk for colorectal cancer (SIR 2.3; 95% CI: 1.7–3.0), but for no other cancer site. In addition, the average age at diagnosis of CRC was older in the families without DNA mismatch-repair deficiency: 61 versus 49 years. This large study concluded that families who fulfill the AC-I should not be counseled as if they have hereditary DNA mismatch-repair defect because the cancer risks are lower and different. The authors proposed use of the term "Lynch Syndrome" to describe families with hereditary DNA mismatch-repair defect and the term "Familial Colorectal Cancer Type X" to signify the other HNPCC-like clusters in which no DNA mismatch-repair defect could be identified. The word "hereditary" was avoided as this remained unproven and the "X" signified the unknown nature of this disorder. A call has been made for retirement of the term HNPCC (Jass 2006).

Familial Colorectal Cancer Type X (FCCTX) is undoubtedly a heterogenous grouping of: (1) random aggregations of a common tumor; (2) aggregations of a tumor related to shared lifestyle factors; (3) polygenic predisposition; (4) some yet-to-be-defined single-gene disorders. In a population-based study of 1,042 CRC probands with verified family histories, Aaltonen et al. (2007) explored how much of familial risk is attributable to Lynch Syndrome or other known genetic syndrome. When known syndromes were excluded from the analysis, 32% of familial risk remains unaccounted for by the known loci. Genetic modeling of the data did not suggest a better explanation than a simple polygenic model. Studies are now beginning to chip away at the genetic causes of FCCTX.

Several additional studies have furthered knowledge regarding FCCTX. In a study of 41 German families, Mueller-Koch et al. (2005) confirmed the older age of onset in FCCTX compared to Lynch Syndrome (55 vs. 41 years) and also noted that two-third of the tumors were left sided which is the inverse of CRC in Lynch Syndrome. Although the Lynch Syndrome group had more synchronous and metachronous CRCs, the FCCTX group had greater adenoma/carcinoma ratio and a tendency toward more adenomas, perhaps suggesting a slower progression of adenomas to carcinomas. This interpretation appeared to be confirmed in a separate study of 97 families with dominant CRC family history in the United Kingdom (Dove-Edwin et al. 2006) in which individuals with Lynch Syndrome and FCCTX had an equal likelihood of having high-risk adenomas but CRC developed only in those with Lynch Syndrome. In 64 Spanish families that met Amsterdam criteria, 40% had normal DNA mismatch repair in tumor tissue, and Valle et al. (2007) also confirmed the older age at CRC diagnosis in the FCCTX compared to Lynch Syndrome (53 vs. 41 years). In addition, the FCCTX cases were less likely to be in the right colon, to have mucinous tumors, and to have fewer multiple primary tumors. Llor et al. (2005) also studied 25 Spanish families meeting Amsterdam criteria: among 100 patients, 40% had normal DNA mismatch repair and, compared to Lynch Syndrome cases, age at diagnosis of CRC in relatives was older (60 vs. 54 years), 89% of tumors were left sided, none showed tumor infiltrating lymphocytes (whereas half of the Lynch Syndrome CRC did so). Table 8.2.1 compares and contrasts Lynch Syndrome with FCCTX.

Caution must be exercised in assigning a family to the FCCTX group. With the studies so far depending heavily upon tumor microsatellite instability results, one must consider the following: the possibility of a phenocopy within a Lynch Syndrome family (i.e., a microsatellite-stable tumor that arose by chance in a family that actually does have a mismatch-repair defect); the fact that not all tumors with germline *MSH6* mutations are MSI-high; and laboratory quality control problems such as the adequacy of the representation of tumor cells in the MSI-assay. In general, the age at diagnosis in the FCCTX families is older than in Lynch families and, in light of the findings to date, families with young average age of onset of

Table 8.2.1 Families that fulfill the pedigree Amsterdam criteria

	Lynch Syndrome	Familial Colorectal Cancer Type X
Colorectal		
Cancer risk	Very high	Modestly increased
Age of onset	~45 years average	50–60s
Usual location	Proximal colon	Distal colon
Polyps	Few	More
Other cancers		
Endometrial risk	Very high	Not very high
Other cancer sites	Many	None known
DNA mismatch-repair (MMR) genes		
Germline	Mutations found	No mutations found
Tumor	Microsatellite instability	No MSI/stable
Tumor staining	Loss of MMR expression	Normal expression

colorectal tumors, or manifesting other classical Lynch Syndrome tumors such as endometrial cancer, should probably not be categorized as having FCCTX. Cancer-screening recommendations have been suggested for FCCTX (Lindor et al. 2005; Hendriks et al. 2006; Dove-Edwin et al. 2006), but it is essential to not miscategorize such families and to continue to search for single-gene syndromes.

References

Bisgaard ML, Jager AC, Myrhoj T, Bernstein I, Nielsen FC. Hereditary non-polyposis colorectal cancer (HNPCC): phenotype-genotype correlation between patients with and without identified mutation. Hum Mutat 2002;20:20–7.

Burke W, Petersen G, Lynch P, Botkin J, Daly M, Garber J, et al. Recommendations for follow-up care of individuals with an inherited predisposition to cancer. 1. Hereditary Nonpolyposis Colon Cancer. JAMA 1997;277:915–9.

Dove-Edwin I, De Jong AE, Adams J, Mesher D, Lipton L, Sasieni P, Vasen HFA, Thomas HJW. Prospective results of surveillance colonoscopy in dominant familial colorectal cancer with and without Lynch Syndrome. Gastroenterology 2006;130:1995–2000.

Hendriks YMC, deJong AE, Morreaqu H, Tops CMJ, Vasen HF, Wijnen JT, et al. Diagnostic approach and management of Lynch syndrome (hereditary nonpolyposis colorectal carcinoma): a guide for clinicians. CA Cancer J Clin 2006;56:213–25.

Jass JR. Hereditary non-polyposis colorectal cancer: the rise and fall of a confusing term. World J Gastroenterol 2006;12(3):4943–50.

Llor X, Pons E, Xicola RM, Castells A, Alenda C, Pinol V, Andreu M, Castellvi-Bel S, paya A, Jover R, Bessa X, Giros A, Roca A, Gassull MA for the Gastrointestinal Oncology Group of the Spanish Gastroenterological Association. Differential features of colorectal cancers fulfilling Amsterdam criteria without involvement of the mutation pathway. Clin Cancer Res 2005;11(20):7304–10.

Mueller-Koch Y, Vogelsang H, Kopp R, Lohse P, Keller G, Aust D, Muders M, Gross M, Daum J, Schiemann U, Grabowski M, Scholz M, Kerker B, Becker I, Henke G, Holinski-Feder E. Hereditary non-polyposis colorectal cancer: clinical and molecular evidence for a new entity of hereditary colorectal cancer. Gut 2005;54:1733–40.

Valle L, Perea J, Carbonell P, Fernandez V, Dotor AM, Benitez J, Urioste M. Clinicopathologic and pedigree differences in Amsterdam-I-positive hereditary nonpolyposis colorectal cancer families according to tumor microsatellite instability status. JCO 2007;25(7):781–6.

Vasen HFA, Mecklin JP, Meera Khan P, Lynch HT. The International Collaborative Group on Hereditary Non-Polyposis Colorectal Cancer. Dis Colon Rectum 1991;34:424–5.

Wijnen JT, Vasen HFA, Kahn M, Zwinderman AH, van der Klift H, Mulder A, et al. Clinical findings with implications for genetic testing in families with clustering of colorectal cancer. N Engl J Med 1998;339:511–8.

Chapter 8.3
Families with Serrated Neoplasia of the Colon

Joanne P. Young

Introduction

Only one-third of all familial colorectal cancer (CRC) has a well-characterised genetic basis. The two most common hereditary CRC predispositions, familial adenomatous polyposis (FAP) and hereditary non-polyposis colon cancer (HNPCC or Lynch Syndrome) (Jass 2006), arise through the traditional CRC developmental model, the adenoma-carcinoma sequence (Iino et al. 2000; Young et al. 2001) and are discussed in Chaps. 5 and 6. With the recognition of the serrated neoplasia pathway as a major contributor to CRC in the wider population (Jass 2001), an opportunity arose to understand a further subset of familial cases. Familial serrated neoplasia encompasses a spectrum of phenotypes that includes: (1) apparently isolated cases of hyperplastic polyposis syndrome (HPS) which may have an autosomal recessive or co-dominant aetiology (Young and Jass 2006; Young et al. 2007), (2) HPS in a familial setting, and (3) the recently described autosomal dominant CRC predisposition known as the serrated pathway syndrome (SPS) (Young et al. 2005). The single trait that occurs frequently across this range of presentations is the advanced serrated polyp (Torlakovic et al. 2003), a lesion that has increased malignant potential (Goldstein et al. 2003; Jass 2003) and that is rarely seen in families with FAP or Lynch Syndrome (Rijcken et al. 2003). The management of families with serrated neoplasia and individuals with hyperplastic polyposis presents a major challenge, inasmuch as there are no defined guidelines for their screening and management. In addition, recognition of cases and their families that do not fulfil the criteria for HPS has been difficult because this spectrum of disorders lacks a defined clinical perimeter, reflecting its recent description. Finally, the genetic mutation(s) underlying serrated neoplasia predisposition have yet to be discovered.

J.P. Young
Cancer Council, Queensland Institute of Medical Research
e-mail: Joanne.Young@qimr.edu.au

J.D. Potter and N.M Lindor (eds.), *Genetics of Colorectal Cancer*,
DOI: 10.1007/978-0-387-09568-4_11,

Hyperplastic Polyposis Syndrome

HPS has been defined by Burt and Jass as a series of alternative phenotypes. Those who fulfil the criteria demonstrate: (1) at least five histologically diagnosed hyperplastic polyps proximal to the sigmoid colon, two of which are greater than 10-mm in diameter; or (2) any number of hyperplastic polyps occurring proximal to the sigmoid colon in an individual who has a first-degree relative with hyperplastic polyposis; or (3) more than 20 hyperplastic polyps of any size distributed throughout the colon (Burt and Jass 2000). In addition, Higuchi and Jass have suggested that atypical serrated polyps (these include sessile serrated adenomas, serrated adenomas, and mixed polyps) are counted in the total and that the polyp count be cumulative over time (Higuchi and Jass 2004). HPS affects both sexes and is usually diagnosed in the sixth or seventh decade (Leggett et al. 2001; Rubio et al. 2006), though earlier presentations have been reported (Cohen et al. 1981; Bengoechea et al. 1987; Torlakovic and Snover 1996; Keljo et al. 1999). The polyposis involves at least the proximal colon; however, it is frequently pan-colonic (Leggett et al. 2001; Yeoman et al. 2007), and importantly, includes a smaller number of traditional adenomas. HPS is more common in Europeans (Young et al. 2007; Yeoman et al. 2007), shows evidence of genetic predisposition (Young and Jass 2006), and importantly, is a condition now considered to carry a high risk of CRC (Bengoechea et al. 1987; Teoh et al. 1989; Jeevaratnam et al. 1996; Torlakovic and Snover 1996; Azimuddin et al. 2000; Jass et al. 2000; Rashid et al. 2000; Leggett et al. 2001; Lage et al. 2004; Chow et al. 2006). In reports of series where CRC was present, it appeared that the risk of a synchronous CRC was higher in those with atypical or large serrated polyps, and with dysplastic changes (Leggett et al. 2001; Lage et al. 2004). The molecular features of the pathway, namely, somatic *BRAF* mutation and CIMP, demonstrate a high rate of concordance within individual lesions in subjects with HPS (Chan et al. 2002; Beach et al. 2005).

In parallel with the recognition that HPS carries a high risk of CRC development, the notion that there is a genetic predisposition has slowly evolved. Though affected first-degree relatives with HPS are rare and mostly involve sibships (Jeevaratnam et al. 1996; Chow et al. 2006), the presence of a substantial family history of CRC was noted as early as 1980 (Williams et al. 1980). Factors that delayed the recognition of a genetic link include the long-held premise that serrated polyps are of little clinical consequence, and the publication of several HPS case-series where family history of CRC had either not been observed or not been examined (Torlakovic and Snover 1996; Place and Simmang 1999; Jass et al. 2000; Ferrandez et al. 2004; Oberschmid et al. 2004). However, HPS with a family history of CRC has now been reported on multiple occasions (Jeevaratnam et al. 1996; Jass et al. 1997; Azimuddin et al. 2000; Hyman et al. 2004; Lage et al. 2004; Chow et al. 2006). As with individual cases where a synchronous CRC was present, a family history of CRC was more likely to occur when the polyps were found to show dysplastic changes (Azimuddin et al. 2000). A phenotype of multiple serrated polyps, and occasional affected sibships including consanguineous kindreds (Chow et al. 2006) and identical twins, suggest an autosomal recessive or co-dominant mechanism as the most likely mode of inheritance (Young and Jass 2006; Young et al. 2007), though the identification of the specific genetic variant associated with HPS will be necessary in order to analyse the mode of inheritance in an empirical manner. Sequence variants in *MYH* and *EPHB2* have been reported in some HPS cases, though these did not account for the majority of cases (Chow et al. 2006; Kokko et al. 2006).

Phenotypic Dichotomy in Hyperplastic Polyposis Syndrome

In 1996, it was proposed that HPS is a heterogeneous condition (Torlakovic and Snover 1996). This is supported by the apparent dichotomy of phenotypes, namely one sub-type with numerous, and relatively uniform, serrated polyps and a second sub-type with fewer polyps but in which the polyps were more likely to be proximal, to have atypical features and to have diameters greater than 1 cm (Rashid et al. 2000). For example, Azimuddin and colleagues described 16 cases of large atypical hyperplastic polyps from a series of colonoscopies. All but one lesion occurred in the proximal colon, and 9 of 16 cases had a family history of CRC (Azimuddin et al. 2000). It is the second sub-type which is more likely to be associated with a personal and family history of CRC, though the sub-types demonstrate considerable overlap in these features.

Serrated Pathway Syndrome

Familial cancer syndromes associated with *BRAF*-mutation-bearing tumours, and thereby reflecting an origin in serrated polyps, have been described recently in Australia (where 2 of 11 CRC families included cases of HPS) (Young et al. 2005) and Sweden (Vandrovcova et al. 2006). Such families show a pedigree consistent with autosomal dominant inheritance. Individuals with CRC or advanced serrated polyps are present across several generations and both sexes are affected. The features which characterise these families include: a relatively high frequency of *BRAF* mutation (18–70%); increased levels of methylation in the CpG island marker *MINT31*; a background of advanced serrated polyps; increased glandular serration within CRCs; and variable levels of tumour MSI. It is currently not known whether these families represent a single penetrant co-dominant allele of an HPS gene with an overlapping phenotype, or a distinct syndrome.

Conclusion

An understanding of the genetic basis for a predisposition to CRC contributes greatly to the management of families, allowing for pre-symptomatic genetic testing and screening of those where a high level of risk is suspected (Aaltonen et al. 2007). The implications of a genetic predisposition to serrated neoplasia may also have implications for the wider population (Young and Jass 2006). If HPS is a co-dominantly inherited condition, then carriers of a single allele will be several orders of magnitude more common than bi-allelic mutation carriers and will be at increased risk for colorectal cancer. It has been shown, at a population level, that individuals with microsatellite stable serrated-pathway CRC are more likely to have a family history of CRC, evidence which supports this proposition (Samowitz et al. 2005). Issues remain to be addressed in families with serrated neoplasia, such as the establishment of a clinicopathological definition of the syndrome and recognition of its genetic basis. Establishing these will allow the development of recommendations for frequency and method of screening in those most at risk for the development of advanced serrated polyps, particularly individuals with HPS and their families. To this end, the importance of the serrated polyp in screening programs has been recently highlighted (Winawer et al. 2006).

References

Aaltonen, L., L. Johns, et al. (2007). "Explaining the familial colorectal cancer risk associated with mismatch repair (MMR)-deficient and MMR-stable tumors." *Clin Cancer Res* **13**(1): 356–61.

Azimuddin, K., J. J. Stasik, et al. (2000). "Hyperplastic polyps: "more than meets the eye?" Report of sixteen cases." *Dis Colon Rectum* **43**(9): 1309–13.

Beach, R., A. O. Chan, et al. (2005). "BRAF mutations in aberrant crypt foci and hyperplastic polyposis." *Am J Pathol* **166**(4): 1069–75.

Bengoechea, O., J. M. Martinez-Penuela, et al. (1987). "Hyperplastic polyposis of the colorectum and adenocarcinoma in a 24-year-old man." *Am J Surg Pathol* **11**(4): 323–7.

Burt, R. and J. R. Jass (2000). Hyperplastic polyposis. In S. R. Hamilton and L. A. Aaltonen (eds.), *Pathology and Genetics of Tumours of the Digestive System*. Lyon: IARC Press, pp. 135–6.

Chan, A. O., J. P. Issa, et al. (2002). "Concordant CpG island methylation in hyperplastic polyposis." *Am J Pathol* **160**(2): 529–36.

Chow, E., L. Lipton, et al. (2006). "Hyperplastic polyposis syndrome: phenotypic presentations and the role of MBD4 and MYH." *Gastroenterology* **131**(1): 30–9.

Cohen, S. M., L. Brown, et al. (1981). "Multiple metaplastic (hyperplastic) polyposis of the colon." *Gastrointest Radiol* **6**(4): 333–5.

Ferrandez, A., W. Samowitz, et al. (2004). "Phenotypic characteristics and risk of cancer development in hyperplastic polyposis: case series and literature review." *Am J Gastroenterol* **99**(10): 2012–8.

Goldstein, N. S., P. Bhanot, et al. (2003). "Hyperplastic-like colon polyps that preceded microsatellite-unstable adenocarcinomas." *Am J Clin Pathol* **119**(6): 778–96.

Higuchi, T. and J. Jass (2004). "Serrated polyps of the colorectum." *J Clin Pathol* **57**: 675–81.

Hyman, N. H., P. Anderson, et al. (2004). "Hyperplastic polyposis and the risk of colorectal cancer." *Dis Colon Rectum* **47**(12): 2101–4.

Iino, H., L. Simms, et al. (2000). "DNA microsatellite instability and mismatch repair protein loss in adenomas presenting in hereditary non-polyposis colorectal cancer." *Gut* **47**(1): 37–42.

Jass, J. R. (2001). "Serrated route to colorectal cancer: back street or super highway?" *J Pathol* **193**(3): 283–5.

Jass, J. R. (2003). "Hyperplastic-like polyps as precursors of microsatellite-unstable colorectal cancer." *Am J Clin Pathol* **119**(6): 773–5.

Jass, J. R. (2006). "Hereditary Non-Polyposis Colorectal Cancer: the rise and fall of a confusing term." *World J Gastroenterol* **12**(31): 4943–50.

Jass, J. R., D. S. Cottier, et al. (1997). "Mixed epithelial polyps in association with hereditary non-polyposis colorectal cancer providing an alternative pathway of cancer histogenesis." *Pathology* **29**(1): 28–33.

Jass, J. R., H. Iino, et al. (2000). "Neoplastic progression occurs through mutator pathways in hyperplastic polyposis of the colorectum." *Gut* **47**(1): 43–9.

Jeevaratnam, P., D. S. Cottier, et al. (1996). "Familial giant hyperplastic polyposis predisposing to colorectal cancer: a new hereditary bowel cancer syndrome." *J Pathol* **179**(1): 20–5.

Keljo, D. J., A. G. Weinberg, et al. (1999). "Rectal cancer in an 11-year-old girl with hyperplastic polyposis." *J Pediatr Gastroenterol Nutr* **28**(3): 327–32.

Kokko, A., P. Laiho, et al. (2006). "EPHB2 germline variants in patients with colorectal cancer or hyperplastic polyposis." *BMC Cancer* **6**: 145.

Lage, P., M. Cravo, et al. (2004). "Management of Portuguese patients with hyperplastic polyposis and screening of at-risk first-degree relatives: a contribution for future guidelines based on a clinical study." *Am J Gastroenterol* **99**(9): 1779–84.

Leggett, B. A., B. Devereaux, et al. (2001). "Hyperplastic polyposis: association with colorectal cancer." *Am J Surg Pathol* **25**(2): 177–84.

Oberschmid, B. I., W. Dietmaier, et al. (2004). "Distinct secreted Frizzled receptor protein 1 staining pattern in patients with hyperplastic polyposis coli syndrome." *Arch Pathol Lab Med* **128**(9): 967–73.

Place, R. J. and C. L. Simmang (1999). "Hyperplastic-adenomatous polyposis syndrome." *J Am Coll Surg* **188**(5): 503–7.

Rashid, A., P. S. Houlihan, et al. (2000). "Phenotypic and molecular characteristics of hyperplastic polyposis." *Gastroenterology* **119**(2): 323–32.

Rijcken, F. E., T. van der Sluis, et al. (2003). "Hyperplastic polyps in hereditary nonpolyposis colorectal cancer." *Am J Gastroenterol* **98**(10): 2306–11.

Rubio, C. A., S. Stemme, et al. (2006). "Hyperplastic polyposis coli syndrome and colorectal carcinoma." *Endoscopy* **38**(3): 266–70.

Samowitz, W. S., C. Sweeney, et al. (2005). "Poor survival associated with the BRAF V600E mutation in microsatellite-stable colon cancers." *Cancer Res* **65**(14): 6063–9.

Teoh, H. H., B. Delahunt, et al. (1989). "Dysplastic and malignant areas in hyperplastic polyps of the large intestine." *Pathology* **21**(2): 138–42.

Torlakovic, E. and D. C. Snover (1996). "Serrated adenomatous polyposis in humans." *Gastroenterology* **110**(3): 748–55.

Torlakovic, E., E. Skovlund, et al. (2003). "Morphologic reappraisal of serrated colorectal polyps." *Am J Surg Pathol* **27**(1): 65–81.

Vandrovcova, J., K. Lagerstedt-Robinsson, et al. (2006). "Somatic BRAF-V600E Mutations in Familial Colorectal Cancer." *Cancer Epidemiol Biomarkers Prev* **15**(11): 2270–3.

Williams, G. T., J. F. Arthur, et al. (1980). "Metaplastic polyps and polyposis of the colorectum." *Histopathology* **4**(2): 155–70.

Winawer, S. J., A. G. Zauber, et al. (2006). "Guidelines for colonoscopy surveillance after polypectomy: a consensus update by the US Multi-Society Task Force on Colorectal Cancer and the American Cancer Society." *Gastroenterology* **130**(6): 1872–85.

Yeoman, A., J. Young, et al. (2007). "Hyperplastic polyposis in the New Zealand population: a condition associated with increased colorectal cancer risk and European ancestry." *NZ Med J* **120**(1266):U2827.

Young, J. and J. R. Jass (2006). "The case for a genetic predisposition to serrated neoplasia in the colorectum: hypothesis and review of the literature." *Cancer Epidemiol Biomarkers Prev* **15**(10): 1778–84.

Young, J., L. A. Simms, et al. (2001). "Features of colorectal cancers with high-level microsatellite instability occurring in familial and sporadic settings: parallel pathways of tumorigenesis." *Am J Pathol* **159**(6): 2107–16.

Young, J., M. A. Barker, et al. (2005). "Evidence for BRAF mutation and variable levels of microsatellite instability in a syndrome of familial colorectal cancer." *Clin Gastroenterol Hepatol* **3**(3): 254–63.

Young, J. P., M. A. Jenkins, et al. (2007). "Serrated pathway colorectal cancer in the population: an alternative to the adenoma-carcinoma sequence." *Gut* **56**: 1453–1459.

Chapter 8.4
Peutz–Jeghers Syndrome

Douglas L. Riegert-Johnson and Lisa A. Boardman

Background

Peutz–Jeghers syndrome (PJS) is a rare autosomal dominant disorder characterized by melanotic macules, hamartomatous intestinal polyps, and an increased risk for several cancers, including colon cancer. The earliest reported cases consistent with PJS were a pair of identical twins described by Hutchinson in 1896 (Keller et al. 2001). Peutz reported a family in 1921, and an additional ten cases from several families were reported by Jeghers et al. in 1949 (Peutz 1921; Jeghers et al. 1949).

Estimates of the incidence of PJS range from 1 in 8,300 to 1 in 200,000 (Mallory and Stough 1987; Burt 2002). PJS has been reported in populations worldwide and occurs equally in males and females (Anyanwu 1999; Yoon et al. 2000).

Manifestations

Almost all individuals with PJS are thought to express the three features of the disease. There is variability among patients in the degree to which they are affected and at what age they manifest the disease (Westerman et al. 1999).

Melanotic macules are the cardinal feature of PJS (Banse-Kupin and Douglass 1986). They develop on the lips and perioral skin by the end of the first year of life and are almost always present by 5 years. The macules are 1–5 mm in diameter and vary from dark chocolate to latte in color. They may fade in puberty and adulthood.

PJS intestinal polyps are hamartomata with hypertrophied disorganized normal epithelium over an underlying smooth-muscle core (Jansen et al. 2006). The smooth-muscle core is unique to PJS hamartoma. PJS hamartomatous polyps arise most frequently in the small intestine, less so in the colon, and least frequently in the stomach. Between the ages of 9 and 14 years, most PJS patients develop episodes of abdominal pain caused by intermittent intussusception of small bowel polyp(s).

D.L. Riegert-Johnson (✉) and L.A. Boardman
Division of Gastroenterology and Hepatology, Mayo Clinic
e-mail: riegertjohnson.douglas@mayo.edu

J.D. Potter and N.M Lindor (eds.), *Genetics of Colorectal Cancer*,
DOI: 10.1007/978-0-387-09568-4_12,

PJS is associated with an increased risk of breast, colon, small bowel, pancreas, gastric, and other cancers. The most current and complete data on cancer risk are from a multicenter, collaborative series of 416 PJS patients (Hearle et al. 2006). The cumulative risk of any cancer was 85% by age 70. The cumulative risk specifically for colon cancer was 3% (40 years), 5% (50 years), 15% (60 years), and 39% (70 years).

STK11

The only gene associated with PJS to date is *STK11*. There is evidence that there may be a second, as yet unknown, gene associated with PJS in a minority of families (Olschwang et al. 1998). In 1997, the PJS locus was identified at 19p13.3 using comparative genome hybridization, loss of heterozygosity, and targeted linkage analysis (Hemminki et al. 1997). One year later, mutations in the *STK11* gene at that locus were identified in PJS patients (Hemminki et al. 1998; Jenne et al. 1998).

STK11 is the official gene designation of the Human Genome Organization. *LKB1* is sometimes used. The mouse homolog of *STK11* is designated *lkb1*.

STK11 consists of ten exons covering 22.6kb of genomic DNA located at 19p13.3. Nine exons are coding, one is noncoding. Only one transcript is known. *STK11* codes for a 433-amino acid protein that is ubiquitously expressed and present primarily in the cytoplasm and to a lesser extent the nucleus (Rowan et al. 2000; Boudeau et al. 2003). *STK11* is highly conserved with approximately 88 and 84% homology, respectively, with its mouse (*lkb1*) and Xenopous homologs (*XEET1*) (Hemminki et al. 1998).

Functions of *STK11*

STK11 is a tyrosine kinase. It is the only tyrosine kinase known to function as a tumor suppressor. *STK11*'s primary function is energy homeostasis and it is the primary kinase of AMP Kinase (*AMPK*) (Shaw et al. 2004). *STK11*, *AMPK*, and the downstream mTOR pathway allow cells to integrate available energy, amino acid supplies, and growth-factor inputs and to adjust energy expenditure and protein production accordingly (Hay and Sonenberg 2004). In situations of low ATP levels – hypoxia, low glucose – *AMPK* is phosphorylated by *STK11*. *AMPK*, in turn, downregulates the mTOR (mammalian target of rapamycin) pathway through the *TSC1/TSC2* complex. Protein synthesis and energy expenditure are then decreased by downregulation of ribosomal RNA and ribosome synthesis (Høyer-Hansen and Jäättelä 2007).

STK11 also plays a role in the *VEGF* pathway and cellular polarity. *Lkb1*$^{-/-}$ mice have high levels of *VEGF* and vascular malformations (Ylikorkala et al. 2001). *STK11* homologs in *C. elegans* and *D. melanogaster* are involved in embryonic polarity (Watts et al. 2000; Martin et al. 2003). In human epithelial cell lines, activation of *STK11* results in polarization of cells and formation of an apical brush border (Baas et al. 2004).

Mouse Model

Lkb1$^{-/+}$ knockout mice have a phenotype similar to PJS patients. They develop polyps at the junction of the stomach and duodenum between 10 and 14 months of life. They rarely develop small bowel polyps and do not have colon polyps. Connective tissue arborizes through the polyps, in a manner similar to the smooth-muscle arborization seen in human PJS polyps. After 50 weeks of life, some *lkb1*$^{-/+}$ mice develop hepatocellular carcinoma (Nakau et al. 2002). *Lkb1*$^{-/-}$ mice die in midgestation.

Carcinogenesis in PJS

Hamartoma → adenoma → carcinoma is the putative pathway for colon cancer in PJS. Hamartomatous polyps are thought to have a very low malignant potential, and it was unclear whether PJS-associated hamartomas were the premalignant lesions associated with cancer in PJS. Molecular and histological studies have confirmed that hamartomatous polyps are premalignant in PJS (Flageole et al. 1994).

It is not known whether inactivation of the second *STK11* allele is necessary for carcinogenesis or if a 50% decrease in protein expression (haploinsufficiency) is enough for carcinogenesis. Supporting the haploinsufficiency hypothesis are studies of polyps from *lkb1*$^{-/+}$ mice showing 50% levels of *lkb1* mRNA transcripts and protein (Jishage et al. 2002). Supporting the two-hit hypothesis are studies of hepatocellular carcinomas from *lkbIb1*$^{-/+}$ mice that show LOH at the *lkb1* locus (Nakau et al. 2002). Further, hypomorphic *lkb1* mice, *lkb1*$^{fl/fl}$, have been created that have *lkb1* protein levels that are 10% of normal; these mice do not develop polyps or tumors, suggesting that complete loss of *lkb1* function is necessary for polyp and tumor growth (Alessi et al. 2006).

Data from human PJS patients also show that LOH of *STK11* is variably present in polyps and cancers (Table 8.4.1). Neither PJS-associated hamartomas nor carcinomas exhibit many of the genetic events to the same degree as that seen in

Table 8.4.1 Features of PJS-associated hamartomatous polyps and cancers

Characteristic	PJS Hamartomatous polyp	PJS Cancer[a]
LOH *STK11*	7/22	8/11
APC somatic mutation	0/22	2/11
Microsatellite instability	0/22	1/11
Nuclear β-catenin	4/22	5/11
LOH *APC*	0/22	0/11
COX-2 epithelial expression (any)	10/22	8/11
COX-2 stromal expression (any)	12/22	2/7

[a]Cancers studied were colon, small bowel, pancreas, nasopharynx, and lung.
Adapted with permission from De Leng et al. (2003)

nonfamilial colon cancer such as somatic *APC* mutations. Conclusions are limited by the small number of PJS hamartomas and carcinomas that have been studied.

STK11 Loss of Heterozygosity and Somatic Mutations in Sporadic Cancer

Twenty to thirty percent of nonfamilial colon cancers have LOH at the *STK11* locus. In studies of colon cancers with *STK11* LOH, somatic mutations have been identified in only a few cases. Somatic *STK11* mutations are rare in colon cancer and in cancers other than lung cancer where they have been reported in 30% of cases (Forster et al. 2000; Sanchez-Cespedes et al. 2002).

Cyclooxygenase-2

COX-2 (cyclooxygenase-2) is overexpressed in the hamartomatous polyps of the *lkb1*$^{+/-}$ mouse (Rossi et al. 2002). Crossing *lkb1*$^{+/-}$ mice onto a COX-2$^{-/+}$ or COX-2$^{-/-}$ background decreases polyp burden (Udd et al. 2004). In a study comparing COX-2 expression in PJS hamartomatous polyps and carcinomas, 24% of polyps, compared to 64% of carcinomas, had moderate or strong COX-2 expression (De Leng et al. 2003). *Lkb1*$^{+/-}$ mice treated with the COX-2 inhibitor, celecoxib, had a decrease in both the formation of new polyps and the size of pre-existing polyps. When six PJS patients were treated with celecoxib, two had a decrease in gastric polyps (Udd et al. 2004).

References

Alessi, D. R., Sakamoto, K., and Bayascas, J. R. 2006. LKB1-dependent signaling pathways. Annu. Rev. Biochem. 75:137–63.

Anyanwu, S. N. 1999. Sporadic Peutz–Jeghers syndrome in a Nigerian. Cent. Afr. J. Med. 45(7):182–4.

Baas, A. F., Kuipers, J., van der Wel, N. N., Batlle, E., Koerten, H. K., Peters, P. J., and Clevers, H. C. 2004. Complete polarization of single intestinal epithelial cells upon activation of LKB1 by STRAD. Cell 116(3):457–66.

Banse-Kupin, L. A., and Douglass, M. C. 1986. Localization of Peutz–Jeghers macules to psoriatic plaques. Arch. Dermatol. 122(6):679–83.

Boudeau, J., Baas, A. F., Deak, M., Morrice, N. A., Kieloch, A., Schutkowski, M., Prescott, A. R., Clevers, H. C., and Alessi, D. R. 2003. MO25alpha/beta interact with STRAD alpha/beta enhancing their ability to bind, activate and localize LKB1 in the cytoplasm. EMBO J. 22(19):5102–14.

Burt, R. W. 2002. Polyposis syndromes. Clin. Perspect. Gastroenterol. 51–9.

De Leng, W. W., Westerman, A. M., Weterman, M. A., De Rooij, F. W., Dekken, Hv. H., De Goeij, A. F., Gruber, S. B., Wilson, J. H., Offerhaus, G. J., Giardiello, F. M., and Keller, J. J. 2003. Cyclooxygenase 2 expression and molecular alterations in Peutz–Jeghers hamartomas and carcinomas. Clin. Cancer Res. 9(8):3065–72.

Flageole, H., Raptis, S., Trudel, J. L., and Lough, J. O. 1994. Progression toward malignancy of hamartomas in a patient with Peutz–Jeghers syndrome: case report and literature. Can. J. Surg. 37(3):231–6.

Forster, L. F., Defres, S., Goudie, D. R., Baty, D. U., and Carey, F. A. J. 2000. An investigation of the Peutz–Jeghers gene (LKB1) in sporadic breast and colon cancers. Clin. Pathol. 53(10):791–3.

Hay, N., and Sonenberg, N. 2004. Upstream and downstream of mTOR. Genes Dev. 18: 1926–45.

Hearle, N., Schumacher, V., Menko, F. H., Olschwang, S., Boardman, L. A., Gille, J. J., Keller, J. J., Westerman, A. M., Scott, R. J., Lim, W., Trimbath, J. D., Giardiello, F. M., Gruber, S. B., Offerhaus, G. J., de Rooij, F. W., Wilson, J. H., Hansmann, A., Moslein, G., Royer-Pokora, B., Vogel, T., Phillips, R. K., Spigelman, A. D., and Houlston, R. S. 2006. Frequency and spectrum of cancers in the Peutz–Jeghers syndrome. Cancer Res. 12(10):3209–15.

Hemminki, A., Tomlinson, I., Markie, D., Jarvinen, H., Sistonen, P., Bjorkqvist, A. M., Knuutila, S., Salovaara, R., Bodmer, W., Shibata, D., de la Chapelle, A., and Aaltonen, L. A. 1997. Localization of a susceptibility locus for Peutz–Jeghers syndrome to 19p using comparative genomic hybridization and targeted linkage analysis. Nat. Genet. 15(1):87–90.

Hemminki, A., Markie, D., Tomlinson, I., Avizienyte, E., Roth, S., Loukola, A., Bignell, G., Warren, W., Aminoff, M., Hoglund, P., Jarvinen, H., Kristo, P., Pelin, K., Ridanpaa, M., Salovaara, R., Toro, T., Bodmer, W., Olschwang, S., Olsen, A. S., Stratton, M. R., de la Chapelle, A., and Aaltonen, L. A. 1998. A serine/threonine kinase gene defective in Peutz–Jeghers syndrome. Nature 391(6663):184–7.

Høyer-Hansen, M., and Jäättelä, M. 2007. AMP-activated protein kinase: a universal regulator of autophagy? Autophagy 3(4):381–3.

Jansen, M., de Leng, W. W., Baas, A. F., Myoshi, H., Mathus-Vliegen, L., Taketo, M. M., Clevers, H., Giardiello, F. M., and Offerhaus, G. J. 2006. Mucosal prolapse in the pathogenesis of Peutz–Jeghers polyposis. Gut 55(1):1–5.

Jenne, D. E., Reimann, H., Nezu, J., Friedel, W., Loff, S., Jeschke, R., Muller, O., Back, W., and Zimmer, M. 1998. Peutz–Jeghers syndrome is caused by mutations in a novel serine threonine kinase. Nat. Genet. 18:38–43.

Jeghers, H., McKusick, V. A., and Katz, K. H. 1949. Generalized intestinal polyposis and melanin spots of the oral mucosa, lips and digits. N. Engl. J. Med. 241:993–1005, 1031–6.

Jishage, K., Nezu, J., Kawase, Y., Iwata, T., Watanabe, M., Miyoshi, A., Ose, A., Habu, K., Kake, T., Kamada, N., Ueda, O., Kinoshita, M., Jenne, D. E., Shimane, M., and Suzuki, H. 2002. Role of Lkb1, the causative gene of Peutz–Jegher's syndrome, in embryogenesis and polyposis. Proc. Natl Acad. Sci. U. S. A. 99(13):8903–8.

Keller, J. J., Offerhaus, G. J. A., Giardello, F. M., and Menko, F. H. 2001. Jan Peutz, Harold Jeghers and a remarkable combination of polyposis and pigmentation of the skin and mucus membranes. Fam. Cancer 1:181–5.

Mallory, S. B., and Stough, D. B. 1987. Genodermatoses with malignant potential. Dermatol. Clin. 5(1):221–30.

Martin, S. G., and St. Johnston, D. 2003. A role for Drosophila LKB1 in anterior–posterior axis formation and epithelial polarity. Nature 421:379–84.

Nakau, M., Miyoshi, H., Seldin, M. F., Imamura, M., Oshima, M., and Taketo, M. M. 2002. Hepatocellular carcinoma caused by loss of heterozygosity in Lkbl gene knockout mice. Cancer Res. 62:4549–53.

Olschwang, S., Markie, D., Seal, S., Neale, K., Phillips, R., Cottrell, S., Ellis, I., Hodgson, S., Zauber, P., Spigelman, A., Iwama, T., Loff, S., McKeown, C., Marchese, C., Sampson, J., Davies, S., Talbot, I., Wyke, J., Thomas, G., Bodmer, W., Hemminki, A., Avizienyte, E., de la Chapelle, A., Aaltonen, L., Tomlinson, I., et al. 1998. Peutz–Jeghers disease: most, but not all, families are compatible with linkage to 19p13.3. J. Med. Genet. 35(1):42–4.

Peutz, J. L. A. 1921. Very remarkable case of familial polyposis of mucous membrane of intestinal tract and nasopharynx accompanied by peculiar pigmentations of skin and mucous membrane (Dutch). Nederl. Maandschr. Geneesk. 10:134–46.

Rossi, D. J., Ylikorkala, A., Korsisaari, N., Salovaara, R., Luukko, K., Launonen, V., Henkemeyer, M., Ristimaki, A., Aaltonen, L. A., and Makela, T. P. 2002. Induction of cyclooxygenase-2 in a mouse model of Peutz–Jeghers polyposis. Proc. Natl Acad. Sci. U. S. A. 99(19):12327–32.

Rowan, A., Churchman, M., Jefferey, R., Hanby, A., Poulsom, R., and Tomlinson, I. 2000. In situ analysis of LKB1/STK11 mRNA expression in human normal tissues and tumours. J. Pathol. 192(2):203–6.

Sanchez-Cespedes, M., Parrella, P., Esteller, M., Nomoto, S., Trink, B., Engles, J. M., Westra, W. H., Herman, J. G., and Sidransky, D. 2002. Inactivation of LKB1/STK11 is a common event in adenocarcinomas of the lung. Cancer Res. 62:3659–62.

Shaw, R. J., Kosmatka, M., Bardeesy, N., Hurley, R. L., Witters, L. A., DePinho, R. A., and Cantley, L. C. 2004. The tumor suppressor LKB1 kinase directly activates AMP-activated kinase and regulates apoptosis in response to energy stress. Proc. Natl Acad. Sci. U.S.A. 101(10):3329–35.

Udd, L., Katajisto, P., Rossi, D. J., Lepisto, A., Lahesmaa, A. M., Ylikorkala, A., Jarvinen, H. J., Ristimaki, A. P., and Makela, T. P. 2004. Suppression of Peutz–Jeghers polyposis by inhibition of cyclooxygenase-2. Gastroenterology 127(4):1030–7.

Watts, J. L., Morton, D. G., Bestman, J., and Kemphues, K. J. 2000. The *C. elegans* par-4 gene encodes a putative serine-threonine kinase required for establishing embryonic asymmetry. Development 127:1467–75.

Westerman, A. M., Entius, M. M., de Baar, E., Boor, P. P., Koole, R., van Velthuysen, M. L., Offerhaus, G. J., Lindhout, D., de Rooij, F. W., and Wilson, J. H. 1999. Peutz–Jeghers syndrome: 78-year follow-up of the original family. Lancet 353(9160):1211–5.

Ylikorkala, A., Rossi, D. J., Korsisaari, N., Luukko, K., Alitalo, K., Henkemeyer, M., and Makela, T. P. 2001. Vascular abnormalities and deregulation of VEGF in Lkb1-deficient mice. Science 293:1323–6.

Yoon, K. A., Ku, J. L., Choi, H. S., Heo, S. C., Jeong, S. Y., Park, Y. J., Kim, N. K., Kim, J. C., Jung, P. M., and Park, J. G. 2000. Germline mutations of the *STK11* gene in Korean Peutz–Jeghers syndrome patients. Br. J. Cancer 82(8):1403–6.

Chapter 8.5
Juvenile Polyposis

Kara A. Mensink, Jeremy R. Jass, and Noralane M. Lindor

Juvenile Polyposis (JP) is an autosomal dominant, genetically heterogeneous disorder, characterized by multiple (5–200) (Aaltonen et al. 2000) hamartomatous polyps of the gastrointestinal tract. A variety of extracolonic manifestations have been recorded (McColl et al. 1964; Soper and Kent 1971; Sachatello et al. 1974; Walpole and Cullity 1989; Erkul and Ariyurek 1994; Coburn et al. 1995; Desai et al. 1998; Pacheco et al. 2007), but it is unclear whether patients in earlier reports had JP or some other hamartoma syndrome. No formal study of disease prevalence has been published, but the population incidence of JP is estimated to be 1 in 100,000 (Burt et al. 1990).

The most commonly cited clinical diagnostic criteria for JP were proposed by Jass and colleagues (1988) (Table 8.5.1). Wide variability in intrafamilial and interfamilial expressivity of the clinical phenotype is observed. Affected individuals typically present with symptoms secondary to polyp formation (hematochezia, anemia, melena, abdominal pain) during their first two decades of life. However, one-third of affected individuals remain asymptomatic until adulthood (Coburn et al. 1995). The name, *juvenile*, refers to the histology of the polyps (which resemble sporadic inflammatory polyps of childhood), not to the age at diagnosis.

Juvenile Polyposis shares clinical features with other colonic hamartomatous polyp syndromes (Cowden, Bannayan–Riley–Ruvalcaba, Peutz–Jeghers, Basal Cell Nevus/Gorlin), often leading to misdiagnosis. The benefit of clinical and molecular hindsight has permitted better classification of patients previously diagnosed with JP. Distinction among syndromes is important for both clinical and research purposes. Careful pathologic examination and clinical and family history can help to differentiate. Many hamartomatous polyp syndromes have a characteristic dermatologic or histopathological finding that can be especially helpful in establishing the correct diagnosis (Tables 8.5.2 and 8.5.3). Hamartomatous polyps arise from the disorganized growth of surrounding normal tissue element(s). The term juvenile polyp is sometimes used synonymously with hamartomatous polyp. But, more precisely, a juvenile polyp is a unique type of hamartomatous polyp.

N.M. Lindor (✉)
Department of Medical Genetics, Mayo Clinic College of Medicine
e-mail: nlindor@mayo.edu

J.D. Potter and N.M Lindor (eds.), *Genetics of Colorectal Cancer*,
DOI: 10.1007/978-0-387-09568-4_13,

Table 8.5.1 Clinical diagnostic criteria for JP (Jass et al. 1988)

One or more of the following:
• More than five juvenile polyps of the colorectum
• Juvenile polyps throughout the gastrointestinal tract
• Any number of juvenile polyps with a family history of juvenile polyposis

Table 8.5.2 Dermatologic findings in the hamartomatous polyp syndromes

Syndrome	Inheritance	Dermatologic findings
JP (hereditary)	Autosomal dominant	Multiple CALs (not consistent)
Sporadic juvenile polyp	Sporadic	Multiple CALs (uncertain if associated)
Bannayan–Riley–Ruvalcaba	Autosomal dominant (allelic with Cowden)	Lipomas Pigmented macules of the glans penis Vascular malformations
Cowden	Autosomal dominant	Trichilemmomas Papillomatous papules Acral/planter keratosis Glycogenic acanthosis Multiple CALs
Cronkhite–Canada	Sporadic	Diffuse lentigines Alopecia Nail dystrophy
Gorlin (nevoid basal cell carcinoma syndrome)	Autosomal dominant	Jaw keratocysts Basal cell carcinomas Facial milia Meibomian cysts Sebaceous cysts Dermoid cysts Skin tags (especially on neck)
Neurofibromatosis type 1	Autosomal dominant	Multiple CALs Freckling, axillary, inguinal, and elsewhere Neurofibromas
Peutz–Jeghers	Autosomal dominant	Dark blue/brown macules around mouth, buccal mucosa, eyes, nostrils, perianal area Hyperpigmented macules of the finger

CAL café-au-lait macules

Juvenile polyps range in size from 5 to 50mm. Although most common in the colon, they can also occur in the stomach and small intestine. They are spherical in shape, can be single or multilobulated, and commonly exhibit surface erosion. With the exception of gastric juvenile polyps, which are sessile, most juvenile

Table 8.5.3 Comparing and contrasting the hamartomous gastrointestinal polyps between syndromes

Disorder	Polyp location	Polyp number	Polyp appearance and histology
JP (hereditary)	Stomach, small bowel, colon	Multiple	Expanded lamina propria; hyperplastic, focally crowded, and cystic glands; surface erosions common; not as often pedunculated as sporadic juvenile polyp; dysplasia
Sporadic juvenile polyp	Colon	1	Pedunculated, smooth, round eroded surface; lacks smooth muscle; surface erosions common; rarely has areas of dysplasia
Banayan–Ruvalcaba–Riley	Esophagus, stomach, small bowel, colorectum	Multiple	Spectrum of intestinal findings: hamartomatous polyps, ganglioneuromas, clusters of ganglia within lamina propria, atypical polyps with some features of tubulovillous adenomas
Cowden	Esophagus, stomach, small bowel, colorectum	Multiple	Hamartomatous polyps, lipomatous hamartomas, ganglioneuromatosis
Cronkhite–Canada	Stomach, small bowel, colon	Multiple	Broad sessile base, expanded edematous lamina propria and cystic glands; no dysplasia
Gorlin (nevoid basal cell carcinoma syndrome)	Stomach	Not common	Hamartomas not well described in literature
Neurofibromatosis type I	Colon	Not common	Hamartomas, not well described in literature; neurofibromas, ganglioneuromas
Peutz–Jeghers	Mostly small bowel; can involve stomach and colorectum	Multiple	Smooth muscle forming infrastructure; branching bands of smooth muscle; hyperplasia, elongation, and cystic change of the foveolar epithelium; atrophy of deeper glandular components

polyps are pedunculated (i.e., have a stalk). Histologically, expanded edematous lamina propria with mucinous, dilated glands, abundant stroma, and inflammatory infiltrate are commonly observed. Some juvenile polyps may have areas of dysplasia (Aaltonen 2000; Brosens 2007; Burke and Sobin 1989; Burger et al. 2002; Howe 2004; Jass et al. 1988). Solitary juvenile polyps occur in approximately 2% of the pediatric population. Although histologically quite similar to juvenile polyps observed in JP, these polyps are not associated with either increased risk of malignancy or extracolonic manifestations and are seldom dysplastic (Nugent et al. 1993; Giardiello and Hamilton 1991). Similar polyp pathology is also observed with Cronkhite–Canada syndrome, a sporadic, adult-onset colonic hamartomatous polyp syndrome that is more prevalent among Japanese. Certain pathologic nuances may help to distinguish solitary hamartomatous polyps and the hamartomatous polyps observed as part of Cronkhite–Canada syndrome from those associated with familial JP. For example, Cronkhite–Canada hamartomatous polyps tend to have a broad sessile base and are not pedunculated like those observed in JP (Burke and Sobin 1989). In addition, a frond-like growth pattern, comparatively less stroma, and dilated glands with more proliferative smaller glands are more commonly observed in familial JP rather than solitary, sporadic juvenile polyps (Brosens et al. 2007).

Individuals with familial JP have an increased risk of GI malignancy. A survey of the literature assesses this risk at approximately 50% (Brosens et al. 2007 ; Howe et al. 1998) with a reported range of 9–68% (Jass et al. 1988; Jarvinen and Franssila 1984; Coburn et al. 1995; Howe et al. 1998; Brosens et al. 2007). Cancers of the stomach, duodenum, and pancreas have been described in some patients, but the risk for malignancy is highest (~40%) in the colon (Howe et al. 1998). The mechanism by which malignant transformation occurs remains a subject of research. A juvenile polyp-adenomatous change-dysplasia-carcinoma sequence is suspected following several reported cases of colorectal adenocarcinoma in patients with JP, and a documented correlation between risk of malignancy and a preponderance of juvenile polyps exhibiting dysplastic change (Merg et al. 2005; Roth 1999). It has also been suggested that disruption of TGF-β signaling due to an abnormal microenvironment created largely by abundant stroma is responsible (Kim et al. 2006; Kinzler and Vogelstein 1998). Additional research is required to determine whether individuals with JP are generally predisposed to malignancy separate from a predisposition to polyps, or are predisposed to polyps that, in turn, become malignant, or both.

Considerable heterogeneity complicates the molecular diagnosis of JP. To date, two genes, *BMPR1A* and *MADH4/SMAD4*, and recently a third gene, *ENG*, have been implicated in JP. These three genes encode proteins of the closely related TGF-β- and BMP-signaling pathways. The TGF-β and BMP signal transduction pathways each involve a signaling cascade where ligands bind to type 2 receptors to recruit type 1 receptors, like *BMPR1A*, creating a receptor complex that, in turn, phosphorylates unique SMADs. The TGF-β and BMP pathways converge when the unique, activated SMADs (*SMAD1*, *SMAD5*, *SMAD8* for BMP, and *SMAD2* and *SMAD3* for TGF-β) bind to the common *SMAD4* product, creating a SMAD oligomer. The SMAD oligomer shuttles into the cell nucleus and binds to transcription factors to form a transcriptional complex, thereby regulating gene expression and ultimately cellular homeostasis (Heldin et al. 1997; Howe et al. 2004; Brosens et al. 2007; Merg et al. 2005).

The prevalence of *BMPR1A* and *MADH4/SMAD4* mutations in JP patients is reported to be 11.4–20.8 and 18.2–18.6%, respectively (Sayed et al. 2002; Howe et al. 2004; Pyatt et al. 2006). This low combined detection rate has prompted a continued search for other possible genes/proteins within the TGF-β and BMP pathways.

Comprehensive molecular screening of the following TGF-β type I, type II, and SMAD receptor proteins: *BMPR1B*, *BMPR2*, *ACVR1*, *SMAD2*, *SMAD3*, and *SMAD7* has not identified a single causative mutation among *BMPR1A* and *MADH4/SMAD4* mutation-negative JP patients (Bevan et al. 1999; Howe et al. 2004; Roth et al. 1999). However, mutations in the *ENG* gene that encodes endoglin, a TGF-β accessory receptor protein, have been reported in two patients with JP with onset in infancy (Sweet et al. 2005), and it is notable that neither had clinical features consistent with Hereditary Hemorrhagic Telangiectasia (HHT) syndrome, which is often caused by mutations in the *ENG* gene. The prevalence of *ENG* gene mutations in JP patients without features of HHT has yet to be adequately described (Howe et al. 2007).

HHT is inherited in an autosomal dominant fashion and is clinically characterized by arteriovenous malformations and skin and mucosal telangiectasias. Two genes, *ACVRL1/ALK-1* and *ENG*, are associated with HHT. *ACVRL1/ALK-1* encodes activin A receptor type-like kinase 1 that acts as an alternate TGF-β type I receptor in endothelial cells. The *ENG* gene encodes endoglin, an accessory receptor protein that binds to specific TGF-β proteins depending on the presence of particular type I and type II receptors. The endoglin protein is required for efficient *ALK-1* signaling (Blanco et al. 2005). More than 92% of individuals meeting clinical diagnostic criteria of HHT have mutations in the *ENG* or *ACVRL1* genes (Letteboer et al. 2005). Several patients with features of both JP and HHT have been reported. None of these patients was reported to have mutations in the *ENG* or *ACVRL1* genes; instead, many have mutations in the *MADH4/SMAD4* gene, which is mutated in approximately 10–20% of JP patients without obvious features of HHT (Burger et al. 2002; Gallione et al. 2004, 2006; Howe et al. 2004; Roth et al. 1999). A retrospective review of such patients may show findings consistent with HHT. Some authors have suggested that individuals who exhibit features of both HHT and JP have a distinct syndrome, while others suggest that variable expressivity and age-related penetrance may explain the clinical overlap. The molecular and clinical relationship of JP and HHT continues to be the subject of research.

Efforts to correlate genotype and phenotype have elucidated specific trends that, depending on clinical features, may be helpful guiding molecular testing strategies for affected patients:

- Polyp morphology of *MADH4/SMAD4* mutation-positive patients has been reported to show less prominent stroma than polyps of mutation-negative individuals (Woodford-Richens et al. 2001).
- *MADH4/SMAD4* mutations have been identified in individuals exhibiting features of both HHT and JP.

- A sometimes massive preponderance of gastric polyps, has been observed in *MADH4/SMAD4* mutation-positive patients which distinguishes them from *BMPR1A* mutation-positive and affected mutation-negative JP patients (Sayed et al. 2002; Friedl et al. 2002; Handra-Luca et al. 2005).
- In general, *MADH4/SMAD4* or *BMPR1A* mutation-positive patients are more likely to have: (1) more polyps; (2) a family history of JP; and (3) a family history of GI cancer (Sayed et al. 2002).

Although relatively rare, JP is the most common of the hamartomatous polyp syndromes. Wide clinical variability and genetic heterogeneity complicate the diagnosis of affected individuals, but careful pathologic and clinical examination coupled with appropriate molecular studies can correctly distinguish familial JP from those with nonfamilial hamartomas or with other colonic hamartomatous syndromes. Continued clinical and molecular research will not only improve diagnosis and management of JP patients, but will continue to provide insight into the complicated interrelationship of the TGF-β superfamily proteins and cancer mechanisms, and thus, probably influence progress in cancer research more generally.

References

Aaltonen, L. A., J. R. Jass, and J. R. Howe (2000). Juvenile polyposis. In: Hamilton SR, Aaltonen LA, editors. Pathology and genetics. Tumours of the digestive system. Lyon: IARC Press, pp. 130–132.

Bevan, S., K. Woodford-Richens, et al. (1999). "Screening SMAD1, SMAD2, SMAD3, and SMAD5 for germline mutations in juvenile polyposis syndrome." Gut 45(3): 406–8.

Blanco, F. J., J. F. Santibanez, et al. (2005). "Interaction and functional interplay between endoglin and ALK-1, two components of the endothelial transforming growth factor-beta receptor complex." J Cell Physiol 204(2): 574–84.

Brosens, L. A., W. A. van Hattem, et al. (2007). "Gastrointestinal polyposis syndromes." Curr Mol Med 7(1): 29–46.

Burger, B., S. Uhlhaas, et al. (2002). "Novel de novo mutation of MADH4/SMAD4 in a patient with juvenile polyposis." Am J Med Genet 110(3): 289–91.

Burke, A. P. and L. H. Sobin (1989). "The pathology of Cronkhite–Canada polyps. A comparison to juvenile polyposis." Am J Surg Pathol 13(11): 940–6.

Burt, R. W., D. T. Bishop, et al. (1990). "Risk and surveillance of individuals with heritable factors for colorectal cancer. WHO Collaborating Centre for the Prevention of Colorectal Cancer." Bull World Health Organ 68(5): 655–65.

Coburn, M. C., V. E. Pricolo, et al. (1995). "Malignant potential in intestinal juvenile polyposis syndromes." Ann Surg Oncol 2(5): 386–91.

Desai, D. C., V. Murday, et al. (1998). "A survey of phenotypic features in juvenile polyposis." J Med Genet 35(6): 476–81.

Erkul, P. E., O. M. Ariyurek, et al. (1994). "Colonic hamartomatous polyposis associated with hypertrophic osteoarthropathy." Pediatr Radiol 24(2): 145–6.

Friedl, W., S. Uhlhaas, et al. (2002). "Juvenile polyposis: massive gastric polyposis is more common in MADH4 mutation carriers than in BMPR1A mutation carriers." Hum Genet 111(1): 108–11.

Gallione, C. J., G. M. Repetto, et al. (2004). "A combined syndrome of juvenile polyposis and hereditary haemorrhagic telangiectasia associated with mutations in MADH4 (SMAD4)." Lancet 363(9412): 852–9.

Gallione, C. J., J. A. Richards, et al. (2006). "SMAD4 mutations found in unselected HHT patients." J Med Genet 43(10): 793–7.

Giardiello, F. M., S. R. Hamilton, et al. (1991). "Colorectal neoplasia in juvenile polyposis or juvenile polyps." Arch Dis Child 66(8): 971–5.

Handra-Luca, A., C. Condroyer, et al. (2005). "Vessels' morphology in SMAD4 and BMPR1A-related juvenile polyposis." Am J Med Genet A 138(2): 113–7.

Heldin, C. H., K. Miyazono, et al. (1997). "TGF-beta signalling from cell membrane to nucleus through SMAD proteins." Nature 390(6659): 465–71.

Howe, J. R., F. A. Mitros, et al. (1998). "The risk of gastrointestinal carcinoma in familial juvenile polyposis." Ann Surg Oncol 5(8): 751–6.

Howe, J. R., M. G. Sayed, et al. (2004). "The prevalence of MADH4 and BMPR1A mutations in juvenile polyposis and absence of BMPR2, BMPR1B, and ACVR1 mutations." J Med Genet 41(7): 484–91.

Howe, J. R., J. L. Haidle, et al. (2007). "ENG mutations in MADH4/BMPR1A mutation negative patients with juvenile polyposis." Clin Genet 71(1): 91–2.

Jarvinen, H. and K. O. Franssila (1984). "Familial juvenile polyposis coli; increased risk of colorectal cancer." Gut 25(7): 792–800.

Jass, J. R., C. B. Williams, et al. (1988). "Juvenile polyposis – a precancerous condition." Histopathology 13(6): 619–30.

Kim, B. G., C. Li, et al. (2006). "Smad4 signalling in T cells is required for suppression of gastrointestinal cancer." Nature 441(7096): 1015–9.

Kinzler, K. W. and B. Vogelstein (1998). "Landscaping the cancer terrain." Science 280(5366): 1036–7.

Letteboer, T. G., R. A. Zewald, et al. (2005). "Hereditary hemorrhagic telangiectasia: ENG and ALK-1 mutations in Dutch patients." Hum Genet 116(1–2): 8–16.

McColl, I., H. J. Busxey, et al. (1964). "Juvenile polyposis coli." Proc R Soc Med 57: 896–7.

Merg, A., H. T. Lynch, et al. (2005). "Hereditary colorectal cancer-part II." Curr Probl Surg 42(5): 267–333.

Nugent, K. P., I. C. Talbot, et al. (1993). "Solitary juvenile polyps: not a marker for subsequent malignancy." Gastroenterology 105(3): 698–700.

Pacheco, T. R., L. S. Scatena, et al. (2007). "Cafe au Lait Macules and Juvenile Polyps." Pediatr Dermatol 24(3): E17–E21(1).

Pyatt, R. E., R. Pilarski, et al. (2006). "Mutation screening in juvenile polyposis syndrome." J Mol Diagn 8(1): 84–8.

Roth, S., P. Sistonen, et al. (1999). "SMAD genes in juvenile polyposis." Genes Chromosomes Cancer 26(1): 54–61.

Sachatello, C. R., I. S. Hahn, et al. (1974). "Juvenile gastrointestinal polyposis in a female infant: report of a case and review of the literature of a recently recognized syndrome." Surgery 75(1): 107–14.

Sayed, M. G., A. F. Ahmed, et al. (2002). "Germline SMAD4 or BMPR1A mutations and phenotype of juvenile polyposis." Ann Surg Oncol 9(9): 901–6.

Soper, R. T. and T. H. Kent (1971). "Fatal juvenile polyposis in infancy." Surgery 69(5): 692–8.

Sweet, K., J. Willis, et al. (2005). "Molecular classification of patients with unexplained hamartomatous and hyperplastic polyposis." JAMA 294(19): 2465–73.

Walpole, I. R. and G. Cullity (1989). "Juvenile polyposis: a case with early presentation and death attributable to adenocarcinoma of the pancreas." Am J Med Genet 32(1): 1–8.

Woodford-Richens, K. L., A. J. Rowan, et al. (2001). "Comprehensive analysis of SMAD4 mutations and protein expression in juvenile polyposis: evidence for a distinct genetic pathway and polyp morphology in SMAD4 mutation carriers." Am J Pathol 159(4): 1293–300.

Chapter 8.6
BLM Mutation and Colorectal Cancer Susceptibility

Beatriz Russell and Joanna Groden

Bloom Syndrome and the *BLM* Gene

Bloom syndrome (BS) was first described by Dr. David Bloom in 1954 after he observed a small number of patients of Ashkenazi Jewish origin with erythematous lesions of the face and small stature (Bloom 1954). Approximately 10 years later, the chromosomal instability and cancer predisposition of BS were reported (German et al. 1965). BS is a rare autosomal recessive disorder in which affected individuals show pre- and postnatal growth retardation, sun-sensitive facial erythema, immunodeficiency, and male infertility. Those affected are predisposed to a plethora of cancers, most commonly occurring before the age of 25 (German 1993). Carcinomas are observed with highest frequency, followed by leukemias and lymphomas (German and Ellis 1998). Cytologically, the hallmark of BS is elevated sister chromatid exchange (SCE), approximately 5- to 10-fold higher than in cells from unaffected individuals. SCE is often used as a diagnostic marker of BS, although molecular genetic mutational analysis is available. Other cytological features of BS include quadriradial structures, telomere associations, and chromosome breaks (German 1964; German et al. 1965; Chaganti et al. 1974).

The disease gene for BS, known as *BLM*, maps to chromosome 15q26.1 and encodes a 159-kDa protein that is a member of the recQ family of helicases (Ellis et al. 1995). This family of helicases is highly conserved throughout evolution; multicellular organisms have multiple recQ-like helicase genes in contrast to unicellular organisms that have only one. The *BLM* gene encodes a protein that contains a central helicase domain. Carboxy terminal to the helicase domain is the conserved RQC domain (*recQC*-terminal) that defines this family of helicases, and the HRDC domain (*h*elicase, *R*Nase*D* and *C*-terminal), a domain common to RNA helicases. A nuclear localization signal is also present in the C terminus of most eukaryotic recQ-like helicases. The human recQ-like helicases have very little

B. Russell and J. Groden (✉)
Department of Molecular Virology, Immunology and Medical Genetics, The Ohio State University College of Medicine.
e-mail: joanna.groden@osumc.edu

J.D. Potter and N.M Lindor (eds.), *Genetics of Colorectal Cancer*,
DOI: 10.1007/978-0-387-09568-4_14,

homology outside the central helicase domain; the other regions vary greatly in length. Such differences permit distinct protein binding partners suggesting some unique functions for each helicase (Hickson 2003).

The recQ-like family of helicases in humans consists of five members: BLM, WRN, RecQL1, RecQL4, and RecQL5 (Karow et al. 2000a,). Mutations in *BLM*, *WRN*, and *RecQL4* lead to the chromosome breakage syndromes, Bloom, Werner, and Rothmund–Thompson syndromes, respectively. The functions of RecQL1 and RecQL5 are less clear, as germline mutations of these genes are not associated with a human syndrome. Homozygous mutation of *Recql5* in mice leads to an increase in cancer susceptibility, with lymphomas occurring in more than half of the mice with tumors. Embryonic stem (ES) cells deficient in *Recql5* show elevated levels of spontaneous DNA double-strand breaks (DSB) (Hu et al. 2007). Cells lacking RecQL1 are sensitive to IR induced damage and camptothecin, resulting in high SCE (Sharma and Brosh 2007). Also, RecQL1 depletion by siRNA leads to mitotic cell death in cancer cells, but not in normal cells suggesting that RecQL1 is necessary for the repair of replication induced damage that persists in cancer cells due to lack of checkpoints (Futami et al. 2008).

Since the cloning of *BLM* in 1995, 64 unique mutations have been reported in the Bloom's Syndrome Database (German et al. 2007), a large number of which (54) lead to premature stop codons, and the remaining (10) code for missense mutations. Frameshift mutations caused by insertions and deletions are found throughout the gene. The Blm^{Ash} mutation is a 6-bp deletion and 7-bp insertion found at high frequency in those of Ashkenazi Jewish descent, of whom one in 120 is a carrier (Straughen et al. 1998). One-third of those affected by BS are of Ashkenazi Jewish origin.

Biochemical Properties of BLM

The BLM helicase has ATP- and Mg^{2+}-dependent 3'–5' helicase activity. BLM does not effectively unwind blunt-ended DNA substrates and has poor processivity on dsDNA with a free 3'-end (Karow et al. 1997). This can be enhanced in the presence of the single-stranded binding protein, RPA (Brosh et al. 2000). Preferred in vitro substrates include unusual DNA structures such as duplex DNA containing a bubble and G4 DNA, stable structures that can form in G-rich regions of the genome such as telomeres (Mohaghegh et al. 2001; Brosh et al. 2000). BLM can recognize and promote branch migration of double Holliday junctions, and its interaction with topoisomerase IIIα-BLAP75-BLAP18, known as the BTB complex, is necessary for proper processing of these structures to yield a noncrossover product (Wu and Hickson 2003; Raynard et al. 2006; Singh et al. 2008). BLM can promote reverse branch migration of stalled replication forks thus facilitating repair without initiation of homologous recombination (HR) (Karow et al. 2000b), and can disrupt D-loops, early intermediates in HR (van Brabant et al. 2000). In vitro evidence, therefore, suggests an antirecombinogenic role for BLM.

In vivo, slow progression of DNA replication supports a role for BLM in the resolution of stalled replication forks (Hand and German 1975; Lönn et al. 1990). Additionally, BS cells are highly sensitive to hydroxyurea (HU), a nucleotide analog that halts DNA replication and activates the S-phase cell-cycle checkpoint. BLM colocalizes with proteins involved in replication and DNA repair such as γH2AX, RAD51, RPA, and PCNA (Brosh et al. 2000; Rassool et al. 2003; Yankiwski et al. 2000; Davies et al. 2004). BLM has been implicated in the repair of DSB with putative roles in both HR and nonhomologous end-joining (NHEJ) (Langland et al. 2002; Wu and Hickson 2003).

Blm/BLM Mutation and Colorectal Cancer Susceptibility

The cancer spectrum that results from lack of BLM, albeit wide, is restricted to proliferative tissues where BLM is normally expressed. Of the 238 affected in the Bloom's Syndrome Registry, 25 have reported colorectal tumors. Colorectal tumors have been identified from cecum to rectum, and range from numerous adenomatous polyps to adenocarcinomas (German 1996; Lowy et al. 2001). Although the case numbers are small, almost half of the carcinomas occur in the ascending or transverse colon, rather than the more common descending colon for the general population (Lowy et al. 2001). Genetic analysis of six adenomas in one BS person revealed somatic mutations of *APC*, but no germline mutation (Lowy et al. 2001). Different *APC* mutations were found in four of these adenomas suggesting that each adenoma formed independently. Two adenomas from the same person were positive for microsatellite instability. These analyses suggest a general increase in mutation frequency in the epithelial cells of the colon in BS persons. In another study, Calin et al. (2000) examined 63 colon carcinomas with high microsatellite instability and determined that a small proportion of these tumors carried microsatellite mutations in *BLM* and that such frameshifts were significantly associated with a mucinous histopathology of the tumor. It is unclear from this work whether *BLM* is a genomic or functional target of mutation.

Although BS is a recessive disorder, reports suggest that *BLM* haploinsufficiency leads to an increase in the incidence of intestinal cancers. A study in which mice carrying one *Blm* null allele ($Blm^{Cin/+}$) were challenged with a murine leukemia virus infection showed that the $Blm^{Cin/+}$ mice died earlier from lymphoma than their wild-type littermates (Goss et al. 2002). To determine the effect of *Blm* haploinsufficiency on intestinal tumorigenesis, $Blm^{Cin/+}$ mice were crossed with $Apc^{Min/+}$ mice carrying a premature stop codon in one allele of *Apc*. At 4 months of age, $Apc^{Min/+}$; $Blm^{Cin/+}$ developed twice the number of adenomas compared to $Apc^{Min/+}$; $Blm^{+/+}$ mice. The adenomas in the double heterozygotes were characterized by high-grade dysplasia, rather than the low-grade dysplasia of the $Apc^{Min/+}$ mouse adenomas. Mutational analysis of the adenomas, using genetic markers proximal and distal to *Apc* on mouse chromosome 18, showed that the loss of the second *Apc* allele could be mediated by somatic recombination, rather than just associated with isodisomy. All the adenomas examined remained heterozygous at *Blm*.

Gruber et al. (2002) asked similar questions using a human population to determine if carriers of a *BLM* mutation are at an increased risk of developing colon cancer. One thousand two hundred forty-four Ashkenazi Jews with colorectal cancers were genotyped at *BLM* and determined to have an allele frequency of 1 in 54. Ashkenazi Jews without colorectal cancers were determined to have an allele frequency of 1 in 118. The age of colon cancer diagnosis did not differ between $BLM^{Ash/+}$ and $BLM^{+/+}$ patients. Gruber et al. (2002) concluded that carriers of *BLM* mutation are twice as likely to develop colorectal cancer as noncarriers. All individuals in the study had three of four grandparents of Ashkenazi Jewish origin, and were evaluated by full colonoscopy. Anyone with a history of inflammatory bowel disease or a family history of colon cancer was eliminated from both study groups.

Since these reports were published, three other reports have been unable to demonstrate an association between *BLM* haploinsufficiency and susceptibility to colorectal cancer. Cleary et al. (2003) genotyped 2,333 Jewish individuals to determine the allele frequency of BLM^{Ash}. Four hundred ninety-seven individuals were diagnosed with colorectal cancer, 125 with adenomatous polyps, 767 with noncolorectal cancers, and 944 were cancer-free. They found the allele frequency of BLM^{Ash} mutation did not significantly differ between individuals with colorectal tumors, noncolorectal cancer, or those that were cancer-free (0.80, 0.87, and 0.85%, respectively). Cleary et al. (2003) also found that, among their sample populations, the mean age of colorectal cancer diagnosis for BLM^{Ash} carriers was 74 years compared to 71 years for noncarriers, suggesting that BLM^{Ash} heterozygosity does not markedly alter the mean age for cancer diagnosis. A second study reported an allele frequency of 0.9% for BLM^{Ash} mutation in paraffin-embedded blocks of colorectal tumors from 429 Ashkenazi Jews (Zauber et al. 2005). In a third report, Baris et al. (2007) retrospectively studied three generations of 28 individuals carrying the BLM^{Ash} and 43 non-carriers. They found no significant difference in the prevalence of malignancies (breast and colon) among carriers and non-carriers.

Although the role of *BLM* haploinsufficiency in susceptibility to colorectal cancer still awaits larger human population studies, it is clear, from the mouse-model experiments, that haploinsufficiency affects tumor number, tumor histopathology, and mutational mechanism. It is also clear that individuals with Bloom Syndrome can develop a wide range of cancers at an early age, but seem to develop colon cancer at an unusually high frequency.

Acknowledgments We would like to thank the Bloom Syndrome Foundation, and Lauri and Richard Gladstein for their support of research on BS. This work was also supported by NIH grants CA-063507 and CA-084291.

References

Baris HN, Kedar I, Halpern GJ, Shohat T, Magal N, Ludman MD, Shohart M (2007) Prevalence of breast and colorectal cancer in Ashkenazi Jewish carriers of Fanconi anemia and Bloom syndrome. IMAJ 9:847–850.

Bloom D (1954) Congenital telangiectatic erythema resembling lupus erythematosus in dwarfs. Am J Dis Child 88:754.

Brosh RM, Li JL, Kenny MK, Karow JK, Cooper MP, Kureekattil RP, Hickson ID Bohr VA (2000) Replication protein A physically interacts with the Bloom's syndrome protein and stimulates its helicase activity. J Biol Chem 275:23500–23508.

Calin GA, Gafa R, Tibileti MG, Herlea V, Becheanu G, Cavazzini L, Barbanti-Brodano G, Nenci I, Negrini M, Lanza G (2000) Genetic progression in microsatellite instability high (MSI-H) colon cancers correlates with clinico-pathological parameters: a study of the *TGRbetaRII*, *BAX*, *hMSH3*, *hMSH6*, *IGFIIR* and *BLM* genes. Int J Cancer 89:230–235.

Chaganti RS, Schonberg S, German J (1974) A manifold increase in sister chromatid exchanges in Bloom's syndrome lymphocytes. Proc Natl Acad Sci U S A 71:4508–4512.

Cleary SP, Zhang W, Di Nicola N, Aronson M, Aube J, Steinman A, Haddad R, Redston M, Gallinger S, Narod SA, Gryfe R (2003) Heterozygosity for the BLM^{Ash} mutation and cancer risk. Cancer Res 63:1769–1771.

Davies SL, North PS, Dart A, Larkin ND, Hickson ID (2004) Phosphorylation of the Bloom's syndrome helicase and its role in recovery from S-phase arrest. Mol Cell Biol 24:1279–1291.

Ellis NA, Groden J, Ye T-Z, Straughen J, Lennon DJ, Ciocci S, Proytcheva M, German J (1995) The Bloom's syndrome gene product is homologous to RecQ helicases. Cell 83:655–666.

Futami K, Kumagai E, Makino H, Goto H, Takagi M, Shimamoto A, Furuichi Y. (2008) Induction of mitotic cell death in cancer cells by small interference RNA suppressing the expression of RecQL1 helicase. Cancer Sci 99:71–80.

German J (1964) Cytological evidence for crossing-over in vitro in human lymphoid cells. Science 144:298–301.

German J (1993) Bloom syndrome: a Mendelian prototype of somatic mutational disease. Medicine 72:393–406.

German J (1996) Bloom's syndrome. XX. The first 100 cancers. Cancer Genet Cytogenet 93:100–106.

German J, Archibald R, Bloom D (1965) Chromosomal breakage in a rare and probably genetically determined syndrome of man. Science 148:506–507.

German J, Ellis NA (1998) Bloom syndrome. In: Vogelstein B, Kinzler KW, eds. The Genetic Basis of Human Cancer. McGraw Hill, New York, pp 301–309.

German J, Sanz MM, Ciocci S, Ye TZ, Ellis NA (2007) Syndrome-causing mutations of the *BLM* gene in persons in the Bloom's syndrome registry. Hum Mutat 28:743–753.

Goss KH, Risinger MA, Kordich JJ, Sanz MM, Straughen JE, Slovek LE, Capobianco AJ, German J, Boivin GP, Groden J (2002) Enhanced tumor formation in mice heterozygous for *Blm* mutation. Science 297:2051–2053.

Gruber SB, Ellis NA, Scott KK, Almg R, Kolchana P, Bonner JD, Kirchhoff T, Nafa K Pierce H, Low M, Stagopan J, Rennert H, Huang H, Greenson JK, Groden J, Rapaport B, Shia J, Johnson S, Gregersen PK, Harrris CC, Boyd J, Renert G, Offit K (2002) *BLM* heterozygosity and the risk of colorectal cancer. Science 297:2013.

Hand R, German J (1975) A retarded rate of DNA chain growth in Bloom's syndrome. Proc Natl Acad Sci U S A 72:758–762.

Hickson ID (2003) RecQ helicases: caretakers of the genome. Nat Rev 3:169–178.

Hu Y, Raynard S, Sehorn MG, Lu X, Bussen W, Zheng L, Stark JM, Barnes EL, Chi P, Janscak P, Jasin M, Vogel H, Sung P, Luo G (2007) RECQL5/Recql5 helicase regulates homologous recombination and suppresses tumor formation via disruption of Rad51 presynaptic filaments. Genes Dev 21:3073–3084.

Karow JK, Chakraverty RK, Hickson ID (1997) The Bloom's syndrome gene product is a 3'–5' DNA helicase. J Biol Chem 272:30611–30614.

Karow JK, Constantinou A, Li JL, West SC, Hickson ID (2000a) The Bloom's syndrome gene product promotes branch migration of Holliday junctions. Proc Natl Acad Sci USA 97:6504–6508.

Karow JK, Wu L, Hickson ID (2000b) DNA helicases: roles in cancer and aging. Curr Opin Genet Dev 10:2–8.

Langland G, Elliot J, Li Y, Creaney J, Dixon K, Groden J (2002) The BLM helicase is necessary for normal DNA double-strand break repair. Cancer Res 62:2766–2770.

Lönn U, Lönn S, Nylen U, Winblad G, German J (1990) An abnormal profile of DNA replication intermediates in Bloom's syndrome. Cancer Res 50:3141–3145.

Lowy AM, Kordich JJ, Gismondi V, Varesco L, Blough RI, Groden J (2001) Numerous colonic adenomas in an individual with Bloom's syndrome. Gastroenterology 121:435–439.

Mohaghegh P, Karow JK, Brosh Jr. RM, Bohr VA, Hickson ID (2001) The Bloom's syndrome and Werner's syndrome proteins are DNA structure-specific helicases. Nucleic Acids Res 29:2843–2849.

Rassool FV, North PS, Mufti GJ, Hickson ID (2003) Constitutive DNA damage is linked to DNA replication abnormalities in Bloom's syndrome cells. Oncogene 22:8749–8757.

Raynard S, Bussen W, Sung P (2006) A double Holliday junction dissolvasome comprising BLM, topoisomerase IIIα, and BLAP75. J Biol Chem 281:13861–13864.

Sharma S and Brosh RM Jr. (2007) Human RECQ1 is a DNA damage responsive protein required for genotoxic stress resistance and suppression of sister chromatid exchanges. PLoS ONE 2:e1297.

Singh TR, Ali AM, Busygina V, Raynard S, Fan Q, Du C, Andreassen PR, Sung P, Meetei AR (2008) BLAP18/RM12, a novel OB-fold-containing protein, is an essential component of the Bloom helicase-double Holliday junction dissolvasome. Genes Dev 22:2856–2868.

Straughen JE, Johnson J, McLaren D, Proytcheva M, Ellis N, German J, Groden J (1998) A rapid method for detecting the predominant Ashkenazi Jewish mutation in the Bloom's syndrome gene. Hum Mutat 11:175–178.

van Brabant AJ, Ye T, Sanz M, German J, Ellis NA, Holloman WK (2000) Binding and melting of D-loops by the Bloom syndrome helicase. Biochemistry 39:14617–14625.

Wu L, Hickson ID (2003) The Bloom's syndrome helicase suppresses crossing over during homologous recombination. Nature 426:870–874.

Yankiwski V, Marcinia RA, Guarente L, Neff NF (2000) Nuclear structure in normal and Bloom syndrome cells. Proc Natl Acad Sci U S A 97:5214–5219.

Zauber NP, Sabbath-Solitare M, Marotta S, Zauber A, Foulkes W, Chan M, Turner F, Bishop DT (2005) Clinical and genetic findings in an Ashkenazi Jewish population with colorectal neoplasms. Cancer 104:719–729.

Chapter 8.7
The Role of *p53* in Colorectal Cancer

Serena Masciari and Sapna Syngal

The hereditary colorectal cancer syndromes have been the focus of intense molecular and clinical investigations aimed at formulating models of tumorigenesis and optimizing the diagnosis, management, and genetic counseling of affected families. However, despite the vastly increasing amount of knowledge regarding the genetic basis of inherited colorectal cancer in the past 15 years, there remains a substantial portion of the disease where a genetic basis cannot be identified. Although many of these may be related to environmental or dietary causes and others may reflect an interaction between low- or intermediate-penetrance genes with environmental factors, additional high-penetrance genes may also be responsible for some cases, particularly those diagnosed at unusually young ages or associated with a family history of other cancers. *p53* is one of these high-penetrance genes that underlies the hereditary cancer syndrome known as Li–Fraumeni Syndrome (LFS), the mutation of which confers a predisposition to a variety of tumors including colorectal cancer at early age. Although initial studies focused on the classic tumors associated with LFS, subsequent reports suggested that germline *p53* mutation carriers might have an increased susceptibility to a much broader range of neoplasms (Garber et al. 1991; Birch et al. 1994, 1998; Varley et al. 1995; Kleihues et al. 1997; Hisada et al. 1998; Nichols et al. 2001). These include carcinomas of the colon, lung, stomach, pancreas, ovary, and lymphomas.

LFS is a rare familial cancer syndrome in which cancer susceptibility is dominantly inherited (Li and Fraumeni 1969). LFS is characterized by the occurrence of several cancers at remarkably early ages. The classic syndrome (Table 8.7.1) includes a number of specific tumor types: soft tissue sarcomas and osteosarcomas, brain tumors, adrenocortical carcinoma, leukemias, and breast cancer. In 70% of families with classic LFS and 30% of Li–Fraumeni-Like (LFL) families (more relaxed criteria) (Table 8.7.1), a germline mutation in the *p53* gene can be identified (Kleihues et al. 1997).

S. Masciari and S. Syngal (✉)
Familial Gastrointestinal Cancer Program, Dana-Farber Cancer Institute/Brigham and Women's Cancer Center
e-mail: ssyngal@partners.org

J.D. Potter and N.M Lindor (eds.), *Genetics of Colorectal Cancer*,
DOI: 10.1007/978-0-387-09568-4_15,

Table 8.7.1 Li–Fraumeni Syndrome (LFS) and Li–Fraumeni-like (LFL) criteria

Li–Fraumeni classic criteria

- Proband diagnosed with sarcoma before 45 years of age, and
- A first-degree relative with any cancer before 45 years of age, and
- Another first-or second-degree relative in the lineage with any cancer before age 45 years or sarcoma at any age

Li–Fraumeni-like (LFL) criteria

Birch's Definition:

- Proband with any childhood cancer or sarcoma, brain tumor, or adrenal cortical tumor before 45 years of age, and
- First- or second-degree relative with a typical LFS tumor (sarcoma, brain tumor, breast cancer, adrenal cortical tumor or leukemia) at any age, and
- First- or second-degree relative with any cancer before 60 years of age

Eeles' Definition:

- Two first- or second-degree relatives with any LFS-related malignancies at any age

The *p53* gene, first identified in 1979, is located on chromosome 17p (Malkin et al. 1990) and encodes for a 53-kDa nuclear phosphoprotein that binds DNA sequences and functions as a negative regulator of cell growth and proliferation in the setting of DNA damage. Often considered as the "guardian of the genome" (Lane 1992), the *p53* protein recognizes damaged cells and functions as a "checkpoint" by delaying the progression of the cell cycle so that damaged DNA can be repaired or apoptosis (programmed cell death) can be ensured (Fisher 2001). It accomplishes these tasks either by: repairing the DNA via the transcriptional activation of the downstream genes (*p21*, *MDM-2*, *GADD45*, Bax, *IGF-BP*, and cyclin-G); or directly signaling a "sensor" molecule that confirms the DNA damage and proceeds with apoptosis. *p53* not only mediates the proper activation of the *RB* pathway (Levine 1997), which is essential to arresting the cell cycle, but also may directly aid in the DNA repair process (Varley et al. 1997). Inactivation of the *p53* gene or disruption of the *p53* protein product can determine the persistence of damaged DNA and the possible development of malignant cells. Most of the germline *p53* mutations are missense mutations involving the binding domain of *p53* and are localized between exon 4 through exon 9. Although *p53* germline deletions are very rare, they do need to be considered in patients with clear LFS features in the absence of detectable missense mutations (Bougeard et al. 2003; Walsh et al. 2006).

LFS is a rare syndrome, with estimates of the frequency of germline *p53* mutations in the range of 1:8,000 in the general population (Nichols et al. 2001), or one-tenth the frequency of mutated germline *BRCA1* and *BRCA2* mutations. In LFS, the risk of developing cancer is 50% by age 30 and 90% by age 70 years. The rate of multiple primary cancers is also markedly elevated in LFS individuals who survive a first cancer diagnosis (Hwang et al. 2003).

Analysis of 45 LFS families and 140 other affected cases within the literature performed by Nichols et al. showed that carriers of a *p53* mutation had significantly earlier age of diagnosis (median age: 33 years) of colorectal cancer (CRC) than the general population (median age: 72 years) (Nichols et al. 2001). This unusually early age of presentation is characteristic of hereditary cancers, and suggested that CRC, among other neoplasms, may also be associated with LFS. The prevalence of early-onset colon cancer, defined as CRC diagnosed at or below age 50, was subsequently evaluated in 397 patients from 64 families with LFS (Wong et al. 2006) who are part of the Dana-Farber Cancer Institute LFS Family registry assembled by Dr. Frederick Li and Joseph Fraumeni. The goal of this analysis was to determine whether CRC is associated with LFS and, therefore, to determine if LFS should be considered in patients with early-onset CRC. Of the total families, 12.5% had individuals with a germline *p53* mutation and CRC diagnosed at age less than 50 years. The mean age at diagnosis was 33 years with a median age of 41 years (range: 9–50 years). From this group, three patients developed colon cancer before age 20 (27% of 11 patients with early-onset CRC) and one patient (9.1%) between age 20 and 34 (Table 8.7.2). The results of this study demonstrated a high rate of CRC in LFS families, often occurring at very young ages (less than age 20). These results are in contrast with the incidence rate of CRC in the general population from the Surveillance, Epidemiology, and End Results (SEER) database, which showed that 0.2% of all colorectal cancers were diagnosed before age 20; 2.2% between ages 20 and 34; 7.6% between ages 35 and 44; and 22.1% between ages 45 and 54.

Table 8.7.2 Classic LFS patients with early-onset colorectal cancer – age of diagnosis, method of conformation, and pathology report. Table data reprinted with permission from Gastroenterology (Wong et al. 2006)

Patient	Age dx	Method of confirmation	Tumor type	Location	Grade	Lymph nodes	Metastases
1	9	Pathology	AdenoCa	L colon	Mod well diff	No	Omentum, peritoneum
2	11	Pathology	AdenoCa	Transverse colon	No report	Yes	Lungs, liver, adrenal, thymus
3	15	Death certificate					
4	20	Pathology	AdenoCa	R colon	Mod undiff	Yes	No
5	41	Pathology	AdenoCa	L colon	Well diff	Yes	Omentum, liver, peritoneum
6	41	Pathology	AdenoCa	L colon, Rectum	Mod diff	No	No
7	41	Verbal report					
8a	41	Pathology	AdenoCa	L colon, rectum	Well diff	No	No
9	43	Verbal report					
10	49	Pathology	AdenoCa	L colon	Mod diff	No	Mesocolon
8b	50	Pathology	AdenoCa	Rectum	Well diff	No	No

AdenoCa adenocarcinoma; *mod diff* moderately differentiated; *undiff* undifferentiated

An additional study by Olivier et al. also confirmed the early age of onset of CRC among individuals in another Li–Fraumeni database (Olivier et al. 2003). The frequency of CRC in *p53*-confirmed or obligate carriers with a family history of LFS or LFL was 1.8% with a median age of onset of 34 years and a slightly higher incidence (57%) in male *p53* carriers versus sporadic CRC cases (50%).

The findings of early-onset CRC in *p53* carriers, if confirmed, will result in the inclusion of LFS in risk assessment models and genetic counseling as well as the consideration of LFS as a possible alternative etiology of early-onset CRC when the other hereditary conditions (Lynch Syndrome, FAP) have been excluded (Lynch and de la Chapelle 1999, 2003).

Additional studies are required to confirm these preliminary findings about the role of germline *p53* in inherited CRC, including further assessment of the prevalence of germline *p53* mutations in individuals with young-onset CRC (<40 years); determination of the phenotypic characteristics of CRC associated with germline *p53* mutations compared with non-familial cases; and determination of genotype-phenotype correlations with CRC phenotype and other associated tumors.

References

Birch, J. M., A. L. Hartley, et al. (1994). "Prevalence and diversity of constitutional mutations in the p53 gene among 21 Li–Fraumeni families." *Cancer Res* **54**(5): 1298–304.

Birch, J. M., V. Blair, et al. (1998). "Cancer phenotype correlates with constitutional TP53 genotype in families with the Li–Fraumeni syndrome." *Oncogene* **17**(9): 1061–8.

Bougeard, G., L. Brugieres, et al. (2003). "Screening for TP53 rearrangements in families with the Li–Fraumeni syndrome reveals a complete deletion of the TP53 gene." *Oncogene* **22**(6): 840–6.

Fisher, D. E. (2001). "The p53 tumor suppressor: critical regulator of life & death in cancer." *Apoptosis* **6**(1–2): 7–15.

Garber, J. E., A. M. Goldstein, et al. (1991). "Follow-up study of twenty-four families with Li–Fraumeni syndrome." *Cancer Res* **51**(22): 6094–7.

Hisada, M., J. E. Garber, et al. (1998). "Multiple primary cancers in families with Li–Fraumeni syndrome." *J Natl Cancer Inst* **90**(8): 606–11.

Hwang, S. J., G. Lozano, et al. (2003). "Germline p53 mutations in a cohort with childhood sarcoma: sex differences in cancer risk." *Am J Hum Genet* **72**(4): 975–83.

Kleihues, P., B. Schauble, et al. (1997). "Tumors associated with p53 germline mutations: a synopsis of 91 families." *Am J Pathol* **150**(1): 1–13.

Lane, D. P. (1992). "Cancer. p53, guardian of the genome." *Nature* 358(6381): 15–6.

Levine, A. J. (1997). "p53, the cellular gatekeeper for growth and division." *Cell* **88**(3): 323–31.

Li, F. P. and J. F. Fraumeni, Jr. (1969). "Soft-tissue sarcomas, breast cancer, and other neoplasms. A familial syndrome?" *Ann Intern Med* **71**(4): 747–52.

Lynch, H. T. and A. de la Chapelle (1999). "Genetic susceptibility to non-polyposis colorectal cancer." *J Med Genet* **36**(11): 801–18.

Lynch, H. T. and A. de la Chapelle (2003). "Hereditary colorectal cancer." *N Engl J Med* **348**(10): 919–32.

Malkin, D., F. P. Li, et al. (1990). "Germ line p53 mutations in a familial syndrome of breast cancer, sarcomas, and other neoplasms." *Science* **250**(4985): 1233–8.

Nichols, K. E., D. Malkin, et al. (2001). "Germ-line p53 mutations predispose to a wide spectrum of early-onset cancers." *Cancer Epidemiol Biomarkers Prev* **10**(2): 83–7.

Olivier, M., D. E. Goldgar, et al. (2003). "Li–Fraumeni and related syndromes: correlation between tumor type, family structure, and TP53 genotype." *Cancer Res* **63**(20): 6643–50.

Varley, J. M., G. McGown, et al. (1995). "An extended Li–Fraumeni kindred with gastric carcinoma and a codon 175 mutation in TP53." *J Med Genet* **32**(12): 942–5.

Varley, J. M., D. G. Evans, et al. (1997). "Li–Fraumeni syndrome – a molecular and clinical review." *Br J Cancer* **76**(1): 1–14.

Walsh, T., S. Casadei, et al. (2006). "Spectrum of mutations in BRCA1, BRCA2, CHEK2, and TP53 in families at high risk of breast cancer." *JAMA* **295**(12): 1379–88.

Wong, P., S. J. Verselis, et al. (2006). "Prevalence of early onset colorectal cancer in 397 patients with classic Li–Fraumeni syndrome." *Gastroenterology* **130**(1): 73–9.

Chapter 8.8
Chromosomes 8q24 and 9p24: Associations with Colorectal Cancer

John D. Potter

At least five studies have recently identified single nucleotide polymorphism variants associated with prostate cancer risk at chromosome 8q24 (Amundadottir et al. 2006; Freedman et al. 2006; Gudmundsson et al. 2007; Haiman et al. 2007b; Yeager et al. 2007). A comparable series of studies has now identified variants in the same region, including rs6983267 and rs10505477, as being associated with colorectal cancer risk (Gruber et al. 2007; Haiman et al. 2007a; Poynter et al. 2007; Tomlinson et al. 2007; Zanke et al. 2007), although the mode of inheritance has been inconsistent across studies: two observed a log-linear per-allele effect (Tomlinson et al. 2007; Zanke et al. 2007), one reported a statistically significant per-allele association and a nonmultiplicative risk (Haiman et al. 2007a), one observed a dominant effect for the association (Gruber et al. 2007), the fifth found a deviation from additivity in a dominant model (Poynter et al. 2007). There is some inconsistency in whether heterogeneity exists by age or family history (Gruber et al. 2007; Haiman et al. 2007a; Poynter et al. 2007; Tomlinson et al. 2007). Microsatellite instability has been included in two studies: one showed no heterogeneity (Tomlinson et al. 2007); the other suggested a stronger association for MSI-H cases (Poynter et al. 2007). The Poynter et al. study observed an association in the population-based, but not the clinic-based, families between colorectal cancer risk and rs10505477 (OR = 1.38, 95% CI: 1.09–1.75 for heterozygous carriers of the T allele, OR = 1.15, 95% CI: 0.85–1.55 for homozygous carriers of the T allele $p = 0.005$), with, thus, no evidence of a per-allele association (Poynter et al. 2007).

Zanke et al. (2007) and Poynter et al. (2007) have also identified another susceptibility allele for colorectal cancer at 9p24 (rs719725). In the Poynter et al. study, the OR per A allele was 1.21 (95% CI: 1.03–1.42, $p = 0.02$), with no heterogeneity by MSI status, age, or family history. The Poynter et al. findings also suggested that there was no statistical interaction between variants at the 8q24 and the 9p24 loci (Poynter et al. 2007).

J.D. Potter,
Cancer Prevention Research Program, Division of Public Health Sciences, Fred Hutchinson Cancer Research Center
e-mail: jpotter@fhcrc.org

J.D. Potter and N.M Lindor (eds.), *Genetics of Colorectal Cancer*,
DOI: 10.1007/978-0-387-09568-4_16,

The 8q24 locus is not in a known gene but is 400- to 500-kb telomeric of the *MYC* oncogene, which has a known role in colon cancer biology. Zanke et al. failed to show any difference in immunohistochemical expression by 8q24 genotype (Zanke et al. 2007). The 9p24 locus is also in a gene desert; the gene most proximal (37-kb telomeric) is protein kinase NYD-SP25 isoform 3 (TPD52L3) (Boutros et al. 2004). Two other neighboring genes include IL33 (124-kb telomeric) (Schmitz et al. 2005) and ubiquitin-like PHD and RING finger domain-containing protein (UHRF2, 47-kb centromeric) (Li et al. 2004). None of these has an established relationship to colorectal cancer risk.

The associations between these loci and colorectal cancer are weak. The consistency of the finding, especially for 8q24, and the relatively high frequency of the variants in the population, nonetheless, mean that the importance to overall disease burden is high.

References

Amundadottir, L. T., P. Sulem, J. Gudmundsson, et al. (2006). A common variant associated with prostate cancer in European and African populations. *Nat Genet* 38(6): 652–8.

Boutros, R., S. Fanayan, M. Shehata, et al. (2004). The tumor protein D52 family: many pieces, many puzzles. *Biochem Biophys Res Commun* 325(4): 1115–21.

Freedman, M. L., C. A. Haiman, N. Patterson, et al. (2006). Admixture mapping identifies 8q24 as a prostate cancer risk locus in African-American men. *Proc Natl Acad Sci U S A* 103(38): 14068–73.

Gruber, S. B., V. Moreno, L. S. Rozek, et al. (2007). Genetic variation in 8q24 associated with risk of colorectal cancer. *Cancer Biol Ther* 6(7): 1143–1147.

Gudmundsson, J., P. Sulem, A. Manolescu, et al. (2007). Genome-wide association study identifies a second prostate cancer susceptibility variant at 8q24. *Nat Genet* 39(5): 631–7.

Haiman, C. A., L. Le Marchand, J. Yamamato, et al. (2007a). A common genetic risk factor for colorectal and prostate cancer. *Nat Genet* 39(8): 954–6.

Haiman, C. A., N. Patterson, M. L. Freedman, et al. (2007b). Multiple regions within 8q24 independently affect risk for prostate cancer. *Nat Genet* 39(5): 638–44.

Li, Y., T. Mori, H. Hata, et al. (2004). NIRF induces G1 arrest and associates with Cdk2. *Biochem Biophys Res Commun* 319(2): 464–8.

Poynter, J. N., J. C. Figueiredo, D. V. Conti, et al. (2007). Variants on 9p24 and 8q24 are associated with risk of colorectal cancer: results from the Colon Cancer Family Registry. *Cancer Res* 67(23): 11128–32.

Schmitz, J., A. Owyang, E. Oldham, et al. (2005). IL-33, an interleukin-1-like cytokine that signals via the IL-1 receptor-related protein ST2 and induces T helper type 2-associated cytokines. *Immunity* 23(5): 479–90.

Tomlinson, I., E. Webb, L. Carvajal-Carmona, et al. (2007). A genome-wide association scan of tag SNPs identifies a susceptibility variant for colorectal cancer at 8q24.21. *Nat Genet* 39(8): 984–8.

Yeager, M., N. Orr, R. B. Hayes, et al. (2007). Genome-wide association study of prostate cancer identifies a second risk locus at 8q24. *Nat Genet* 39(5): 645–9.

Zanke, B. W., C. M. T. Greenwood, J. Rangrej, et al. (2007). Genome-wide association scan identifies a colorectal cancer susceptibility locus on chromosome 8q24. *Nat Genet* 39(8): 989–4.

Section IV
Germline Susceptibility – Gene-Environment Interactions

Chapter 9
Genetic Variability in Folate-Mediated One-Carbon Metabolism and Risk of Colorectal Neoplasia

Amy Y. Liu and Cornelia M. Ulrich

Introduction

Folate is a necessary micronutrient in humans, essential for transferring single-carbon units for important biochemical reactions such as the biosynthesis of methionine, thymidylate, purines, and glycine, and in the metabolism of serine, formate, and histidine. The principal reactions of folate-mediated one-carbon metabolism (FOCM) in the cytosol and the major transport mechanisms of folate into the cell are illustrated in Fig. 9.1.

Experiments in animal models and epidemiologic studies investigating dietary intakes or biomarkers of folate status have provided evidence linking FOCM to colorectal cancer risk. It has been shown in animal studies that a methyl-group-deficient diet can enhance colon carcinogenesis, with potentially different effects depending on the transformation state of the cell (Kim 2004). The results from epidemiologic studies on both colorectal adenomas and colorectal cancer show a strong inverse association between folate status and colorectal neoplasia: either high dietary folate intakes or high biomarkers of folate intake are consistently associated with a decreased risk (Benito et al. 1990; Freudenheim et al. 1991; Giovannucci et al. 1993, 1995; Boutron-Ruault et al. 1996; Glynn et al. 1996; Slattery et al. 1997; Kato et al. 1999; Giovannucci 2002; Konings et al. 2002).

These associations are not explained merely by some other truly causal factors correlating with a high folate intake or blood level; rather, genetic studies provide good evidence of a causal relationship between aspects of FOCM and an altered risk of cancer. In order for FOCM to function, several nutrients, such as vitamin B12, B6, and B2, are necessary as cofactors of various enzymes. In addition to critical cofactors, many feedback mechanisms and other regulatory processes ensure the robustness of the complex metabolic pathways that constitute folate metabolism (Wagner 1995; Nijhout et al. 2004). Consequently, for phenotypic effects to occur, multiple disturbances within the pathway, or stress on the system as a whole (e.g., low intakes of

C.M. Ulrich (✉)
Division of Public Health Sciences, Fred Hutchinson Cancer Research Center
email: nulrich@fhcrc.org

J.D. Potter and N.M Lindor (eds.), *Genetics of Colorectal Cancer*,
DOI: 10.1007/978-0-387-09568-4_17,

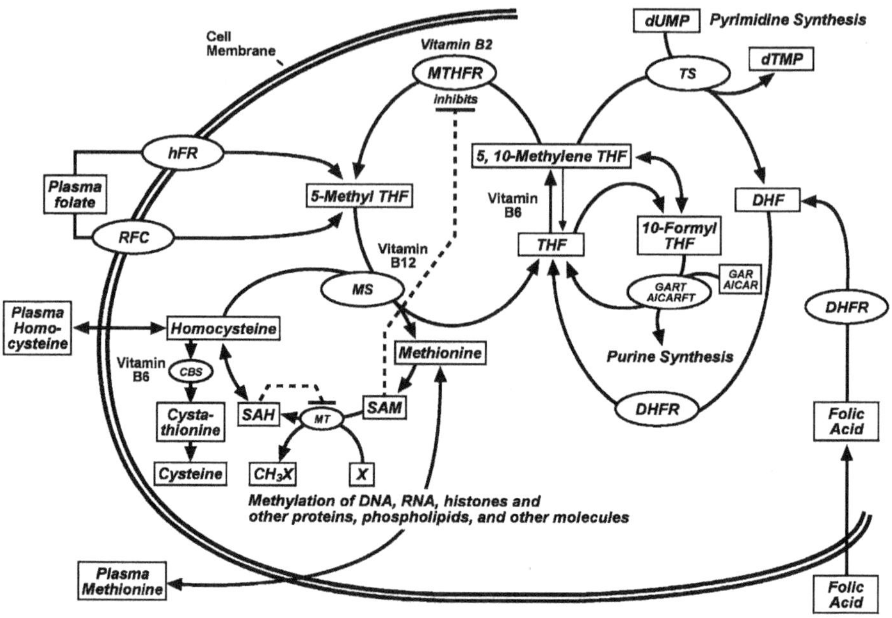

Fig. 9.1 Overview of folate-mediated one-carbon metabolism (simplified), links to methylation reactions and nucleotide synthesis (modified with permission from (Ulrich et al. 2003)). *THF* tetrahydrofolate; *DHF* dihydrofolate; *RFC* reduced folate carrier; *hFR* human folate receptor; *MTHFR* 5,10-methylenetetrahydrofolate reductase; *DHFR* dihydrofolate reductase; *GART* glycinamide ribonucleotide transformylase; *AICARFT* 5-amino-imidazole-4-carboxamide ribonucleotide transformylase; *AICAR* 5-aminoimidazole-4-carboxamide ribonucleotide; *GAR* glycinamide ribonucleotide; *SAM (AdoMet) S*-adenosylmethionine; *SAH (AdoHcy) S*-adenosylhomocysteine; *dUMP* deoxyuridine monophosphate; *dTMP* deoxythymidine monophosphate; *MS* methionine synthase; *TS* thymidylate synthase; *MT* methyltransferases; *X* a variety of substrates for methylation

folate or other nutrients involved in FOCM), are required. For example, the association between genetic polymorphisms in *MTHFR* and biomarkers, such as homocysteine concentrations, are strongest under low folate conditions. Thus, examining gene–gene and gene–nutrient interactions within this complex system is a critical step in understanding the relationship between folate metabolism and cancer risk.

Investigating Genetic Polymorphisms in Epidemiologic Studies

A wealth of information regarding inherited human genetic variability has become accessible as part of the Human Genome Project. Studies investigating the associations between genetic susceptibility and disease are important because they can (a) provide evidence for a causal relationship between an environmental factor (e.g., folate) and disease outcomes; (b) help to elucidate whether increasing or reducing intakes of specific nutrients may benefit certain people or population groups depending on genetic susceptibility; and (c) assist in discovering biologic mechanisms connecting specific biologic pathways to disease outcomes.

Early epidemiologic studies focused mainly on specific candidate polymorphisms in proteins that appear essential in a biologic pathway (for the discussion here, folate metabolism) or are functionally relevant. However, this approach is limited because it omits either other genetic variants within the same biologic pathway or other genetic polymorphisms within the same gene. Consequently, performing comprehensive investigations covering genetic variability in numerous biologically interrelated proteins will become more and more critical in future studies. However, only if suitable statistical methods are available for pathway-based analyses and if sample sizes are sufficient to provide stable estimates for gene–gene and gene–environment interactions can a full integration of biologic relationships into the epidemiologic data analysis be achieved. In this chapter, many of the studies discussed were limited by sample size; therefore, studies in other populations are needed in order to confirm the reported gene–gene or gene–nutrient interactions. Furthermore, by utilizing a candidate polymorphism approach, as noted earlier, additional genetic variation within the same gene or in adjacent genomic regions may be overlooked. The candidate polymorphism may not be the causal variant; rather, another polymorphism in linkage disequilibrium (LD) with the candidate may be causing the observed associations. In order to address these concerns, investigations of gene-wide haplotypes or combination of haplotypes (i.e., diplotypes) (as well as methods development) are underway for integrating these approaches in epidemiologic analyses (Johnson et al. 2001; Patil et al. 2001; Dawson et al. 2002; Gabriel et al. 2002).

We here focus on the studies to date of those polymorphisms that are expected to have phenotypic effects (as indicated by biomarker measurements) or have been associated with disease endpoints.

Polymorphisms in One-Carbon Metabolism and Their Functional Impact

Thymidylate Synthase

5,10-Methylenetetrahydrofolate (5,10-methylene-THF) is a crucial substrate in folate metabolism with connections to three biosynthetic pathways: thymidine synthesis, purine synthesis, or methionine synthesis via 5,10-methylenetetrahydrofolate reductase (MTHFR). Thymidylate synthase (TS) catalyzes the transfer of a methyl group from 5,10-methylene-THF to deoxyuridine monophosphate, creating deoxythymidine monophosphate and dihydrofolate. Additionally, TS is an important drug target for chemotherapeutic agents (Ulrich et al. 2002b, 2003). A polymorphic 28-bp tandem repeat is located in the 5′-UTR of the *TS* gene and functions as a cis-acting transcriptional enhancer element (Kaneda et al. 1987). Two and three repeats occur most commonly; however, among populations of African descent, the rarer four and nine tandem repeats have been observed (Marsh et al. 2000). Individuals with the triple repeat have approximately 2–4 times greater gene expression than individuals with the double repeat (Horie et al. 1995; Pullarkat et al. 2001). The 3R/3R genotype has been found to be associated with reduced plasma folate and, furthermore, with increased plasma homocysteine concentrations among individuals with low folate intake (Trinh et al. 2002). Additionally, within the second repeat of 3R alleles, a G > C polymorphism has been identified that alters the transcriptional activation of *TS* gene (Mandola et al. 2003). Within the 3′-UTR of the *TS* gene is a third functionally relevant polymorphism – a 6-bp deletion (1494del6) which has been associated with reduced mRNA stability (Ulrich et al. 2000; Mandola et al. 2004).

5,10-Methylenetetrahydrofolate Reductase

5,10-Methylenetetrahydrofolate reductase (MTHFR) catalyzes 5,10-methylene-THF to 5-methyltetrahydrofolate (5-methyl-THF). MTHFR plays a critical role in FOCM because it balances the DNA methylation and DNA synthesis pathways. Two common polymorphisms have been investigated extensively within *MTHFR*. First, *MTHFR* C677T was identified as reducing MTHFR activity, with the thermolabile variant 677TT reported to decrease normal enzyme activity by as much as 70% (Frosst et al. 1995). Epidemiologic studies have consistently observed that the 677TT genotype is associated with higher plasma homocysteine concentrations than wild type, with this relationship being strongest under a low folate status (Jacques et al. 1996; Girelli et al. 1998). Low levels of vitamin B2 lead to higher homocysteine concentrations as well, but only in 677TT individuals (Hustad et al. 2000). *MTHFR* C677T provides a classic example of a gene–nutrient interaction with 677TT, 677CT, and 677CC having highest, intermediate, and lowest homocysteine levels, respectively, under low folate levels; yet, in the presence of normal folate status, there were no differences among the different genotypes. Furthermore, it has been observed that under low folate levels, 677TT individuals have decreased levels of genomic DNA methylation (Friso and Choi 2002; Friso et al. 2002a).

The variant genotype of the second common polymorphism, *MTHFR* A1298C, also confers reduced enzyme activity compared to wild type (van der Put et al. 1998). Initial studies reporting on the association between A1298C and homocysteine levels were inconsistent (van der Put et al. 1998; Weisberg et al. 1998; Friedman et al. 1999; Chango et al. 2000a; Friso et al. 2002b): observing no association (Friso et al. 2002b); lower (Friedman et al. 1999); or higher (Chango et al. 2000a; Ulvik et al. 2007) homocysteine levels in individuals with the 1298CC genotype. A recent large-scale study by Ulvik et al. with 10,034 study participants established, nonetheless, that the 1298C allele does have functional impact, independent of the C677T polymorphism: with each additional 1298C allele, homocysteine increased statistically significantly and serum folate levels decreased (Ulvik et al. 2007). In relation to DNA methylation, studies showed that individuals with the 677TT and 1298AA genotypes, compared to the other genotypes, had reduced genomic DNA methylation in the presence of low folate (Stern et al. 2000; Friso et al. 2002a, 2005). The two most common *MTHFR* variants studied also appear to interact with each other: individuals heterozygous for both polymorphisms had decreased MTHFR enzyme activity, increased homocysteine concentrations, and reduced plasma folate levels (van der Put et al. 1998). The very large study of Ulvik et al. had sufficient statistical power to establish that individuals with the 677TT/1298AA genotype had the lowest serum folate and highest homocysteine concentrations (Ulvik et al. 2007). Homocysteine concentration changes were greatest in the presence of low folate. This result can be explained by the model of MTHFR as an enzyme dimer, in which its six main configurations are sensitive to low folate levels (Ulvik et al. 2007).

Methionine Synthase

Methionine synthase (MTR) catalyzes the methylation of homocysteine to methionine while simultaneously converting 5-methyl-THF to tetrahydrofolate (THF). Studies have identified a variant in the *MTR* gene (A2756G, Asp919Gly) (Leclerc et al. 1998) that may affect plasma homocysteine concentrations. For example, some studies (Harmon et al. 1999; Chen et al. 2001), although not all (van der Put et al. 1997; Ma et al. 1999; Jacques et al. 2003), observed that, across genotypes, homocysteine concentrations tend to decrease linearly, with the AA genotype associated with the highest homocysteine concentrations.

Methionine Synthase Reductase

Methionine synthase reductase (MTRR) is responsible for the reductive methylation of the cobalamin cofactor of MTR (Leclerc et al. 1998). A variety of disorders of folate or cobalamin metabolism have been described in individuals who lack this enzyme activity (Watkins and Rosenblatt 1989). An association between the A66G (Ile22Met) polymorphism and homocysteine concentrations has been reported (Wilson et al. 1999; Gaughan et al. 2001); however, the functional impact of the variant is not well defined (Jacques et al. 2003). Furthermore, investigators have proposed a relationship between the A66G polymorphism and risk of developmental defects (Wilson et al. 1999; O'Leary et al. 2002).

Serine Hydroxymethyltransferase

Serine hydroxymethyltransferase (SHMT), with the cofactor pyridoxal phosphate (vitamin B6), catalyzes the conversion of THF to 5,10-methylene-THF in a reversible reaction. As mentioned previously, 5,10-methylene-THF is a crucial substrate in folate metabolism that links to three biosynthetic pathways: thymidine synthesis, purine synthesis, or methionine synthesis via MTHFR. Within FOCM, one of the critical regulatory mechanisms concerns the synthesis and fate of 5,10-methylene-THF. Cytosolic SHMT (cSHMT) activity increases when glycine concentrations increase; consequently, more 5,10-methylene-THF is committed to serine synthesis and less 5-methyl-THF is produced (Herbig et al. 2002). However, the function of cSHMT in FOCM is not clearly characterized because, in mammals, mitochondrial SHMT also provides single-carbon units for the cytosolic metabolism (Garrow et al. 1993; Wagner 1995; Stover et al. 1997; Liu et al. 2001). Mean red blood cell and plasma folate concentrations were higher in individuals with a T-allele compared to CC homozygotes of the C1420T (Leu474Phe) polymorphism in the *cSHMT* gene (Heil et al. 2001).

Cystathionine β-Synthase

Cystathionine β-synthase (CBS) catalyzes the trans-sulfuration of homocysteine to cystathionine, with a deficiency in this enzyme leading to classical homocystinuria (Mudd et al. 1995). Vitamin B6 is necessary as a cofactor for this reaction, thereby providing a potential motivation for studying gene–nutrient interactions with nutrients other than folate. Within *CBS*, several polymorphisms have been described that are in linkage disequilibrium (a 68-bp insertion in exon 8; C699T; C1080T; and C1985T) (De Stefano et al. 1998; Kraus et al. 1998). These variants may modify homocysteine levels (De Stefano et al. 1998; Kruger et al. 2000), change postmethionine-load homocysteine concentrations (Aras et al. 2000), and influence coronary artery disease risk (Kruger et al. 2000). Furthermore, postmethionine-load homocysteine concentrations may also be affected by a 31-bp variable number tandem repeat that spans the exon12/intron12 boundary (Sebastio et al. 1995; Yang et al. 2000; Lievers et al. 2001).

Reduced Folate Carrier

The reduced folate carrier (RFC) actively transports 5-methyl-THF from the plasma to the cytosol. The polymorphism G80A (Arg27His) in the *RFC* gene may result in better carrier function or higher affinity for folate (Chango et al. 2000b). Chango et al. observed that the variant A-allele was consistently and linearly associated with increasing plasma folate concentrations; however, these differences were not statistically significant (Chango et al. 2000b). Further supporting the idea that differential carrier activity exists among those with the variant allele is the discovery that plasma concentrations of the chemotherapeutic agent, methotrexate, 24–48 h after administration were higher among children with the AA genotype (Laverdiere et al. 2002).

Other Genes

Candidate polymorphisms have been described in virtually all proteins relevant to FOCM. Thus far, there have not been any epidemiologic studies on colorectal cancer risk and polymorphisms in enzymes responsible for polyglutamation or cleavage of γ-glutamyl groups, or other key enzymes, such as those involved in transcobalamin transport. Some of these candidate polymorphisms are briefly described here.

Polyglutamation of folate molecules via folylpolyglutamyl synthase (FPGS) and cleavage of these glutamyl groups by γ-glutamyl hydrolase (GGH) are implicated in the regulation of intracellular folate concentrations. Several variants have been reported in the *GGH* gene: C-401T, G-354T, T-124G, and C452T (Chave et al. 2003; Cheng et al. 2004). The polymorphism, C452CT in exon 5, is associated with GGH activity and reduces GGH hydrolysis of long-chain methotrexate polyglutamates in leukemia patients treated with high-dose methotrexate (Cheng et al. 2004).

In order for folic acid to enter the FOCM pathway, dihydrofolate reductase (DHFR) is vital. Several polymorphisms reported among Japanese (Goto et al. 2001) have not been seen by our group in a North American population (unpublished results). A 19-bp deletion polymorphism within intron1 of *DHFR* may be associated with an increased risk of spina bifida (Johnson et al. 2004).

Transcobalamin II (TCII) is a serum protein that transports vitamin B12 to tissues. As shown in Fig. 9.1, the conversion of homocysteine to methionine by MTR requires vitamin B12 as a cofactor. Variants, C776G, A67G, G280A, C1043T, and G1196A, have been reported in *TCII* (Afman et al. 2001, 2002; Lievers et al. 2002). The most common *TCII* polymorphism is C776G, but studies investigating this polymorphism and homocysteine and vitamin B12 concentrations have been inconsistent (Afman et al. 2001, 2002; Lievers et al. 2002; Miller et al. 2002; Zetterberg et al. 2002, 2003; Fodinger et al. 2003; Geisel et al. 2003; Wans et al. 2003; Anello et al. 2004; Winkelmayer et al. 2004). However, two studies have found that among those with the 776GG polymorphism, methylmalonic acid is higher (Miller et al. 2002; Geisel et al. 2003). The other variants of *TCII*, A67G, G280A, C1043T, and G1196A, may be associated with homocysteine and vitamin B12 concentrations (Afman et al. 2002; Lievers et al. 2002). Given the numerous polymorphisms in *TCII*, evaluating the genetic variability within this gene in a comprehensive manner (e.g., by haplotype or diplotype analyses) is essential.

Polymorphisms in One-Carbon Metabolism and Colorectal Cancer Risk

Because colorectal cancer is common (Sandler et al. 2002), numerous research groups have conducted large case-control and intermediate-size prospective cohort studies. Although the case-control studies were usually much larger and thereby provided more statistical power to assess gene–gene or gene–nutrient interactions, such studies rely on questionnaire data to evaluate folate status because biomarkers, such as serum folate, may be altered by the presence of a tumor. Prospective cohort studies are not similarly limited. Colorectal adenoma, an established precursor of colorectal cancer, has also been examined in both case-control and prospective studies (Winawer et al. 1993); studies largely based on screening by colonoscopy or sigmoidoscopy. However, as sigmoidoscopy does not detect polyps in the proximal colon, this type of screening can lead to misclassification of individuals as polyp-free. Studies of colorectal adenoma may be especially pertinent to FOCM because genomic hypomethylation is an early stage in colorectal carcinogenesis, perhaps implicating the folate pathway early in progression (Fearon and Vogelstein 1990). Furthermore, although folate may protect against colon carcinogenesis early in the process, increased intakes appear to promote the growth of existing premalignant lesions (Kim 2004; Cole et al. 2007; Ulrich and Potter 2007) creating additional complexity. Nonetheless, it may thus be conjectured that the inverse association between folate and colon neoplasia will be stronger for adenoma than cancer risk.

5,10-Methylenetetrahydrofolate Reductase

Polymorphisms in *MTHFR* and colorectal cancer risk have been investigated extensively. An initial case-control study discovered that 677TT individuals were at a reduced risk for colorectal cancer under high dietary methyl sources, an association not seen among those consuming alcohol (Chen et al. 1996). Later studies observed the same association when investigating folate status: carriers of the 677TT genotype were at a reduced risk at adequate folate levels; however, under low folate, these individuals were at an increased risk for colorectal cancer (Ma et al. 1997; Le Marchand et al. 2005; Koushik et al. 2006; Huang et al. 2007; Hubner et al. 2007). However, some studies did not find a statistically significant association between the C677T polymorphism and colorectal cancer (Matsuo et al. 2005; Otani et al. 2005). Studies on A1298C suggest that the1298AA homozygotes had an increased colorectal cancer risk (Chen et al. 2006; Huang et al. 2007). A study of colon cancer observed that, among 677TT individuals, high intakes of vitamin B6 and B12, in addition to folate, were associated with lower risk (Slattery et al. 1999). However, a recent study found no association between the 677 T-allele and risk of colon cancer, but did report an increased risk for rectal cancer (Wang et al. 2006). A case-control study investigating C > T mutations within the p53 gene reported that individuals with a 677TT genotype were at a reduced risk for these mutations, but only at CpG sites (Ulrich et al. 2005b). The Physicians' Health Study reported a weak association between the 1298CC genotype and reduced colon cancer risk which was unaltered by folate status (Chen et al. 2002). Wang et al. later verified these findings (Wang et al. 2006), and Keku et al. described this relationship among Caucasians, but not African-Americans (Keku et al. 2002).

Adenoma studies have been inconsistent on the association between the C677T polymorphism and risk: some have reported no associations (van den Donk et al.

2005; Mitrou et al. 2006; Huang et al. 2007), although others have reported a reduced risk (Marugame et al. 2003; Hirose et al. 2005; Hubner et al. 2006b). An initial study observed a reduced risk among individuals with the 677TT genotype and a high plasma folate level and an elevated risk for those with low folate (Marugame et al. 2003). Martinez et al. observed that individuals with both the homozygote variant genotypes, 677TT and 1298CC, were at an increased risk for metachronous adenoma, with a higher risk in the presence of low folate (Martinez et al. 2006). Future studies investigating the gene–nutrient interactions between one or more *MTHFR* polymorphisms and folate are needed to elucidate mechanisms.

Thymidylate Synthase

There have been several published studies investigating polymorphisms in thymidylate synthase and the risk of colorectal neoplasia. Ulrich et al. initially reported on 510 cases and 604 polyp-free controls and observed little association between the *TSER* and *TS* 1494del6 polymorphisms and risk of colorectal adenoma (Ulrich et al. 2002a). Hubner et al. also observed no association between *TSER* and *TS* 1494del6 variants and metachronous adenoma risk (Hubner et al. 2006b). However, in the former study, a statistically significant interaction between the *TSER* genotype and folate intake was observed: among individuals with 3R/3R genotypes (corresponding to higher TS expression), persons taking >440 μg per day of folate (highest tertile) were at a 2-fold decreased risk compared to persons taking ≤440 μg per day; concomitantly, among individuals with 2R/2R genotypes, high folate intake was associated with an 1.5-fold increased risk (Ulrich et al. 2002a). A similar trend was found for vitamin B12. However, a study by Chen et al. (2004a) among 373 sigmoidoscopy-detected cases (thus limited to the distal colon) and 720 controls did not find such interactions. In contrast, they reported a statistically significant *TSER*-alcohol interaction: although individuals with the 2rpt/2rpt genotype were not at an increased risk if they had high alcohol consumption, heterozygotes and those with the 3rpt/3rpt genotype showed an elevated risk.

A prospective study reporting of 270 cases of colorectal cancer and 454 controls discovered that individuals with 2rpt/2rpt genotypes (lower TS expression) were at a reduced risk (Chen et al. 2003). A more recent study of 1,600 cases of colon cancer and 1,962 controls observed that the *TSER* variant was associated with a statistically significantly decreased risk among men, and individuals with both *TS* polymorphisms (*TSER* and *TS* 1496del6) were at a reduced risk, with statistically significant results for women (Ulrich et al. 2005a). Additionally, the *TSER* variant was associated with a reduced colon cancer risk in the presence of both low folate and methionine intakes, which is consistent with previous reports. It has been hypothesized that a greater diversion of 5,10-methylene-THF toward thymidine synthesis may explain why the *MTHFR* 677TT genotype is generally associated with a decreased colorectal cancer risk (Choi and Mason 2000). However, the results for *TS* polymorphisms show that individuals with low TS expression, in conjunction with reduced MTHFR activity, also have a decreased risk of colon cancer (Ulrich et al. 2005a); these findings refute the aforementioned hypothesis and suggest that purine synthesis is a more likely mechanism connecting FOCM to colorectal carcinogenesis (Ulrich et al. 2005a). As discussed earlier, there seem to be at least three genetic polymorphisms within the *TS* gene that influence gene expression or protein stability (Horie et al. 1995; Ulrich et al. 2000; Mandola et al. 2003). As a result, future studies exploring diplotype analyses should genotype all of these genetic variants and ensure that sample sizes are appropriate.

Methionine Synthase

The A2756G (Asp919Gly) polymorphism in *MTR* has been investigated by several groups (Chen et al. 1998; Ma et al. 1999; Le Marchand et al. 2002; Goode et al. 2004; Ulvik et al. 2004). With a minor allele frequency of approximately 0.20, few studies investigating this less common variant had a large enough sample size to explore risks or gene–diet interactions associated with the homozygous variant genotype, which makes up approximately 3% of the Caucasian populations. A statistically significantly reduced risk of colorectal cancer with the variant *MTR* genotype (2756GG) has been reported by both the Physician's Health Study and a very large Norwegian cohort of more than 2,000 case-control pairs (Ma et al. 1999; Ulvik et al. 2004). Conversely, Le Marchand and colleagues did not observe any associations between this polymorphism and the risk of colorectal cancer in 727 cases of Japanese, Caucasian, and Native Hawaiian origin and 727 ethnicity-matched controls. A case-control study by Matsuo et al. also observed no statistically significant association between colorectal cancer risk and A2756G; but, the 2756GG genotype was associated with an increased risk among drinkers. However, a nested case-control study with 140 colorectal patients and 343 controls conducted in China observed that the 2756 G-allele was associated with an increased risk of colorectal cancer, and patients with *MTHFR* 1298AA and either *MTR* 2756AG or *MTR* 2756GG genotypes were at an increased risk (Chen et al. 2006). These results need to be further evaluated with larger sample studies and studies evaluating different ethnic groups. Ulrich et al. did not observe any association between the D919G polymorphism and colon cancer risk (Ulrich et al. 2005a).

There are conflicting findings from studies of colorectal adenoma, showing either a trend toward higher risk (Goode et al. 2004) or decreased risk (Chen et al. 1998). The 2756GG variant genotype may be associated with lower risk only in the presence of low alcohol intake (Ma et al. 1999; Goode et al. 2004) or high methionine intake (Goode et al. 2004). Additionally, several interactions with *MTHFR* or *TS* have been proposed (Le Marchand et al. 2002; Goode et al. 2004). However, in the large Norwegian JANUS cohort (Ulvik et al. 2004), the only biomarker measured was homocysteine, which may not be an ideal biomarker of folate status, as it could be influenced by *MTR* or *CBS* genotypes. Limited statistical power in studies of gene–nutrient or gene–gene interactions may be responsible for these discrepancies. Accordingly, a comprehensive evaluation of the association of the *MTR* variant and the risk of colorectal neoplasia under specific dietary conditions is necessary. Furthermore, the phenotype associated with the *MTR* G2756A polymorphism has not been investigated; it is uncertain if the associations reported here are attributable to the variant itself or another polymorphism with which it is in LD. In order to help answer this question, gene-wide haplotype studies of *MTR*, along with biochemical evaluations, are needed.

Methionine Synthase Reductase

One case-control study of colorectal cancer examining *MTRR* observed an elevated risk among Caucasians, but not among other ethnic groups (Le Marchand et al. 2002). A nested case-control study examining Ser(284)Thr and Art(415)Cys polymorphisms observed that individuals who were carriers of both variants had an increased risk of colorectal cancer (Koushik et al. 2006). Otani et al. did not observe a statistically significant association between the A66G polymorphism and colorectal cancer risk; however, A66G may modify the associations of folate or vitamin B6 with colorectal cancer (Otani et al. 2005). Although a statistically significant interaction was discovered between folate and the A66G polymorphism, there was no linear trend within each stratum. In wild-type *MTRR* individuals, higher vitamin B6 intake was associated with a decreased risk of colorectal cancer. Individuals who were heterozygous for the A66G polymorphism were at a decreased risk for colorectal adenoma recurrence (Hubner et al. 2006a). Additionally, a gene–diet interaction suggested that A66G heterozygotes had a reduced risk for recurrence if they received folate supplement (Hubner et al. 2006a).

Other Genes

Few studies have been conducted on several other FOCM enzymes. There has been only one study investigating colorectal cancer risk and polymorphisms in each of the following enzymes: cSHMT, CBS, RFC, methylenetetrahydrofolate dehydrogenase, and glutamate carboxypeptidase, with no statistically significant associations or gene–gene or gene–diet interactions discovered (Le Marchand et al. 2002; Chen et al. 2004b). However, one study evaluating the Arg(239)Gln variant in betaine-homocysteine methyltransferase (*BHMT*) observed that carriers of this polymorphism were at an increased risk for colorectal cancer (Koushik et al. 2006). These studies were all limited by small sample sizes, especially for evaluating gene–gene or gene–diet interactions.

Several studies have investigated DNA methyltransferases (*DNMT*s), exploring the connection between FOCM and the provision of *S*-adenosylmethionine (SAM), the only human methyl-group donor. In preliminary studies, we evaluated three polymorphisms (C-149T, T-283C, and G-579T) in the promoter region of the DNA methyltransferase-3b (*DNMT3b*) gene and discovered that, among individuals with low folate and low methionine intakes, those with the -149TT genotype were at an increased risk for colorectal adenoma (Jung et al. 2008). Furthermore, results from this study are consistent with previous research (Giovannucci et al. 2003; Tiemersma et al. 2003) suggesting that alcohol dehydrogenase (*ADH*) may also be an important factor in the gene–diet interactions that play a role in colorectal carcinogenesis (Jung et al. 2008). Alcohol inhibits both the absorption of folate and several enzymes in the pathway and, thus, is also an important dietary factor to consider.

Final Thoughts

FOCM is highly pertinent for cancer prevention because this pathway is vital for both nucleotide synthesis and the provision of SAM for methylation reactions. Recent studies have begun to investigate enzymes other than MTHFR. There are numerous genetic polymorphisms, some with strong evidence for phenotypic impact *in vitro*. There is also evidence that these variants may modify cancer risk. Because of the intricacies of the pathway, the array of genetic variability, and the many regulatory mechanisms, this area is in need of a thorough assessment of the associations within epidemiologic studies of sufficient size to consider multiple exposures and polymorphisms using a statistical analytic approach that incorporates biologic information. A tool for integrating biology into statistical analysis may arise from the development of a mathematical model of folate metabolism (Reed et al. 2006). Preliminary results from the model are promising, with predictions matching experimental data, extending our knowledge on the role of FOCM in methylation and in mitochondrial metabolism, and suggesting avenues for the application of this information in statistical analyses (Nijhout et al. 2006a, b; Reed et al. 2006; Ulrich et al. 2006).

References

Afman, L. A., Van Der Put, N. M., Thomas, C. M., Trijbels, J. M., and Blom, H. J. 2001. Reduced vitamin B12 binding by transcobalamin II increases the risk of neural tube defects. QJM 94:159–66.

Afman, L. A., Lievers, K. J., van der Put, N. M., Trijbels, F. J., and Blom, H. J. 2002. Single nucleotide polymorphisms in the transcobalamin gene: relationship with transcobalamin concentrations and risk for neural tube defects. Eur J Hum Genet 10:433–8.

Anello, G., Gueant-Rodriguez, R. M., Bosco, P., Gueant, J. L., Romano, A., Namour, B., Spada, R., Caraci, F., Pourie, G., Daval, J. L., and Ferri, R. 2004. Homocysteine and methylenetetrahydrofolate reductase polymorphism in Alzheimer's disease. Neuroreport 15:859–61.

Aras, O., Hanson, N. Q., Yang, F., and Tsai, M. Y. 2000. Influence of 699C → T and 1080C → T polymorphisms of the cystathionine beta-synthase gene on plasma homocysteine levels. Clin Genet 58:455–9.

Benito, E., Obrador, A., Stiggelbout, A., Bosch, F. X., Mulet, M., Munoz, N., and Kaldor, J. 1990. A population-based case-control study of colorectal cancer in Majorca. I. Dietary factors. Int J Cancer 45:69–76.

Boutron-Ruault, M. C., Senesse, P., Faivre, J., Couillault, C., and Belghiti, C. 1996. Folate and alcohol intakes: related or independent roles in the adenoma-carcinoma sequence? Nutr Cancer 26:337–46.

Chango, A., Boisson, F., Barbe, F., Quilliot, D., Droesch, S., Pfister, M., Fillon-Emery, N., Lambert, D., Fremont, S., Rosenblatt, D. S., and Nicolas, J. P. 2000a. The effect of 677C → T and 1298A → C mutations on plasma homocysteine and 5,10-methylenetetrahydrofolate reductase activity in healthy subjects. Br J Nutr 83:593–6.

Chango, A., Emery-Fillon, N., de Courcy, G. P., Lambert, D., Pfister, M., Rosenblatt, D. S., and Nicolas, J. P. 2000b. A polymorphism (80G → A) in the reduced folate carrier gene and its associations with folate status and homocysteinemia. Mol Genet Metab 70:310–5.

Chave, K. J., Ryan, T. J., Chmura, S. E., and Galivan, J. 2003. Identification of single nucleotide polymorphisms in the human gamma-glutamyl hydrolase gene and characterization of promoter polymorphisms. Gene 319:167–75.

Chen, J., Giovannucci, E., Kelsey, K., Rimm, E. B., Stampfer, M. J., Colditz, G. A., Spiegelman, D., Willett, W. C., and Hunter, D. J. 1996. A methylenetetrahydrofolate reductase polymorphism and the risk of colorectal cancer. Cancer Res 56:4862–4.

Chen, J., Giovannucci, E., Hankinson, S. E., Ma, J., Willett, W. C., Spiegelman, D., Kelsey, K. T., and Hunter, D. J. 1998. A prospective study of methylenetetrahydrofolate reductase and methionine synthase gene polymorphisms, and risk of colorectal adenoma. Carcinogenesis 19:2129–32.

Chen, J., Stampfer, M. J., Ma, J., Selhub, J., Malinow, M. R., Hennekens, C. H., and Hunter, D. J. 2001. Influence of a methionine synthase (D919G) polymorphism on plasma homocysteine and folate levels and relation to risk of myocardial infarction. Atherosclerosis 154:667–72.

Chen, J., Ma, J., Stampfer, M. J., Palomeque, C., Selhub, J., and Hunter, D. J. 2002. Linkage disequilibrium between the 677C > T and 1298A > C polymorphisms in human methylenetetrahydrofolate reductase gene and their contributions to risk of colorectal cancer. Pharmacogenetics 12:339–42.

Chen, J., Hunter, D. J., Stampfer, M. J., Kyte, C., Chan, W., Wetmur, J. G., Mosig, R., Selhub, J., and Ma, J. 2003. Polymorphism in the thymidylate synthase promoter enhancer region modifies the risk and survival of colorectal cancer. Cancer Epidemiol Biomarkers Prev 12:958–62.

Chen, J., Kyte, C., Chan, W., Wetmur, J. G., Fuchs, C. S., and Giovannucci, E. 2004a. Polymorphism in the thymidylate synthase promoter enhancer region and risk of colorectal adenomas. Cancer Epidemiol Biomarkers Prev 13:2247–50.

Chen, J., Kyte, C., Valcin, M., Chan, W., Wetmur, J. G., Selhub, J., Hunter, D. J., and Ma, J. 2004b. Polymorphisms in the one-carbon metabolic pathway, plasma folate levels and colorectal cancer in a prospective study. Int J Cancer 110:617–20.

Chen, K., Song, L., Jin, M. J., Fan, C. H., Jiang, Q. T., and Yu, W. P. 2006. [Association between genetic polymorphisms in folate metabolic enzyme genes and colorectal cancer: a nested case-control study]. Zhonghua Zhong Liu Za Zhi 28:429–32.

Cheng, Q., Wu, B., Kager, L., Panetta, J. C., Zheng, J., Pui, C. H., Relling, M. V., and Evans, W. E. 2004. A substrate specific functional polymorphism of human gamma-glutamyl hydrolase alters catalytic activity and methotrexate polyglutamate accumulation in acute lymphoblastic leukaemia cells. Pharmacogenetics 14:557–67.

Choi, S. W., and Mason, J. B. 2000. Folate and carcinogenesis: an integrated scheme. J Nutr 130:129–32.

Cole, B. F., Baron, J. A., Sandler, R. S., Haile, R. W., Ahnen, D. J., Bresalier, R. S., McKeown-Eyssen, G., Summers, R. W., Rothstein, R. I., Burke, C. A., Snover, D. C., Church, T. R., Allen, J. I., Robertson, D. J., Beck, G. J., Bond, J. H., Byers, T., Mandel, J. S., Mott, L. A., Pearson, L. H., Barry, E. L., Rees, J. R., Marcon, N., Saibil, F., Ueland, P. M., and Greenberg, E. R. 2007. Folic acid for the prevention of colorectal adenomas: a randomized clinical trial. JAMA 297:2351–9.

Dawson, E., Abecasis, G. R., Bumpstead, S., Chen, Y., Hunt, S., Beare, D. M., Pabial, J., Dibling, T., Tinsley, E., Kirby, S., Carter, D., Papaspyridonos, M., Livingstone, S., Ganske, R., Lohmussaar, E., Zernant, J., Tonisson, N., Remm, M., Magi, R., Puurand, T., Vilo, J., Kurg, A., Rice, K., Deloukas, P., Mott, R., Metspalu, A., Bentley, D. R., Cardon, L. R., and Dunham, I. 2002. A first-generation linkage disequilibrium map of human chromosome 22. Nature 418: 544–8.

De Stefano, V., Dekou, V., Nicaud, V., Chasse, J. F., London, J., Stansbie, D., Humphries, S. E., and Gudnason, V. 1998. Linkage disequilibrium at the cystathionine beta synthase (CBS) locus and the association between genetic variation at the CBS locus and plasma levels of homocysteine. The Ears II Group. European Atherosclerosis Research Study. Ann Hum Genet 62:481–90.

Fearon, E. R., and Vogelstein, B. 1990. A genetic model for colorectal tumorigenesis. Cell 61:759–67.

Fodinger, M., Veitl, M., Skoupy, S., Wojcik, J., Rohrer, C., Hagen, W., Puttinger, H., Hauser, A. C., Vychytil, A., and Sunder-Plassmann, G. 2003. Effect of TCN2 776C > G on vitamin B12 cellular availability in end-stage renal disease patients. Kidney Int 64:1095–100.

Freudenheim, J. L., Graham, S., Marshall, J. R., Haughey, B. P., Cholewinski, S., and Wilkinson, G. 1991. Folate intake and carcinogenesis of the colon and rectum. Int J Epidemiol 20:368–74.

Friedman, G., Goldschmidt, N., Friedlander, Y., Ben-Yehuda, A., Selhub, J., Babaey, S., Mendel, M., Kidron, M., and Bar-On, H. 1999. A common mutation A1298C in human methylenetetrahydrofolate reductase gene: association with plasma total homocysteine and folate concentrations. J Nutr 129:1656–61.

Friso, S., and Choi, S. W. 2002. Gene–nutrient interactions and DNA methylation. J Nutr 132:2382S–7S.

Friso, S., Choi, S. W., Girelli, D., Mason, J. B., Dolnikowski, G. G., Bagley, P. J., Olivieri, O., Jacques, P. F., Rosenberg, I. H., Corrocher, R., and Selhub, J. 2002a. A common mutation in the 5,10-methylenetetrahydrofolate reductase gene affects genomic DNA methylation through an interaction with folate status. Proc Natl Acad Sci USA 99:5606–11.

Friso, S., Girelli, D., Trabetti, E., Stranieri, C., Olivieri, O., Tinazzi, E., Martinelli, N., Faccini, G., Pignatti, P. F., and Corrocher, R. 2002b. A1298C methylenetetrahydrofolate reductase mutation and coronary artery disease: relationships with C677T polymorphism and homocysteine/folate metabolism. Clin Exp Med 2:7–12.

Friso, S., Girelli, D., Trabetti, E., Olivieri, O., Guarini, P., Pignatti, P. F., Corrocher, R., and Choi, S. W. 2005. The MTHFR 1298A > C polymorphism and genomic DNA methylation in human lymphocytes. Cancer Epidemiol Biomarkers Prev 14:938–43.

Frosst, P., Blom, H. J., Milos, R., Goyette, P., Sheppard, C. A., Matthews, R. G., Boers, G. J., den Heijer, M., Kluijtmans, L. A., van den Heuvel, L. P., and Rozen, R. 1995. A candidate genetic risk factor for vascular disease: a common mutation in methylenetetrahydrofolate reductase [letter]. Nat Genet 10:111–3.

Gabriel, S. B., Schaffner, S. F., Nguyen, H., Moore, J. M., Roy, J., Blumenstiel, B., Higgins, J., DeFelice, M., Lochner, A., Faggart, M., Liu-Cordero, S. N., Rotimi, C., Adeyemo, A., Cooper, R., Ward, R., Lander, E. S., Daly, M. J., and Altshuler, D. 2002. The structure of haplotype blocks in the human genome. Science 296:2225–9.

Garrow, T. A., Brenner, A. A., Whitehead, V. M., Chen, X. N., Duncan, R. G., Korenberg, J. R., and Shane, B. 1993. Cloning of human cDNAs encoding mitochondrial and cytosolic serine hydroxymethyltransferases and chromosomal localization. J Biol Chem 268:11910–6.

Gaughan, D. J., Kluijtmans, L. A., Barbaux, S., McMaster, D., Young, I. S., Yarnell, J. W., Evans, A., and Whitehead, A. S. 2001. The methionine synthase reductase (MTRR) A66G polymorphism is a novel genetic determinant of plasma homocysteine concentrations. Atherosclerosis 157:451–6.

Geisel, J., Hubner, U., Bodis, M., Schorr, H., Knapp, J. P., Obeid, R., and Herrmann, W. 2003. The role of genetic factors in the development of hyperhomocysteinemia. Clin Chem Lab Med 41:1427–34.

Giovannucci, E. 2002. Epidemiologic studies of folate and colorectal neoplasia: a review. J Nutr 132:2350S–5S.

Giovannucci, E., Stampfer, M. J., Colditz, G. A., Rimm, E. B., Trichopoulos, D., Rosner, B. A., Speizer, F. E., and Willett, W. C. 1993. Folate, methionine, and alcohol intake and risk of colorectal adenoma. J Natl Cancer Inst 85:875–84.

Giovannucci, E., Rimm, E. B., Ascherio, A., Stampfer, M. J., Colditz, G. A., and Willett, W. C. 1995. Alcohol, low-methionine – low-folate diets, and risk of colon cancer in men. J Natl Cancer Inst 87:265–73.

Giovannucci, E., Chen, J., Smith-Warner, S. A., Rimm, E. B., Fuchs, C. S., Palomeque, C., Willett, W. C., and Hunter, D. J. 2003. Methylenetetrahydrofolate reductase, alcohol dehydrogenase, diet, and risk of colorectal adenomas. Cancer Epidemiol Biomarkers Prev 12:970–9.

Girelli, D., Friso, S., Trabetti, E., Olivieri, O., Russo, C., Pessotto, R., Faccini, G., Pignatti, P. F., Mazzucco, A., and Corrocher, R. 1998. Methylenetetrahydrofolate reductase C677T mutation, plasma homocysteine, and folate in subjects from northern Italy with or without angiographically documented severe coronary atherosclerotic disease: evidence for an important genetic–environmental interaction. Blood 91:4158–63.

Glynn, S. A., Albanes, D., Pietinen, P., Brown, C. C., Rautalahti, M., Tangrea, J. A., Gunter, E. W., Barrett, M. J., Virtamo, J., and Taylor, P. R. 1996. Colorectal cancer and folate status: a nested case-control study among male smokers. Cancer Epidemiol Biomarkers Prev 5:487–94.

Goode, E. L., Potter, J. D., Bigler, J., and Ulrich, C. M. 2004. Methionine synthase D919G polymorphism, folate metabolism, and colorectal adenoma risk. Cancer Epidemiol Biomarkers Prev 13:157–62.

Goto, Y., Yue, L., Yokoi, A., Nishimura, R., Uehara, T., Koizumi, S., and Saikawa, Y. 2001. A novel single-nucleotide polymorphism in the 3′-untranslated region of the human dihydrofolate reductase gene with enhanced expression. Clin Cancer Res 7:1952–6.

Harmon, D., Shields, D., Woodside, J., McMaster, D., Yarnell, J., Young, I., Peng, K., Shane, B., Evans, A., and Whitehead, A. 1999. Methionine synthase D919G polymorphism is a significant but modest determinant of circulating homocysteine concentrations. Genet Epidemiol 17:298–309.

Heil, S. G., Van der Put, N. M., Waas, E. T., den Heijer, M., Trijbels, F. J., and Blom, H. J. 2001. Is mutated serine hydroxymethyltransferase (SHMT) involved in the etiology of neural tube defects? Mol Genet Metab 73:164–72.

Herbig, K., Chiang, E. P., Lee, L. R., Hills, J., Shane, B., and Stover, P. J. 2002. Cytoplasmic serine hydroxymethyltransferase mediates competition between folate-dependent deoxyribonucleotide and *S*-adenosylmethionine biosyntheses. J Biol Chem 277:38381–9.

Hirose, M., Kono, S., Tabata, S., Ogawa, S., Yamaguchi, K., Mineshita, M., Hagiwara, T., Yin, G., Lee, K. Y., Tsuji, A., and Ikeda, N. 2005. Genetic polymorphisms of methylenetetrahydrofolate reductase and aldehyde dehydrogenase 2, alcohol use and risk of colorectal adenomas: Self-Defense Forces Health Study. Cancer Sci 96:513–8.

Horie, N., Aiba, H., Oguro, K., Hojo, H., and Takeishi, K. 1995. Functional analysis and DNA polymorphism of the tandemly repeated sequences in the 5′-terminal regulatory region of the human gene for thymidylate synthase. Cell Struct Funct 20:191–7.

Huang, Y., Han, S., Li, Y., Mao, Y., and Xie, Y. 2007. Different roles of MTHFR C677T and A1298C polymorphisms in colorectal adenoma and colorectal cancer: a meta-analysis. J Hum Genet 52:73–85.

Hubner, R. A., Muir, K. R., Liu, J. F., Logan, R. F., Grainge, M., Armitage, N., Shepherd, V., Popat, S., and Houlston, R. S. 2006a. Genetic variants of UGT1A6 influence risk of colorectal adenoma recurrence. Clin Cancer Res 12:6585–9.

Hubner, R. A., Muir, K. R., Liu, J. F., Sellick, G. S., Logan, R. F., Grainge, M., Armitage, N., Chau, I., and Houlston, R. S. 2006b. Folate metabolism polymorphisms influence risk of colorectal adenoma recurrence. Cancer Epidemiol Biomarkers Prev 15:1607–13.

Hubner, R. A., Lubbe, S., Chandler, I., and Houlston, R. S. 2007. MTHFR C677T has differential influence on risk of MSI and MSS colorectal cancer. Hum Mol Genet 16:1072–7.

Hustad, S., Ueland, P. M., Vollset, S. E., Zhang, Y., Bjorke-Monsen, A. L., and Schneede, J. 2000. Riboflavin as a determinant of plasma total homocysteine: effect modification by the methylenetetrahydrofolate reductase C677T polymorphism. Clin Chem 46:1065–71.

Jacques, P. F., Bostom, A. G., Williams, R. R., Ellison, R. C., Eckfeldt, J. H., Rosenberg, I. H., Selhub, J., and Rozen, R. 1996. Relation between folate status, a common mutation in methylenetetrahydrofolate reductase, and plasma homocysteine concentrations [see comments]. Circulation 93:7–9.

Jacques, P. F., Bostom, A. G., Selhub, J., Rich, S., Curtis Ellison, R., Eckfeldt, J. H., Gravel, R. A., and Rozen, R. 2003. Effects of polymorphisms of methionine synthase and methionine synthase reductase on total plasma homocysteine in the NHLBI Family Heart Study. Atherosclerosis 166:49–55.

Johnson, G. C., Esposito, L., Barratt, B. J., Smith, A. N., Heward, J., Di Genova, G., Ueda, H., Cordell, H. J., Eaves, I. A., Dudbridge, F., Twells, R. C., Payne, F., Hughes, W., Nutland, S., Stevens, H., Carr, P., Tuomilehto-Wolf, E., Tuomilehto, J., Gough, S. C., Clayton, D. G., and Todd, J. A. 2001. Haplotype tagging for the identification of common disease genes. Nat Genet 29:233–7.

Johnson, W. G., Stenroos, E. S., Spychala, J. R., Chatkupt, S., Ming, S. X., and Buyske, S. 2004. New 19 bp deletion polymorphism in intron-1 of dihydrofolate reductase (DHFR): a risk factor for spina bifida acting in mothers during pregnancy? Am J Med Genet 124A:339–45.

Jung, A., Poole, E. M., Bigler, J., Whitton, J., Potter, J. D., and Ulrich, C. M. 2008. DNA methyltransferase and alcohol dehydrogenase: gene–nutrient interactions in relation to risk of colorectal polyps. Cancer Epidemiol Biomarkers Prev 17(2):330–8.

Kaneda, S., Takeishi, K., Ayusawa, D., Shimizu, K., Seno, T., and Altman, S. 1987. Role in translation of a triple tandemly repeated sequence in the 5′-untranslated region of human thymidylate synthase mRNA. Nucleic Acids Res 15:1259–70.

Kato, I., Dnistrian, A. M., Schwartz, M., Toniolo, P., Koenig, K., Shore, R. E., Akhmedkhanov, A., Zeleniuch-Jacquotte, A., and Riboli, E. 1999. Serum folate, homocysteine and colorectal cancer risk in women: a nested case-control study. Br J Cancer 79:1917–22.

Keku, T., Millikan, R., Worley, K., Winkel, S., Eaton, A., Biscocho, L., Martin, C., and Sandler, R. 2002. 5,10-Methylenetetrahydrofolate reductase codon 677 and 1298 polymorphisms and colon cancer in African Americans and whites. Cancer Epidemiol Biomarkers Prev 11: 1611–21.

Kim, Y. I. 2004. Will mandatory folic acid fortification prevent or promote cancer? Am J Clin Nutr 80:1123–8.

Konings, E. J., Goldbohm, R. A., Brants, H. A., Saris, W. H., and van den Brandt, P. A. 2002. Intake of dietary folate vitamers and risk of colorectal carcinoma: results from The Netherlands Cohort Study. Cancer 95:1421–33.

Koushik, A., Kraft, P., Fuchs, C. S., Hankinson, S. E., Willett, W. C., Giovannucci, E. L., and Hunter, D. J. 2006. Nonsynonymous polymorphisms in genes in the one-carbon metabolism pathway and associations with colorectal cancer. Cancer Epidemiol Biomarkers Prev 15:2408–17.

Kraus, J. P., Oliveriusova, J., Sokolova, J., Kraus, E., Vlcek, C., de Franchis, R., Maclean, K. N., Bao, L., Bukovsk, Patterson, D., Paces, V., Ansorge, W., and Kozich, V. 1998. The human cystathionine beta-synthase (CBS) gene: complete sequence, alternative splicing, and polymorphisms. Genomics 52:312–24.

Kruger, W. D., Evans, A. A., Wang, L., Malinow, M. R., Duell, P. B., Anderson, P. H., Block, P. C., Hess, D. L., Graf, E. E., and Upson, B. 2000. Polymorphisms in the CBS gene associated with decreased risk of coronary artery disease and increased responsiveness to total homocysteine lowering by folic acid. Mol Genet Metab 70:53–60.

Laverdiere, C., Chiasson, S., Costea, I., Moghrabi, A., and Krajinovic, M. 2002. Polymorphism G80A in the reduced folate carrier gene and its relationship to methotrexate plasma levels and outcome of childhood acute lymphoblastic leukemia. Blood 100:3832–4.

Le Marchand, L., Donlon, T., Hankin, J. H., Kolonel, L. N., Wilkens, L. R., and Seifried, A. 2002. B-vitamin intake, metabolic genes, and colorectal cancer risk (United States). Cancer Causes Control 13:239–48.

Le Marchand, L., Wilkens, L. R., Kolonel, L. N., and Henderson, B. E. 2005. The MTHFR C677T polymorphism and colorectal cancer: the multiethnic cohort study. Cancer Epidemiol Biomarkers Prev 14:1198–203.

Leclerc, D., Wilson, A., Dumas, R., Gafuik, C., Song, D., Watkins, D., Heng, H. H., Rommens, J. M., Scherer, S. W., Rosenblatt, D. S., and Gravel, R. A. 1998. Cloning and mapping of a cDNA for methionine synthase reductase, a flavoprotein defective in patients with homocystinuria. Proc Natl Acad Sci USA 95:3059–64.

Lievers, K. J., Kluijtmans, L. A., Heil, S. G., Boers, G. H., Verhoef, P., van Oppenraay-Emmerzaal, D., den Heijer, M., Trijbels, F. J., and Blom, H. J. 2001. A 31 bp VNTR in the cystathionine beta-synthase (CBS) gene is associated with reduced CBS activity and elevated post-load homocysteine levels. Eur J Hum Genet 9:583–9.

Lievers, K. J., Afman, L. A., Kluijtmans, L. A., Boers, G. H., Verhoef, P., den Heijer, M., Trijbels, F. J., and Blom, H. J. 2002. Polymorphisms in the transcobalamin gene: association with plasma homocysteine in healthy individuals and vascular disease patients. Clin Chem 48:1383–9.

Liu, X., Szebenyi, D. M., Anguera, M. C., Thiel, D. J., and Stover, P. J. 2001. Lack of catalytic activity of a murine mRNA cytoplasmic serine hydroxymethyltransferase splice variant: evidence against alternative splicing as a regulatory mechanism. Biochemistry 40:4932–9.

Ma, J., Stampfer, M. J., Giovannucci, E., Artigas, C., Hunter, D. J., Fuchs, C., Willett, W. C., Selhub, J., Hennekens, C. H., and Rozen, R. 1997. Methylenetetrahydrofolate reductase polymorphism, dietary interactions, and risk of colorectal cancer. Cancer Res 57:1098–102.

Ma, J., Stampfer, M. J., Christensen, B., Giovannucci, E., Hunter, D. J., Chen, J., Willett, W. C., Selhub, J., Hennekens, C. H., Gravel, R., and Rozen, R. 1999. A polymorphism of the methionine synthase gene: association with plasma folate, vitamin B12, homocyst(e)ine, and colorectal cancer risk. Cancer Epidemiol Biomarkers Prev 8:825–9.

Mandola, M. V., Stoehlmacher, J., Muller-Weeks, S., Cesarone, G., Yu, M. C., Lenz, H. J., and Ladner, R. D. 2003. A novel single nucleotide polymorphism within the 5′ tandem repeat polymorphism of the thymidylate synthase gene abolishes USF-1 binding and alters transcriptional activity. Cancer Res 63:2898–904.

Mandola, M. V., Stoehlmacher, J., Zhang, W., Groshen, S., Yu, M. C., Iqbal, S., Lenz, H. J., and Ladner, R. D. 2004. A 6 bp polymorphism in the thymidylate synthase gene causes message instability and is associated with decreased intratumoral TS mRNA levels. Pharmacogenetics 14:319–27.

Marsh, S., Ameyaw, M. M., Githang'a, J., Indalo, A., Ofori-Adjei, D., and McLeod, H. L. 2000. Novel thymidylate synthase enhancer region alleles in African populations. Hum Mutat 16:528.

Martinez, M. E., Thompson, P., Jacobs, E. T., Giovannucci, E., Jiang, R., Klimecki, W., and Alberts, D. S. 2006. Dietary factors and biomarkers involved in the methylenetetrahydrofolate reductase genotype-colorectal adenoma pathway. Gastroenterology 131:1706–16.

Marugame, T., Tsuji, E., Kiyohara, C., Eguchi, H., Oda, T., Shinchi, K., and Kono, S. 2003. Relation of plasma folate and methylenetetrahydrofolate reductase C677T polymorphism to colorectal adenomas [comment]. Int J Epidemiol 32:64–6.

Matsuo, K., Ito, H., Wakai, K., Hirose, K., Saito, T., Suzuki, T., Kato, T., Hirai, T., Kanemitsu, Y., Hamajima, H., and Tajima, K. 2005. One-carbon metabolism related gene polymorphisms interact with alcohol drinking to influence the risk of colorectal cancer in Japan. Carcinogenesis 26:2164–71.

Miller, J. W., Ramos, M. I., Garrod, M. G., Flynn, M. A., and Green, R. 2002. Transcobalamin II 775G > C polymorphism and indices of vitamin B12 status in healthy older adults. Blood 100:718–20.

Mitrou, P. N., Watson, M. A., Loktionov, A. S., Cardwell, C., Gunter, M. J., Atkin, W. S., Macklin, C. P., Cecil, T., Bishop, T. D., Primrose, J., and Bingham, S. A. 2006. MTHFR (C677T and A1298C) polymorphisms and risk of sporadic distal colorectal adenoma in the UK Flexible Sigmoidoscopy Screening Trial (United Kingdom). Cancer Causes Control 17:793–801.

Mudd, S. H., Levy, H. L., and Skovby, F. 1995. Disorders of transsulfuration. In: The Metabolic and Molecular Basis of Inherited Disease, eds. C. R. Scriver, A. L. Beaudet, W. S. Sly and D. Valle, pp. 1279–327, New York: McGraw-Hill.

Nijhout, H. F., Reed, M. C., Budu, P., and Ulrich, C. M. 2004. A mathematical model of the folate cycle: new insights into folate homeostasis. J Biol Chem 279:55008–16.
Nijhout, H. F., Reed, M., Anderson, D., Mattingly, J., James, S. J., and Ulrich, C. M. 2006a. Long-range allosteric interactions between the folate and methionine cycles stabilize DNA methylation rate. Epigenetics 1:81–7.
Nijhout, H. F., Reed, M. C., Lam, S. L., Shane, B., Gregory, J. F., III, and Ulrich, C. M. 2006b. In silico experimentation with a model of hepatic mitochondrial folate metabolism. Theor Biol Med Model 3:40.
O'Leary, V. B., Parle-McDermott, A., Molloy, A. M., Kirke, P. N., Johnson, Z., Conley, M., Scott, J. M., and Mills, J. L. 2002. MTRR and MTHFR polymorphism: link to Down syndrome? Am J Med Genet 107:151–5.
Otani, T., Iwasaki, M., Hanaoka, T., Kobayashi, M., Ishihara, J., Natsukawa, S., Shaura, K., Koizumi, Y., Kasuga, Y., Yoshimura, K., Yoshida, T., and Tsugane, S. 2005. Folate, vitamin B6, vitamin B12, and vitamin B2 intake, genetic polymorphisms of related enzymes, and risk of colorectal cancer in a hospital-based case-control study in Japan. Nutr Cancer 53:42–50.
Patil, N., Berno, A. J., Hinds, D. A., Barrett, W. A., Doshi, J. M., Hacker, C. R., Kautzer, C. R., Lee, D. H., Marjoribanks, C., McDonough, D. P., Nguyen, B. T., Norris, M. C., Sheehan, J. B., Shen, N., Stern, D., Stokowski, R. P., Thomas, D. J., Trulson, M. O., Vyas, K. R., Frazer, K. A., Fodor, S. P., and Cox, D. R. 2001. Blocks of limited haplotype diversity revealed by high-resolution scanning of human chromosome 21. Science 294:1719–23.
Pullarkat, S. T., Stoehlmacher, J., Ghaderi, V., Xiong, Y.-P., Ingles, S. A., Sherrod, A., Warren, R., Tsao-Wei, D., Groshen, S., and Lenz, H.-J. 2001. Thymidylate synthase gene polymorphism determines response and toxicity of 5-FU chemotherapy. Pharmacogenomics J 1:65–70.
Reed, M. C., Nijhout, H. F., Neuhouser, M. L., Gregory, J. F., III, Shane, B., James, S. J., Boynton, A., and Ulrich, C. M. 2006. A mathematical model gives insights into nutritional and genetic aspects of folate-mediated one-carbon metabolism. J Nutr 136:2653–61.
Sandler, R. S., Everhart, J. E., Donowitz, M., Adams, E., Cronin, K., Goodman, C., Gemmen, E., Shah, S., Avdic, A., and Rubin, R. 2002. The burden of selected digestive diseases in the United States. Gastroenterology 122:1500–11.
Sebastio, G., Sperandeo, M. P., Panico, M., de Franchis, R., Kraus, J. P., and Andria, G. 1995. The molecular basis of homocystinuria due to cystathionine beta-synthase deficiency in Italian families, and report of four novel mutations. Am J Hum Genet 56:1324–33.
Slattery, M. L., Schaffer, D., Edwards, S. L., Ma, K. N., and Potter, J. D. 1997. Are dietary factors involved in DNA methylation associated with colon cancer? Nutr Cancer 28:52–62.
Slattery, M. L., Potter, J. D., Samowitz, W., Schaffer, D., and Leppert, M. 1999. Methylenetetrahydrofolate reductase, diet, and risk of colon cancer. Cancer Epidemiol Biomarkers Prev 8:513–8.
Stern, L. L., Mason, J. B., Selhub, J., and Choi, S. W. 2000. Genomic DNA hypomethylation, a characteristic of most cancers, is present in peripheral leukocytes of individuals who are homozygous for the C677T polymorphism in the methylenetetrahydrofolate reductase gene. Cancer Epidemiol Biomarkers Prev 9:849–53.
Stover, P. J., Chen, L. H., Suh, J. R., Stover, D. M., Keyomarsi, K., and Shane, B. 1997. Molecular cloning, characterization, and regulation of the human mitochondrial serine hydroxymethyltransferase gene. J Biol Chem 272:1842–8.
Tiemersma, E. W., Wark, P. A., Ocke, M. C., Bunschoten, A., Otten, M. H., Kok, F. J., and Kampman, E. 2003. Alcohol consumption, alcohol dehydrogenase 3 polymorphism, and colorectal adenomas. Cancer Epidemiol Biomarkers Prev 12:419–25.
Trinh, B. N., Ong, C.-N., Coetzee, G. A., Yu, M. C., and Laird, P. W. 2002. Thymidylate synthase: a novel genetic determinant of plasma homocysteine and folate levels. Hum Genet 111:299–302.
Ulrich, C. M., and Potter, J. D. 2007. Folate and cancer – timing is everything. JAMA 297:2408–9.
Ulrich, C. M., Bigler, J., Velicer, C., Greene, E., Farin, F., and Potter, J. 2000. Searching expressed sequence tag databases: discovery and confirmation of a common polymorphism in the thymidylate synthase gene. Cancer Epidemiol Biomarkers Prev 9:1381–5.

Ulrich, C. M., Bigler, J., Bostick, R., Fosdick, L., and Potter, J. D. 2002a. *Thymidylate synthase* promoter polymorphism, interaction with folate intake, and risk of colorectal adenomas. Cancer Res 62:3361–4.
Ulrich, C. M., Robien, K., and Sparks, R. 2002b. Pharmacogenetics and folate metabolism – a promising direction. Pharmacogenomics 3:299–313.
Ulrich, C. M., Robien, K., and McLeod, H. L. 2003. Cancer pharmacogenetics: polymorphisms, pathways and beyond. Nat Rev Cancer 3:912–20.
Ulrich, C. M., Curtin, K., Potter, J. D., Bigler, J., Caan, B., and Slattery, M. L. 2005a. Polymorphisms in the reduced folate carrier, thymidylate synthase, or methionine synthase and risk of colon cancer. Cancer Epidemiol Biomarkers Prev 14:2509–16.
Ulrich, C. M., Curtin, K., Samowitz, W., Bigler, J., Potter, J. D., Caan, B., and Slattery, M. L. 2005b. MTHFR variants reduce the risk of G:C → A:T transition mutations within the p53 tumor suppressor gene in colon tumors. J Nutr 135:2462–7.
Ulrich, C. M., Nijhout, H. F., and Reed, M. C. 2006. Mathematical modeling: epidemiology meets systems biology. Cancer Epidemiol Biomarkers Prev 15:827–9.
Ulvik, A., Vollset, S. E., Hansen, S., Gislefoss, R., Jellum, E., and Ueland, P. M. 2004. Colorectal cancer and the methylenetetrahydrofolate reductase 677C → T and methionine synthase 2756A → G polymorphisms: a study of 2,168 case-control pairs from the JANUS cohort. Cancer Epidemiol Biomarkers Prev 13:2175–80.
Ulvik, A., Ueland, P. M., Fredriksen, A., Meyer, K., Vollset, S. E., Hoff, G., and Schneede, J. 2007. Functional inference of the methylenetetrahydrofolate reductase 677 C > T and 1298A > C polymorphisms from a large-scale epidemiological study. Hum Genet 121:57–64.
van den Donk, M., Buijsse, B., van den Berg, S. W., Ocke, M. C., Harryvan, J. L., Nagengast, F. M., Kok, F. J., and Kampman, E. 2005. Dietary intake of folate and riboflavin, MTHFR C677T genotype, and colorectal adenoma risk: a Dutch case-control study. Cancer Epidemiol Biomarkers Prev 14:1562–6.
van der Put, N. M., van der Molen, E. F., Kluijtmans, L. A., Heil, S. G., Trijbels, J. M., Eskes, T. K., Van Oppenraaij-Emmerzaal, D., Banerjee, R., and Blom, H. J. 1997. Sequence analysis of the coding region of human methionine synthase: relevance to hyperhomocysteinaemia in neural-tube defects and vascular disease. QJM 90:511–7.
van der Put, N. M., Gabreels, F., Stevens, E. M., Smeitink, J. A., Trijbels, F. J., Eskes, T. K., van den Heuvel, L. P., and Blom, H. J. 1998. A second common mutation in the methylenetetrahydrofolate reductase gene: an additional risk factor for neural-tube defects? Am J Hum Genet 62:1044–51.
Wagner, C. 1995. Biochemical role of folate in cellular metabolism. New York: Marcel Dekker.
Wang, J., Gajalakshmi, V., Jiang, J., Kuriki, K., Suzuki, S., Nagaya, T., Nakamura, S., Akasaka, S., Ishikawa, H., and Tokudome, S. 2006. Associations between 5,10-methylenetetrahydrofolate reductase codon 677 and 1298 genetic polymorphisms and environmental factors with reference to susceptibility to colorectal cancer: a case-control study in an Indian population. Int J Cancer 118:991–7.
Wans, S., Schuttler, K., Jakubiczka, S., Muller, A., Luley, C., and Dierkes, J. 2003. Analysis of the transcobalamin II 776C > G (259P > R) single nucleotide polymorphism by denaturing HPLC in healthy elderly: associations with cobalamin, homocysteine and holo-transcobalamin II. Clin Chem Lab Med 41:1532–6.
Watkins, D., and Rosenblatt, D. S. 1989. Functional methionine synthase deficiency (cblE and cblG): clinical and biochemical heterogeneity. Am J Med Genet 34:427–34.
Weisberg, I., Tran, P., Christensen, B., Sibani, S., and Rozen, R. 1998. A second genetic polymorphism in methylenetetrahydrofolate reductase (MTHFR) associated with decreased enzyme activity. Mol Genet Metab 64:169–72.
Wilson, A., Platt, R., Wu, Q., Leclerc, D., Christensen, B., Yang, H., Gravel, R. A., and Rozen, R. 1999. A common variant in methionine synthase reductase combined with low cobalamin (vitamin B12) increases risk for spina bifida. Mol Genet Metab 67:317–23.
Winawer, S. J., Zauber, A. G., Ho, M. N., O'Brien, M. J., Gottlieb, L. S., Sternberg, S. S., Waye, J. D., Schapiro, M., Bond, J. H., Panish, J. F., Ackroyd, F., Shike, M., Kurtz, R. C., Hornsby-Lewis, L.,

Gerdes, H., Stewart, E. T., and The National Polyp Study Workgroup. 1993. Prevention of colorectal cancer by colonoscopic polypectomy. The National Polyp Study Workgroup [see comments]. N Engl J Med 329:1977–81.

Winkelmayer, W. C., Skoupy, S., Eberle, C., Fodinger, M., and Sunder-Plassmann, G. 2004. Effects of TCN2 776C > G on vitamin B, folate, and total homocysteine levels in kidney transplant patients. Kidney Int 65:1877–81.

Yang, F., Hanson, N. Q., Schwichtenberg, K., and Tsai, M. Y. 2000. Variable number tandem repeat in exon/intron border of the cystathionine beta-synthase gene: a single nucleotide substitution in the second repeat prevents multiple alternate splicing. Am J Med Genet 95:385–90.

Zetterberg, H., Coppola, A., D'Angelo, A., Palmer, M., Rymo, L., and Blennow, K. 2002. No association between the MTHFR A1298C and transcobalamin C776G genetic polymorphisms and hyperhomocysteinemia in thrombotic disease. Thromb Res 108:127–31.

Zetterberg, H., Nexo, E., Regland, B., Minthon, L., Boson, R., Palmer, M., Rymo, L., and Blennow, K. 2003. The transcobalamin (TC) codon 259 genetic polymorphism influences holo-TC concentration in cerebrospinal fluid from patients with Alzheimer disease. Clin Chem 49:1195–8.

Chapter 10
Genetic Variability in NSAID Targets and NSAID-Metabolizing Enzymes and Colorectal Neoplasia

Elizabeth M. Poole, James T. Cross, John D. Potter, and Cornelia M. Ulrich

Introduction

Inflammation is a known or suspected risk factor for several cancer types (Coussens and Werb 2002), including colorectal cancer. Nonsteroidal anti-inflammatory drugs (NSAIDs), including aspirin (acetylsalicylic acid), indomethacin, piroxicam, sulindac, and ibuprofen, have several functions, including reduction of inflammation, fever, and pain. Results of epidemiologic studies, intervention trials, and animal studies suggest that aspirin and other NSAIDs inhibit colorectal carcinogenesis (Giovannucci 1999; Brown and DuBois 2005). In observational studies, regular aspirin use has been associated with an approximate halving of risk of colorectal cancer compared with nonusers (Thun et al. 1991; Suh et al. 1993; Giovannucci et al. 1994, 1995; Muscat et al. 1994; Peleg et al. 1994; Schreinemachers and Everson 1994; La Vecchia et al. 1997; Freedman et al. 1998; Chan et al. 2005a; Bigler et al. 2001). Studies of adenomatous polyps, precursors of colorectal cancer, have shown similar results (Greenberg et al. 1993; Logan et al. 1993; Suh et al. 1993; Martinez et al. 1995). Recently, two randomized placebo-controlled trials (RCTs) of aspirin for the prevention of recurrent adenomatous polyps have shown risk reductions of 19–35% (Baron et al. 2003; Sandler et al. 2003). Two other RCTs of the COX-2 specific NSAID (coxib) celecoxib showed a 33–36% risk reduction and even greater reduction in the risk of advanced adenoma (Arber et al. 2006; Bertagnolli et al. 2006).

About 25% of chronic NSAID users experience toxicities, in particular gastrointestinal bleeding and renal toxicity (Murray and Brater 1993; Davies 1995). The traditional NSAIDs, such as aspirin and ibuprofen, are associated with gastrointestinal bleeding, whereas coxibs may result in cardiovascular toxicity. The celecoxib randomized control trials were stopped early due to adverse cardiovascular events among the group receiving celecoxib (Bresalier et al. 2005; Nussmeier et al. 2005; Solomon et al. 2005). NSAID-associated side effects have prompted research

C.M. Ulrich (✉)
Division of Public Health Sciences, Fred Hutchinson Cancer Research Center
email: nulrich@fhcrc.org

J.D. Potter and N.M Lindor (eds.), *Genetics of Colorectal Cancer*,
DOI: 10.1007/978-0-387-09568-4_18,

to identify genetic variations that may predispose toward adverse events and help identify population groups most likely to benefit from NSAID use (Ulrich et al. 2006). This chapter is a summary of the known genetic variability in the genes that target and metabolize NSAIDs and the results of epidemiologic studies of these genetic variants. It is likely that polymorphisms in these pathways may interact with NSAID use to alter an individual's colorectal cancer risk/benefit profile.

Genetic Variability in NSAID Targets

The anti-inflammatory activity of NSAIDs is due to their ability to inhibit the cyclooxygenase activity of the prostaglandin H synthase enzymes (also known as the COX enzymes) (Vane 1971), thus decreasing the production of prostaglandins. The COX enzymes have two isoforms: COX-1, which is constitutively expressed in many tissues and is responsible for tissue homeostatic functions, and COX-2, which is inducible and has a role in many inflammatory and proliferative reactions (Taketo 1998; Smith et al. 2000; Gupta and Dubois 2001). The COX enzymes are essential for the production of prostaglandins and have two functions: first, a cyclooxygenase (COX) reaction which converts arachidonic acid (an omega-6 fatty acid) to prostaglandin G2 (PGG_2); and second, a peroxidase reaction which converts PGG_2 to PGH_2.

Prostaglandins are in the eicosanoid family of oxygenated-lipid signaling molecules, formed from arachidonic acid and some other highly unsaturated fatty acids. They are widely produced in the human body and have important roles in inflammation and other physiologic processes, such as blood clotting, wound healing, immune responses, bone metabolism, and nerve growth and development (Mead et al. 1986). Cancer tissues contain high concentrations of prostaglandins (Jaffe 1974; Bennett et al. 1977; Rigas et al. 1993; Pugh and Thomas 1994; Gupta and Dubois 2001). Prostaglandin E_2, specifically, appears to be central to carcinogenesis: activation of prostaglandin E_2 receptors triggers other signaling pathways known to contribute to cancer progression, such as the epidermal growth factor receptor pathway (Pai et al. 2002; Buchanan et al. 2003; Han and Wu 2005). Additionally, disruption of these receptors in mouse models reduces tumor formation (Watanabe et al. 1999, 2000; Mutoh et al. 2002).

Studies of colorectal tissue have shown COX-2 expression in up to 90% of colorectal carcinomas and 40% of adenomas, with no expression in normal colorectal mucosa (Eberhart et al. 1994; Kutchera et al. 1996; Chapple et al. 2000). In a recent study, aspirin use was associated with a deficit of tumors that overexpressed COX-2; in individuals whose tumors expressed little COX-2, there was no reduction of risk with regular aspirin use (Chan et al. 2007). Given its role in inflammation, many of the initial studies focused on COX-2 as the likely target in COX-induced colorectal carcinogenesis. However, COX-1 has been more recently implicated in colorectal cancer development (Oshima et al. 1996; Chulada et al. 2000), and in NSAID pharmacokinetics (Fries et al. 2006). Thus, studies of variability in prostaglandin synthesis in relation to colorectal neoplasia should consider both enzymes.

Cyclooxygenase-1 (COX-1)

COX-1 has been systematically screened for polymorphisms in African Americans and Caucasians by our group (Ulrich et al. 2002) and by the University of Washington-Fred Hutchinson Cancer Research Center Variation Discovery Resource (UW-FHCRC-VDR) and SNP500Cancer. Additionally, several other groups have reported SNPs in *COX-1* (Scott et al. 2002; Halushka et al. 2003; Hillarp et al. 2003); one of these, P17L, has been associated with functional effects (Scott et al. 2002; Halushka et al. 2003; Fries et al. 2006) and two others, R8W and L237M, are predicted to have functional impact by in silico programs such as SIFT (Ng and Henikoff 2003) and PolyPhen (Ramensky et al. 2002).

Cyclooxygenase-2 (COX-2)

Like *COX-1*, *COX-2* has also been screened by the UW-FHCRC-VDR, the University of Washington National Institute of Environmental Health Environmental Genome Project (NIEHS-EGP), and SNP500Cancer and several groups have also reported polymorphisms. Unlike *COX-1*, however, nonsynonymous polymorphisms in *COX-2* are very rare [see dbSNP at http://www.ncbi.nlm.nih.gov]. Thus almost all studies of genetic variability in *COX-2* have focused on intronic or UTR polymorphisms. Of these, the −765G > C polymorphism has been consistently associated with differential expression (Papafili et al. 2002; Cipollone et al. 2004; Zhang et al. 2005; Orbe et al. 2006); however, other UTR polymorphisms have also been reported to alter mRNA expression (Hu et al. 2005; Zhang et al. 2005) or changes in transcription-factor-binding sites (Panguluri et al. 2004).

Other Targets

Although COX-1 and COX-2 have been the focus of much research into colorectal carcinogenesis, inquiries into many targets downstream in this pathway and, in competing pathways, are beginning to show promising results. There are four prostaglandin synthases that act downstream of the COX enzymes that may be of interest for colorectal carcinogenesis (see Fig. 10.1). Specifically, prostaglandin E_2 synthase (PGES) may be of particular interest due to the known activities of PGE_2 in colorectal cancer. This gene has been completely resequenced by the UW-FHCRC-VDR and partially (the exons and UTRs) by our group (Bigler et al. 2007). Although the reported nonsynonymous polymorphisms are rare, two of them are likely to be functional, based on predictions from SIFT and PolyPhen. PGE_2 has four cell-surface receptors that may also be of interest, given that recent evidence has shown that these receptors crosstalk with the EGFR pathway (Pai et al. 2002;

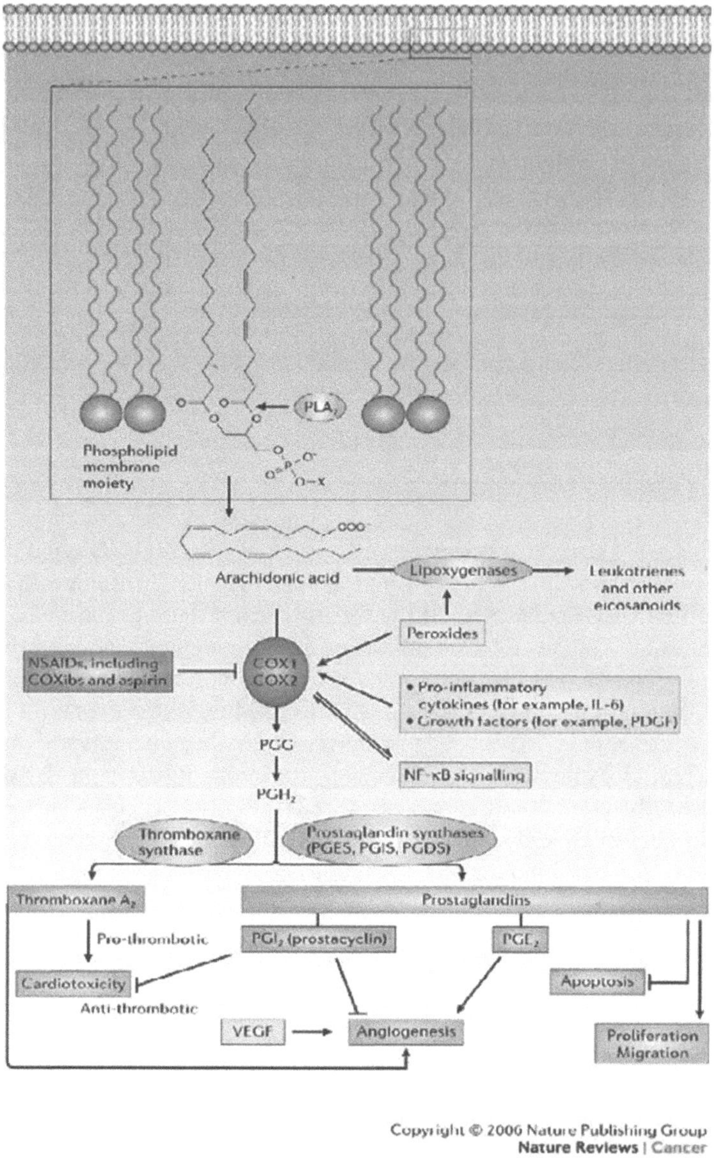

Fig. 10.1 NSAIDs, COX inhibition, and prostaglandins. The main targets of coxibs and NSAIDs are COX-1 and COX-2, enzymes central to the metabolism of arachidonic acid to prostaglandins (PGs); an alternative pathway involves the lipoxygenases. Both pathways are regulated by peroxide concentrations, with COX-2 being influenced at lower concentrations than COX-1. PGs influence angiogenesis, apoptosis, cell proliferation, and migration. The balance between pro- and antithrombotic factors is probably relevant to the cardiovascular toxicity of coxibs. *IL6* interleukin 6; *NF-κB* nuclear factor-κB; *PDGF* platelet-derived growth factor; *PLA_2* phospholipase A; *VEGF* vascular endothelial growth factor

Buchanan et al. 2003; Han and Wu 2005). Like *PGES*, two of these receptors, *EP2* and *EP4*, have been completely resequenced by the UW-FHCRC-VDR and partially by our group (Bigler et al. 2007). As in the case of *PGES*, several of the reported nonsynonymous polymorphisms were predicted to have functional consequences (Bigler et al. 2007).

Another downstream prostaglandin synthase of potential interest is prostaglandin I_2 synthase or prostacyclin synthase (PGIS). PGIS has been of recent interest because of the cardiotoxicity of the coxibs. PGIS has antithrombotic activity and inhibits platelet aggregation, vasoconstriction, and vascular proliferation (Marcus et al. 2002). Because COX-2 is the major source of PGIS (Fitzgerald 2004), targeted inhibition of COX-2, and thus inhibition of PGIS, may be the source of the cardiovascular side effects of the coxibs. *PGIS* has been screened for polymorphisms by SNP500Cancer, and there is a known variable number tandem repeat (VNTR) polymorphism [−3(CCAGCCCCG)3–8] in the promoter region; having fewer than the wild-type number of alleles (6R) has been associated with reduced promoter activity (Iwai et al. 1999; Chevalier et al. 2001).

The arachidonate lipoxygenases (ALOXs) are of interest because they compete with the COXs for their substrate, arachidonic acid. There are three major lipoxygenases: ALOX5, ALOX12, and ALOX15. ALOX5 and -12 are thought to be procarcinogenic, whereas ALOX15 may have anticarcinogenic properties (Shureiqi and Lippman 2001). *ALOX12* and *ALOX15* have been resequenced by the UW-FHCRC-VDR, and all three lipoxygenases have been resequenced by SNP500Cancer. There is a promoter VNTR polymorphism [−176(GGGCGG)2–8] in *ALOX5*, in which having fewer repeats than wild type (5R) has been associated with a decrease in promoter activity, although the effect of having more repeats than wild type has not been resolved (In et al. 1997; Silverman and Drazen 2000).

15-Hydroxyprostaglandin dehydrogenase (PGDH) breaks down PGE_2 into 15-keto PGE, which has greatly reduced biological activity (Ensor and Tai 1995). Recently, PGDH expression was reported to be greatly reduced in colon cancer, providing another mechanism by which cancer cells enhance the production of prostaglandins (Yan et al. 2004; Backlund et al. 2005). *PGDH* has been systematically screened for polymorphisms by the UW-FHCRC-VDR.

Ornithine Decarboxylase

Ornithine decarboxylase (ODC1) may play a role in the development of colorectal polyps and cancer. ODC1 is inhibited by NSAIDs, including celecoxib (Ostrowski et al. 2003); however, this inhibition is through a prostaglandin-independent pathway. ODC1 catalyzes the synthesis of polyamines, which have several carcinogenic actions, including increased cell division, upregulation of genes involved in metastasis and tumor invasion, and downregulation of apoptosis (Babbar et al. 2003; Gerner and Meyskens 2004). Increased concentrations of intracellular polyamines have been positively associated with cancer risk (Janne et al. 1978; Gerner and

Meyskens 2004), including colorectal (Kingsnorth et al. 1984), and have been inversely associated with apoptosis (Scornioni 2001). Additionally, ODC1 has been shown to be overexpressed in cancerous colon epithelium compared to normal colon expression levels (LaMuraglia et al. 1986; Porter et al. 1987; Koo et al. 1988; Gerner and Meyskens 2004; Wolter et al. 2004). Thus, although the main action of NSAIDs is through inhibition of prostaglandin synthesis, ODC1 inhibition may be another pathway by which NSAIDs exert chemopreventive properties in colorectal cancer (Carbone et al. 1998; Turchanowa et al. 2001; Martinez et al. 2003).

ODC1 has been screened by the NIEHS-EGP. However, to date, research into the role of genetic variability has been limited to the 315G > A polymorphism. This polymorphism is located near transcription-factor-binding sites and was associated with differential RNA expression in one study (Guo et al. 2000). In another study, aspirin did not affect the promoter activity of the *ODC1* gene (Martinez et al. 2003), indicating that the role of this polymorphism requires further investigation.

Genetic Variability in NSAID-Metabolizing Enzymes

NSAIDs are metabolized by two main mechanisms: glucuronidation and hydroxylation. Glucuronidation of NSAIDs is accomplished primarily through the UDP-glucuronosyltransferases (UGTs), specifically UGT1A6 (Kuehl et al. 2005); hydroxylation occurs via the cytochrome P450 2C enzymes, specifically CYP2C9 (Miners and Birkett 1998), although other UGTs and CYPs may also contribute. Both the *UGT1A6* and the *CYP2C9* genes have known genetic polymorphisms that are associated with slower metabolism. These functional polymorphisms may interact with NSAID use to affect risk of colorectal neoplasia.

UGT1A6

UGT1A6 has been systematically screened for polymorphisms by SNP500Cancer. Two known variant alleles in *UGT1A6* have been associated with decreased enzyme activity; the first is characterized by amino acid changes at amino acids 181 and 184 (T181A + R184S, also known as *UGT1A6*2*) and the second by R184S (also known as *UGT1A6*4*) alone (Ciotti et al. 1997; Lampe et al. 1999). These genotypes are associated with a 30–50% reduction in enzyme activity (Ciotti et al. 1997).

CYP2C9

CYP2C9 has been screened for genetic variation by the NIEHS-EGP and by SNP500Cancer; there are more than 100 SNPs reported in dbSNP. However, similar to *UGT1A6*, there are two well-studied polymorphisms in *CYP2C9*, R144C (also known as *2) and I359L (also known as *3). Like *UGT1A6*, these polymorphisms have known functional effects, namely, a 5–30% reduction in enzyme activity compared to native CYP2C9 (Rettie et al. 1994; Takahashi et al. 1998).

Genetic Variability in NSAID Targets and NSAID-Metabolizing Enzymes and Risk of Colorectal Neoplasia

Investigations of potential gene–NSAID interactions in relation to risk of colorectal neoplasia have recently been comprehensively summarized (Cross et al. 2008). Following is a review of main-effect associations and gene–NSAID interactions published to date. Evaluation of these studies is complicated by the lack of a consistent definition of NSAID use. Definitions are inconsistent with respect to amount, type of NSAID, duration of use, and recency of use. Given that the benefits of NSAIDs probably vary by all of these, comparisons of studies that used markedly differing definitions may obscure the true relationships among genetic variability in NSAID-related genes, NSAID use, and risk of colorectal neoplasia.

NSAID Targets

COX-1

Studies of *COX-1* are a relatively recent focus of cancer research. To date, four studies have evaluated four *COX-1* polymorphisms (R8W, L15-L16del, P17L, and L237M) for a main-effect association with colorectal neoplasia risk or an interaction with NSAID exposure (Goodman et al. 2004; Ulrich et al. 2004; Siezen et al. 2005, 2006b). Only the L15-L16del polymorphism has been independently associated with increased risk of colorectal adenomatous polyps (OR: 3.6; 95% CI: 1.2–11.2) (Ulrich et al. 2004). In the same study, current, regular NSAID use (more than once per week) was associated with a reduction of adenoma risk among those with the 17PP (wild type) genotype compared to wild type nonusers (OR: 0.6; 95% CI: 0.5–0.8; $p = 0.03$) (Ulrich et al. 2004). This interaction has not been investigated in any other studies. The functional effects of this polymorphism are unclear, because the P17L polymorphism is found in the signal peptide portion of COX-1 and is cleaved from the mature protein. However, Halushka et al. reported that the P17L polymorphism is in complete linkage disequilibrium with a polymorphism in the promoter region of *COX-1*, −842A > G, which may have effects on transcription-factor-binding sites (Halushka et al. 2003). No associations or interactions with the R8W or L237M polymorphisms have been reported for colorectal neoplasia. However, due to the rarity of these polymorphisms, larger studies may be required for adequate power to detect such associations.

COX-2

In eight studies of colorectal neoplasia risk, eleven *COX-2* polymorphisms have been tested for main-effect associations or interaction with NSAID use (Lin et al. 2002; Cox et al. 2004; Goodman et al. 2004; Koh et al. 2004; Ulrich et al. 2005; Sansbury et al. 2006; Siezen et al. 2006a, b). Four of these studies included the −765G > C polymorphism and three included V511A, which occurs only in non-Caucasian populations. Main-effect associations with colorectal cancer have been reported for two SNPs: an inverse association with a synonymous SNP in exon 3 (V102V) (Siezen et al. 2006a) and a positive association with an intronic SNP (Cox et al. 2004), but these have not been confirmed in additional studies. In one study, a suggested interaction between the −765G > C polymorphism and current, regular NSAID use (more than once per week) was detected: when stratified on NSAID use, homozygous variant nonusers were at decreased risk of adenoma (OR: 0.26, 95% CI: 0.07–0.89) compared to wild-type nonusers, whereas homozygous-variant NSAID users (OR: 0.82, 95% CI: 0.25–2.73) showed no reduction in risk (Ulrich et al. 2005). However, a smaller study of 337 adenoma cases and 368 controls did not confirm this potential interaction (Siezen et al. 2006b). The −765G > C polymorphism has been shown to suppress *COX-2* promoter activity in one study (Papafili et al. 2002), but not in another (Orbe et al. 2006). Among atherosclerosis patients, the −765CC genotype was associated with lower levels of C-reactive protein and interleukin-6, biomarkers of inflammatory processes (Orbe et al. 2006).

Two studies have examined potential interactions between the *COX-2* V511A polymorphism and NSAID use among African-Americans in modifying colorectal cancer (Lin et al. 2002) and adenoma (Sansbury et al. 2006). The adenoma study (161 cases, 219 controls) reported statistically significant risk reductions among those who used NSAIDs (more than two times per week for at least 2 years) or carried the A allele (or both) compared to those with neither exposure (Lin et al. 2002). However, this study did not evaluate multiplicative interaction. The study of colorectal cancers observed that the risk reduction associated with regular NSAID use (at least 3 times a week for at least 3 months) (OR: 0.66; 95% CI: 0.45–0.95) may be more pronounced among those carrying at least one variant allele (OR: 0.29, 95% CI: 0.08–1.06), indicating that those with the alanine variant may receive greater benefit from regular NSAID use. This interaction, however, was not statistically significant ($p = 0.59$). A functional analysis of COX-2 polymorphisms has suggested that the V511A variant does not modify COX-2 activity; thus, it is less likely that there is a true NSAID interaction with this polymorphism (Fritsche et al. 2001).

A small hospital-based case-control study (292 cases and 274 controls) conducted in Spain examined eight *COX-2* polymorphisms and reported one statistically significant main-effect association: subjects carrying one or more variant allele of the intronic 9850A > G polymorphism had a statistically significantly increased risk of colorectal cancer (OR: 2.49; 95% CI: 1.17–5.32) (Cox et al. 2004). However, no interaction with NSAID use was observed (p-interaction = 0.19). A haplotype analysis of the eight SNPs in this study confirmed the main-effect association with the 9850A > G SNP – the haplotype containing the 9850 variant allele was the only polymorphism associated with a change in colorectal cancer risk. To date, the 9850A > G polymorphism has not been investigated in other studies of colorectal cancer, so this association remains unconfirmed. Because of the small size of this study, statistical power to detect NSAID interactions was limited. Moreover, because it was hospital-based, the control group may have had underlying comorbidities associated with NSAID use; thus, any true association between NSAID use and colorectal cancer may be attenuated, further limiting ability to detect gene–NSAID interactions.

Other Genes Related to Prostaglandin Synthesis

Although several genes described earlier have been suggested as potential targets for NSAID interactions in relation to colorectal neoplasia risk, studies of genetic variability in these genes are limited to date.

In a case-control study of 516 adenoma cases and 618 polyp-free controls, cases with fewer than the wild-type number of *PGIS* [−3(CCAGCCCCG)3–8] alleles (i.e., <6R/<6R) had a nearly twofold increase in risk compared to the wild-type genotype (6R/6R) (OR: 1.90; 95% CI: 1.09–3.31). Additionally, there was a suggested interaction with regular NSAID use (more than one pill per week for at least 1 year) in which the group of regular users who had at least one allele with greater than wild-type number of repeats (>6R/≥6R) was at greatly reduced adenoma risk compared to wild-type nonusers (OR: 0.33; 95% CI: 0.11–0.99; p-interaction = 0.06) (Poole et al. 2006).

In the same case-control study, evidence neither for an association with two *ALOX5* polymorphisms (−1700A > G or [−176(GGGCGG)2–8]) nor for an NSAID interaction was observed (Poole et al. 2006). Two other studies have also shown no association between colorectal neoplasia risk and the −1700A > G polymorphism (Goodman et al. 2004; Gong et al. 2007). Goodman et al. (2004) also reported no association with another promoter polymorphism (−1753G > A) but did not examine NSAID interactions (Goodman et al. 2004). Gong et al. (2007) reported no association with a fourth *ALOX5* polymorphism (21C > T) and no interaction between either of the studied *ALOX5* polymorphisms and NSAID use for colorectal adenoma risk (Gong et al. 2007). These studies suggest that either there is no association of *ALOX5* with colorectal neoplasia risk and no interaction with NSAIDs, or that a causative variant in *ALOX5* is yet to be discovered.

The study by Gong et al. (2007) also investigated an association between a nonsynonymous polymorphism in *ALOX12* (R261Q) and adenoma risk. In 162 adenoma cases and 211 controls, having at least one Q allele was associated with decreased adenoma risk (OR: 0.62; 95% CI: 0.40–0.96). Although the study was small, the authors reported a statistically significant interaction between this polymorphism and nonaspirin NSAID use (at least once per week) in which only the NSAID users with at least one Q allele were at decreased adenoma risk (p-interaction = 0.02) (Gong et al. 2007). Although intriguing, this result requires confirmation in larger studies.

ODC1

ODC1 is a relatively recent gene of interest for colorectal cancer risk; thus, only two studies have investigated a potential interaction between the *ODC1* 315G > A polymorphism and NSAID use in modifying colorectal neoplasia risk. In a randomized trial of wheat bran to prevent adenoma recurrence, Martinez and colleagues reported in 341 cases and 347 controls that the 315AA genotype was associated with reduced risk of metachronous adenoma (OR: 0.48; 95% CI: 0.24–0.99) (Martinez et al. 2003). Additionally, among the group that used aspirin (use not further defined), those with the 315AA genotype were at greatly reduced risk of metachronous adenoma compared to those with the wild-type (GG) genotype who did not use aspirin (OR: 0.10; 95% CI: 0.02–0.66). The 315AA genotype was not associated with risk reduction among aspirin nonusers (OR: 0.68; 95% CI: 0.30–1.51); however, the interaction was not statistically significant (p-interaction = 0.13). Subsequently, Barry et al. conducted a study of the *ODC1* 315G > A polymorphism in 972 participants in the Aspirin/Folate Polyp Prevention Trial and detected no main-effect association between this polymorphism and risk of metachronous adenoma. However, a statistically significant interaction between 315G > A genotype and aspirin use on metachronous adenoma was detected (Barry et al. 2006): among those with at least one variant 315A allele, those receiving aspirin (either 81 or 300 mg daily) were at statistically significantly reduced risk of metachronous adenoma (RR: 0.77; 95% CI: 0.63–0.95; p-interaction = 0.04) and advanced adenoma (RR: 0.51; 95% CI: 0.29–0.90; p-interaction = 0.02) compared to those on placebo. No risk reduction associated with aspirin use was observed among those with the wild-type genotype. The results of these two studies suggest that aspirin use, in the context of the *ODC1* 315 variant genotype, may be associated with greater protection against colorectal carcinogenesis; however, these results require confirmation in larger studies.

NSAID-Metabolizing Enzymes

UGT1A6

Potential associations with the functional *UGT1A6* polymorphisms (T181Ala + R184S or R184S alone) and risk of colorectal neoplasia or interactions with NSAID use have been investigated in four studies, with conflicting results. In a case-control study of 441 adenoma cases and 488 controls, regular aspirin use was associated with reduced adenoma risk only among those with at least one variant allele (OR: 0.53, 95% CI: 0.33–0.86, p-interaction not reported) (Bigler et al. 2001). This association was also observed for nonaspirin NSAID use, but was less pronounced. A similar association was observed in a case-control study of 530 women with adenoma and 532 control women participating in the Nurses' Health Study: there was greater risk reduction associated with regular NSAID use among women with any variant *UGT1A6* genotype compared to women with the wild-type alleles (p-interaction = 0.02) (Chan et al. 2005b). Two other studies have reported no interaction between *UGT1A6* genotype and NSAID use (Hubner et al. 2006; Samowitz et al. 2006). In the first, no main-effect association was seen for *UGT1A6* among 1,554 colon cancer cases and 1,939 controls or among 671 rectal cancer cases and 860 controls. The risk reductions associated with regular aspirin or ibuprofen use (at least three times per week for 1 month or more during the 2 years prior to diagnosis or reference date) were similar across genotypes (Samowitz et al. 2006). In the second study of 546 participants in the United Kingdom Colorectal Adenoma Prevention trial, having any variant allele was associated with reduced risk of adenoma recurrence (OR: 0.68, 95% CI: 0.52–0.89). However, no interaction with aspirin use (300 mg daily) was observed ($p = 0.70$). In general, the results of these investigations suggest that *UGT1A6* variants may interact with NSAIDs to affect risk of adenoma (Hubner et al. 2006). However, given that the one study of colorectal cancer, the largest of the four studies, found no interaction with NSAID use, further studies may be required to determine whether *UGT1A6* plays a role in colorectal neoplasia risk, and whether its role differs by stage of progression.

CYP2C9

Three of the studies that investigated genetic variability in *UGT1A6* also examined potential associations between the *2 and *3 polymorphisms in *CYP2C9*, and interactions with NSAID use, on risk of colorectal neoplasia (Bigler et al. 2001; Hubner et al. 2006; Samowitz et al. 2006). In the case-control study by Bigler et al. (2001), no main-effect association was reported, but a statistically significant adenoma risk reduction associated with regular aspirin use (more than once per week vs. less than once per month) was seen in those with the wild-type *CYP2C9* genotype (OR: 0.50, 95% CI 0.32–0.78), but not among those with the variant genotype (p-interaction not reported). However, among nonaspirin NSAID users, a statistically significant risk reduction was observed only among those with any variant (Bigler et al. 2001). The study of 1,554 colon cancer cases and 1,939 controls and among 671 rectal cancer cases and 860 controls similarly found no main-effect association, but noted a statistically significant interaction in which those who were homozygous for the variant alleles had a greater decrease in risk with regular ibuprofen use (at least three times per week for 1 month or more during the 2 years prior to diagnosis or reference date) than those with the wild-type alleles (p-interaction = 0.02). No similar interaction was observed for regular aspirin use (Samowitz et al. 2006). In the United Kingdom Colorectal Adenoma Prevention trial, no interaction between *CYP2C9* genotypes and aspirin treatment (300 mg daily) was reported; however, power to detect interactions was limited by the relatively small sample size (266 patients on aspirin and 280 on placebo) (Hubner et al. 2006). The two larger studies of *CYP2C9* *2 and *3 are generally consistent: both found that the combination of NSAID use and variant genotypes afforded the greatest risk reduction. However, the lack of confirmation in the United Kingdom Colorectal Adenoma Prevention trial (Hubner et al. 2006) and the differing results for aspirin and nonaspirin NSAIDs in the adenoma case-control study (Bigler et al. 2001) suggest that the interaction between *CYP2C9* polymorphisms and NSAID use requires additional clarification.

Summary

Initial studies suggest that genetic variability in NSAID targets and NSAID-metabolizing enzymes may be key to understanding the relationship between regular NSAID use and colorectal cancer chemoprevention. Given the great potential of NSAIDs as preventive agents, particularly for colorectal carcinogenesis, research into these genes is highly relevant and an important area of future research.

References

Arber, N., Eagle, C. J., Spicak, J., Racz, I., Dite, P., Hajer, J., Zavoral, M., Lechuga, M. J., Gerletti, P., Tang, J., Rosenstein, R. B., Macdonald, K., Bhadra, P., Fowler, R., Wittes, J., Zauber, A. G., Solomon, S. D., and Levin, B. 2006. Celecoxib for the prevention of colorectal adenomatous polyps. N Engl J Med 355:885–95.

Babbar, N., Ignatenko, N. A., Casero, R. A., and Gerner, E. W. 2003. Cyclooxygenase-independent induction of apoptosis by sulindac sulfone is mediated by polyamines in colon cancer. J Biol Chem 278:47762–75.

Backlund, M. G., Mann, J. R., Holla, V. R., Buchanan, F. G., Tai, H. H., Musiek, E. S., Milne, G. L., Katkuri, S., and DuBois, R. N. 2005. 15-Hydroxyprostaglandin dehydrogenase is down-regulated in colorectal cancer. J Biol Chem 280:3217–23.

Baron, J. A., Cole, B. F., Sandler, R. S., Haile, R. W., Ahnen, D., Bresalier, R., McKeown-Eyssen, G., Summers, R. W., Rothstein, R., Burke, C. A., Snover, D. C., Church, T. R., Allen, J. I., Beach, M., Beck, G. J., Bond, J. H., Byers, T., Greenberg, E. R., Mandel, J. S., Marcon, N., Mott, L. A., Pearson, L., Saibil, F., and van Stolk, R. U. 2003. A randomized trial of aspirin to prevent colorectal adenomas. N Engl J Med 348:891–9.

Barry, E. L., Baron, J. A., Bhat, S., Grau, M. V., Burke, C. A., Sandler, R. S., Ahnen, D. J., Haile, R. W., and O'Brien, T. G. 2006. Ornithine decarboxylase polymorphism modification of response to aspirin treatment for colorectal adenoma prevention. J Natl Cancer Inst 98:1494–500.

Bennett, A., Tacca, M. D., Stamford, I. F., and Zebro, T. 1977. Prostaglandins from tumours of human large bowel. Br J Cancer 35:881–4.

Bertagnolli, M. M., Eagle, C. J., Zauber, A. G., Redston, M., Solomon, D. H., Kim, K., Tang, J., Rosenstein, R. B., Wittes, J., Corle, D., Hess, T. M., Woloj, G. M., Boisserie, F., Anderson, W. F., Viner, J. L., Bagheri, D., Burn, J., Chung, D. C., Dewar, T., Foley, R., Hoffman, N., Macrae, F., Pruitt, R. E., Saltzman, J. R., Zalzberg, B., Sylwestrowicz, T., Gordon, G. B., and Hawk, E. T. 2006. Celecoxib for the prevention of sporadic colorectal adenomas. N Engl J Med 355:873–4.

Bigler, J., Whitton, J., Lampe, J. W., Fosdick, L., Bostick, R. M., and Potter, J. D. 2001. CYP2C9 and UGT1A6 genotypes modulate the protective effect of aspirin on colon adenoma risk. Cancer Res 61:3566–9.

Bigler, J., Sibert, J. G., Poole, E. M., Carlson, C. S., Potter, J. D., and Ulrich, C. M. 2007. Polymorphisms predicted to alter function in prostaglandin E_2 synthase and prostaglandin E_2 receptors. Pharmacogenet Genomics 17:221–7.

Bresalier, R. S., Sandler, R. S., Quan, H., Bolognese, J. A., Oxenius, B., Horgan, K., Lines, C., Riddell, R., Morton, D., Lanas, A., Konstam, M. A., and Baron, J. A. 2005. Cardiovascular events associated with rofecoxib in a colorectal adenoma chemoprevention trial. N Engl J Med 352:1092–102.

Brown, J. R., and DuBois, R. N. 2005. COX-2: a molecular target for colorectal cancer prevention. J Clin Oncol 23:2840–55.

Buchanan, F. G., Wang, D., Bargiacchi, F., and DuBois, R. N. 2003. Prostaglandin E_2 regulates cell migration via the intracellular activation of the epidermal growth factor receptor. J Biol Chem 278:35451–7.

Carbone, P. P., Douglas, J. A., Larson, P. O., Verma, A. K., Blair, I. A., Pomplun, M., and Tutsch, K. D. 1998. Phase I chemoprevention study of piroxicam and alpha-difluoromethylornithine. Cancer Epidemiol Biomarkers Prev 7:907–12.

Chan, A. T., Giovannucci, E. L., Meyerhardt, J. A., Schernhammer, E. S., Curhan, G. C., and Fuchs, C. S. 2005a. Long-term use of aspirin and nonsteroidal anti-inflammatory drugs and risk of colorectal cancer. JAMA 294:914–23.

Chan, A. T., Tranah, G. J., Giovannucci, E. L., Hunter, D. J., and Fuchs, C. S. 2005b. Genetic variants in the UGT1A6 enzyme, aspirin use, and the risk of colorectal adenoma. J Natl Cancer Inst 97:457–60.

Chan, A. T., Ogino, S., and Fuchs, C. S. 2007. Aspirin and the risk of colorectal cancer in relation to the expression of COX-2. N Engl J Med 356:2131–42.

Chapple, K. S., Cartwright, E. J., Hawcroft, G., Tisbury, A., Bonifer, C., Scott, N., Windsor, A. C., Guillou, P. J., Markham, A. F., Coletta, P. L., and Hull, M. A. 2000. Localization of cyclooxygenase-2 in human sporadic colorectal adenomas. Am J Pathol 156:545–53.

Chevalier, D., Cauffiez, C., Bernard, C., Lo-Guidice, J. M., Allorge, D., Fazio, F., Ferrari, N., Libersa, C., Lhermitte, M., D'Halluin, J. C., and Broly, F. 2001. Characterization of new mutations in the coding sequence and 5′-untranslated region of the human prostacyclin synthase gene (CYP8A1). Hum Genet 108:148–55.

Chulada, P. C., Thompson, M. B., Mahler, J. F., Doyle, C. M., Gaul, B. W., Lee, C., Tiano, H. F., Morham, S. G., Smithies, O., and Langenbach, R. 2000. Genetic disruption of Ptgs-1, as well as Ptgs-2, reduces intestinal tumorigenesis in Min mice. Cancer Res 60:4705–8.

Ciotti, M., Marrone, A., Potter, C., and Owens, I. S. 1997. Genetic polymorphism in the human UGT1A6 (planar phenol) UDP-glucuronosyltransferase: pharmacological implications. Pharmacogenetics 7:485–95.

Cipollone, F., Toniato, E., Martinotti, S., Fazia, M., Iezzi, A., Cuccurullo, C., Pini, B., Ursi, S., Vitullo, G., Averna, M., Arca, M., Montali, A., Campagna, F., Ucchino, S., Spigonardo, F., Taddei, S., Virdis, A., Ciabattoni, G., Notarbartolo, A., Cuccurullo, F., and Mezzetti, A. 2004. A polymorphism in the cyclooxygenase 2 gene as an inherited protective factor against myocardial infarction and stroke. JAMA 291:2221–8.

Coussens, L. M., and Werb, Z. 2002. Inflammation and cancer. Nature 420:860–7.

Cox, D. G., Pontes, C., Guino, E., Navarro, M., Osorio, A., Canzian, F., and Moreno, V. 2004. Polymorphisms in prostaglandin synthase 2/cyclooxygenase 2 (PTGS2/COX2) and risk of colorectal cancer. Br J Cancer 91:339–43.

Cross, J. T., Poole, E. M., and Ulrich, C. M. 2008. A review of gene–drug interactions for non-steroidal anti-inflammatory drug (NSAID) use in preventing colorectal neoplasia. Pharmacogenomics J 8:237–247.

Davies, N. M. 1995. Toxicity of nonsteroidal anti-inflammatory drugs in the large intestine. Dis Colon Rectum 38:1311–21.

Eberhart, C. E., Coffey, R. J., Radhika, A., Giardiello, F. M., Ferrenbach, S., and DuBois, R. N. 1994. Up-regulation of cyclooxygenase 2 gene expression in human colorectal adenomas and adenocarcinomas. Gastroenterology 107:1183–8.

Ensor, C. M., and Tai, H. H. 1995. 15-Hydroxyprostaglandin dehydrogenase. J Lipid Mediat Cell Signal 12:313–9.

Fitzgerald, G. A. 2004. Coxibs and cardiovascular disease. N Engl J Med 351:1709–11.

Freedman, A. N., Michalek, A. M., Weiss, H. A., Zhang, Z. F., Marshall, J. R., Mettlin, C. J., Asirwatham, J. E., Petrelli, N. J., and Caporaso, N. E. 1998. Aspirin use and p53 expression in colorectal cancer. Cancer Detect Prev 22:213–8.

Fries, S., Grosser, T., Price, T. S., Lawson, J. A., Kapoor, S., DeMarco, S., Pletcher, M. T., Wiltshire, T., and FitzGerald, G. A. 2006. Marked interindividual variability in the response to selective inhibitors of cyclooxygenase-2. Gastroenterology 130:55–64.

Fritsche, E., Baek, S. J., King, L. M., Zeldin, D. C., Eling, T. E., and Bell, D. A. 2001. Functional characterization of cyclooxygenase-2 polymorphisms. J Pharmacol Exp Ther 299:468–76.

Gerner, E. W., and Meyskens, F. L., Jr. 2004. Polyamines and cancer: old molecules, new understanding. Nat Rev Cancer 4:781–92.

Giovannucci, E. 1999. The prevention of colorectal cancer by aspirin use. Biomed Pharmacother 53:303–8.

Giovannucci, E., Rimm, E. B., Stampfer, M. J., Colditz, G. A., Ascherio, A., and Willett, W. C. 1994. Aspirin use and the risk for colorectal cancer and adenoma in male health professionals. Ann Intern Med 121:241–6.

Giovannucci, E., Egan, K. M., Hunter, D. J., Stampfer, M. J., Colditz, G. A., Willett, W. C., and Speizer, F. E. 1995. Aspirin and the risk of colorectal cancer in women. N Engl J Med 333:609–14.

Gong, Z., Hebert, J. R., Bostick, R. M., Deng, Z., Hurley, T. G., Dixon, D. A., Nitcheva, D., and Xie, D. 2007. Common polymorphisms in 5-lipoxygenase and 12-lipoxygenase genes and the risk of incident, sporadic colorectal adenoma. Cancer 109:849–57.

Goodman, J. E., Bowman, E. D., Chanock, S. J., Alberg, A. J., and Harris, C. C. 2004. Arachidonate lipoxygenase (ALOX) and cyclooxygenase (COX) polymorphisms and colon cancer risk. Carcinogenesis 25:2467–72.

Greenberg, E. R., Baron, J. A., Freeman, D. H., Jr., Mandel, J. S., and Haile, R. 1993. Reduced risk of large-bowel adenomas among aspirin users. The Polyp Prevention Study Group. J Natl Cancer Inst 85:912–6.

Guo, Y., Harris, R. B., Rosson, D., Boorman, D., and O'Brien, T. G. 2000. Functional analysis of human ornithine decarboxylase alleles. Cancer Res 60:6314–7.

Gupta, R. A., and Dubois, R. N. 2001. Colorectal cancer prevention and treatment by inhibition of cyclooxygenase-2. Nat Rev Cancer 1:11–21.

Halushka, M. K., Walker, L. P., and Halushka, P. V. 2003. Genetic variation in cyclooxygenase 1: effects on response to aspirin. Clin Pharmacol Ther 73:122–30.

Han, C., and Wu, T. 2005. Cyclooxygenase-2-derived prostaglandin E_2 promotes human cholangiocarcinoma cell growth and invasion through EP1 receptor-mediated activation of the epidermal growth factor receptor and Akt. J Biol Chem 280:24053–63.

Hillarp, A., Palmqvist, B., Lethagen, S., Villoutreix, B. O., and Mattiasson, I. 2003. Mutations within the cyclooxygenase-1 gene in aspirin non-responders with recurrence of stroke. Thromb Res 112:275–83.

Hu, Z., Miao, X., Ma, H., Wang, X., Tan, W., Wei, Q., Lin, D., and Shen, H. 2005. A common polymorphism in the 3'YTP of cyclooxygenase 2/prostaglandin synthase 2 gene and risk of lung cancer in a Chinese population. Lung Cancer 48:11–7.

Hubner, R. A., Muir, K. R., Liu, J. F., Logan, R. F., Grainge, M., Armitage, N., Shepherd, V., Popat, S., and Houlston, R. S. 2006. Genetic variants of UGT1A6 influence risk of colorectal adenoma recurrence. Clin Cancer Res 12:6585–9.

In, K. H., Asano, K., Beier, D., Grobholz, J., Finn, P. W., Silverman, E. K., Silverman, E. S., Collins, T., Fischer, A. R., Keith, T. P., Serino, K., Kim, S. W., De Sanctis, G. T., Yandava, C., Pillari, A., Rubin, P., Kemp, J., Israel, E., Busse, W., Ledford, D., Murray, J. J., Segal, A., Tinkleman, D., and Drazen, J. M. 1997. Naturally occurring mutations in the human 5-lipoxygenase gene promoter that modify transcription factor binding and reporter gene transcription. J Clin Invest 99:1130–7.

Iwai, N., Katsuya, T., Ishikawa, K., Mannami, T., Ogata, J., Higaki, J., Ogihara, T., Tanabe, T., and Baba, S. 1999. Human prostacyclin synthase gene and hypertension: the Suita Study. Circulation 100:2231–6.

Jaffe, B. M. 1974. Prostaglandins and cancer: an update. Prostaglandins 6:453–61.

Janne, J., Poso, H., and Raina, A. 1978. Polyamines in rapid growth and cancer. Biochim Biophys Acta 473:241–93.

Kingsnorth, A. N., Lumsden, A. B., and Wallace, H. M. 1984. Polyamines in colorectal cancer. Br J Surg 71:791–4.

Koh, W. P., Yuan, J. M., Van Den Berg, D., Lee, H. P., and Yu, M. C. 2004. Interaction between cyclooxygenase-2 gene polymorphism and dietary *n*-6 polyunsaturated fatty acids on colon cancer risk: The Singapore Chinese Health Study. Br J Cancer 90:1760–4.

Koo, H. B., Sigurdson, E. R., Daly, J. M., Berenson, M., Groshen, S., and Decosse, J. J. 1988. Ornithine decarboxylase levels in the rectal mucosa of patients with colonic neoplasia. J Surg Oncol 38:240–3.

Kuehl, G. E., Lampe, J. W., Potter, J. D., and Bigler, J. 2005. Glucuronidation of nonsteroidal anti-inflammatory drugs (NSAIDs): identifying the enzymes responsible in human liver microsomes. Drug Metabol Dispos 33:1027–35.

Kutchera, W., Jones, D. A., Matsunami, N., Groden, J., McIntyre, T. M., Zimmerman, G. A., White, R. L., and Prescott, S. M. 1996. Prostaglandin H synthase 2 is expressed abnormally in human colon cancer: evidence for a transcriptional effect. Proc Natl Acad Sci USA 93:4816–20.

La Vecchia, C., Negri, E., Franceschi, S., Conti, E., Montella, M., Giacosa, A., Falcini, A., and Decarli, A. 1997. Aspirin and colorectal cancer. Br J Cancer 76:675–7.

Lampe, J. W., Bigler, J., Horner, N. K., and Potter, J. D. 1999. UDP-glucuronosyltransferase (UGT1A1*28 and UGT1A6*2) polymorphisms in Caucasians and Asians: relationships to serum bilirubin concentrations. Pharmacogenetics 9:341–9.

LaMuraglia, G. M., Lacaine, F., and Malt, R. A. 1986. High ornithine decarboxylase activity and polyamine levels in human colorectal neoplasia. Ann Surg 204:89–93.

Lin, H. J., Lakkides, K. M., Keku, T. O., Reddy, S. T., Louie, A. D., Kau, I. H., Zhou, H., Gim, J. S., Ma, H. L., Matthies, C. F., Dai, A., Huang, H. F., Materi, A. M., Lin, J. H., Frankl, H. D., Lee, E. R., Hardy, S. I., Herschman, H. R., Henderson, B. E., Kolonel, L. N., Le Marchand, L., Garavito, R. M., Sandler, R. S., Haile, R. W., and Smith, W. L. 2002. Prostaglandin H synthase 2 variant (Val511Ala) in African Americans may reduce the risk for colorectal neoplasia. Cancer Epidemiol Biomarkers Prev 11:1305–15.

Logan, R. F., Little, J., Hawtin, P. G., and Hardcastle, J. D. 1993. Effect of aspirin and non-steroidal anti-inflammatory drugs on colorectal adenomas: case-control study of subjects participating in the Nottingham faecal occult blood screening programme [see comments]. BMJ 307:285–9.

Marcus, A. J., Broekman, M. J., and Pinsky, D. J. 2002. COX inhibitors and thromboregulation. N Engl J Med 347:1025–6.

Martinez, M. E., McPherson, R. S., Levin, B., and Annegers, J. F. 1995. Aspirin and other nonsteroidal anti-inflammatory drugs and risk of colorectal adenomatous polyps among endoscoped individuals. Cancer Epidemiol Biomarkers Prev 4:703–7.

Martinez, M. E., O'Brien, T. G., Fultz, K. E., Babbar, N., Yerushalmi, H., Qu, N., Guo, Y., Boorman, D., Einspahr, J., Alberts, D. S., and Gerner, E. W. 2003. Pronounced reduction in adenoma recurrence associated with aspirin use and a polymorphism in the ornithine decarboxylase gene. Proc Natl Acad Sci USA 100:7859–64.

Mead, J., Alfin-Slater, R., Howton, D., and Popjak, G. 1986. Prostaglandins, thromboxanes, and prostacyclin. New York: Plenum.

Miners, J. O., and Birkett, D. J. 1998. Cytochrome P4502C9: an enzyme of major importance in human drug metabolism. Br J Clin Pharmacol 45:525–38.

Murray, M. D., and Brater, D. C. 1993. Renal toxicity of the nonsteroidal anti-inflammatory drugs. Annu Rev Pharmacol Toxicol 33:435–65.

Muscat, J. E., Stellman, S. D., and Wynder, E. L. 1994. Nonsteroidal antiinflammatory drugs and colorectal cancer. Cancer 74:1847–54.

Mutoh, M., Watanabe, K., Kitamura, T., Shoji, Y., Takahashi, M., Kawamori, T., Tani, K., Kobayashi, M., Maruyama, T., Kobayashi, K., Ohuchida, S., Sugimoto, Y., Narumiya, S., Sugimura, T., and Wakabayashi, K. 2002. Involvement of prostaglandin E receptor subtype EP(4) in colon carcinogenesis. Cancer Res 62:28–32.

Ng, P. C., and Henikoff, S. 2003. SIFT: Predicting amino acid changes that affect protein function. Nucleic Acids Res 31:3812–4.

Nussmeier, N. A., Whelton, A. A., Brown, M. T., Langford, R. M., Hoeft, A., Parlow, J. L., Boyce, S. W., and Verburg, K. M. 2005. Complications of the COX-2 inhibitors parecoxib and valdecoxib after cardiac surgery. N Engl J Med 352:1081–91.

Orbe, J., Beloqui, O., Rodriguez, J. A., Belzunce, M. S., Roncal, C., and Paramo, J. A. 2006. Protective effect of the G-765C COX-2 polymorphism on subclinical atherosclerosis and inflammatory markers in asymptomatic subjects with cardiovascular risk factors. Clin Chim Acta 368:138–43.

Oshima, M., Dinchuk, J. E., Kargman, S. L., Oshima, H., Hancock, B., Kwong, E., Trzaskos, J. M., Evans, J. F., and Taketo, M. M. 1996. Suppression of intestinal polyposis in Apc delta716 knockout mice by inhibition of cyclooxygenase 2 (COX-2). Cell 87:803–9.

Ostrowski, J., Wocial, T., Skurzak, H., and Bartnik, W. 2003. Do altering in ornithine decarboxylase activity and gene expression contribute to antiproliferative properties of COX inhibitors? Br J Cancer 88:1143–51.

Pai, R., Soreghan, B., Szabo, I. L., Pavelka, M., Baatar, D., and Tarnawski, A. S. 2002. Prostaglandin E_2 transactivates EGF receptor: a novel mechanism for promoting colon cancer growth and gastrointestinal hypertrophy. Nat Med 8:289–93.

Panguluri, R. C. K., Long, L. O., Chen, W., Wang, S., Coulibaly, A., Ukoli, F., Jackson, A., Weinrich, S., Ahaghotu, C., Isaacs, W., and Kittles, R. A. 2004. COX-2 gene promoter haplotypes and prostate cancer risk. Carcinogenesis 25:961–6.

Papafili, A., Hill, M. R., Brull, D. J., McAnulty, R. J., Marshall, R. P., Humphries, S. E., and Laurent, G. J. 2002. Common promoter variant in cyclooxygenase-2 represses gene expression: evidence of role in acute-phase inflammatory response.[comment]. Arterioscler Thromb Vasc Biol 22:1631–6.

Peleg, I. I., Maibach, H. T., Brown, S. H., and Wilcox, C. M. 1994. Aspirin and nonsteroidal anti-inflammatory drug use and the risk of subsequent colorectal cancer [see comments]. Arch Intern Med 154:394–9.

Poole, E. M., Bigler, J., Whitton, J., Sibert, J. G., Potter, J. D., and Ulrich, C. M. 2006. Prostacyclin synthase and arachidonate 5-lipoxygenase polymorphisms and risk of colorectal polyps. Cancer Epidemiol Biomarkers Prev 15:502–8.

Porter, C. W., Herrera-Ornelas, L., Pera, P., Petrelli, N. F., and Mittelman, A. 1987. Polyamine biosynthetic activity in normal and neoplastic human colorectal tissue. Cancer 60:1275–81.

Pugh, S., and Thomas, G. A. 1994. Patients with adenomatous polyps and carcinomas have increased colonic mucosal prostaglandin E_2. Gut 35:675–8.

Ramensky, V., Bork, P., and Sunyaev, S. 2002. Human non-synonymous SNPs: server and survey. Nucleic Acids Res 30:3894–900.

Rettie, A. E., Wienkers, L. C., Gonzalez, F. J., Trager, W. F., and Korzekwa, K. R. 1994. Impaired (*S*)-warfarin metabolism catalysed by the R144C allelic variant of CYP2C9. Pharmacogenetics 4:39–42.

Rigas, B., Goldman, I. S., and Levine, L. 1993. Altered eicosanoid levels in human colon cancer. J Lab Clin Med 122:518–23.

Samowitz, W. S., Wolff, R. K., Curtin, K., Sweeney, C., Ma, K. N., Andersen, K., Levin, T. R., and Slattery, M. L. 2006. Interactions between CYP2C9 and UGT1A6 polymorphisms and nonsteroidal anti-inflammatory drugs in colorectal cancer prevention. Clin Gastroenterol Hepatol 4:894–901.

Sandler, R. S., Halabi, S., Baron, J. A., Budinger, S., Paskett, E., Keresztes, R., Petrelli, N., Pipas, J. M., Karp, D. D., Loprinzi, C. L., Steinbach, G., and Schilsky, R. 2003. A randomized trial of aspirin to prevent colorectal adenomas in patients with previous colorectal cancer. N Engl J Med 348:883–90.

Sansbury, L. B., Millikan, R. C., Schroeder, J. C., North, K. E., Moorman, P. G., Keku, T. O., de Cotret, A. R., Player, J., and Sandler, R. S. 2006. COX-2 polymorphism, use of nonsteroidal anti-inflammatory drugs, and risk of colon cancer in African Americans (United States). Cancer Causes Control 17:257–66.

Schreinemachers, D. M., and Everson, R. B. 1994. Aspirin use and lung, colon, and breast cancer incidence in a prospective study [see comments]. Epidemiology 5:138–46.

Scornioni. 2001. Manipulation of the expression of regulatory genes of polyamine metabolism results in specific alterations of the cell-cycle progression. Biochem J 354:217–23.

Scott, B. T., Hasstedt, S. J., Bovill, E. G., Callas, P. W., Valliere, J. E., Wang, L., Wu, K. K., and Long, G. L. 2002. Characterization of the human prostaglandin H synthase 1 gene (PTGS1): exclusion by genetic linkage analysis as a second modifier gene in familial thrombosis. Blood Coagul Fibrinolysis 13:519–31.

Shureiqi, I., and Lippman, S. M. 2001. Lipoxygenase modulation to reverse carcinogenesis. Cancer Res 61:6307–12.

Siezen, C. L., van Leeuwen, A. I., Kram, N. R., Luken, M. E., van Kranen, H. J., and Kampman, E. 2005. Colorectal adenoma risk is modified by the interplay between polymorphisms in arachidonic acid pathway genes and fish consumption. Carcinogenesis 26:449–57.

Siezen, C. L., Bueno-de-Mesquita, H. B., Peeters, P. H., Kram, N. R., van Doeselaar, M., and van Kranen, H. J. 2006a. Polymorphisms in the genes involved in the arachidonic acid-pathway, fish consumption and the risk of colorectal cancer. Int J Cancer 119:297–303.

Siezen, C. L., Tijhuis, M. J., Kram, N. R., van Soest, E. M., de Jong, D. J., Fodde, R., van Kranen, H. J., and Kampman, E. 2006b. Protective effect of nonsteroidal anti-inflammatory drugs on colorectal adenomas is modified by a polymorphism in peroxisome proliferator-activated receptor delta. Pharmacogenet Genomics 16:43–50.

Silverman, E. S., and Drazen, J. M. 2000. Genetic variations in the 5-lipoxygenase core promoter. Description and functional implications. Am J Respir Crit Care Med 161:S77–80.

Smith, W. L., DeWitt, D. L., and Garavito, R. M. 2000. Cyclooxygenases: structural, cellular, and molecular biology. Annu Rev Biochem 69:145–82.

SNP500Cancer. http://snp500cancer.nci.nih.gov/home_1.cfm.

Solomon, S. D., McMurray, J. J., Pfeffer, M. A., Wittes, J., Fowler, R., Finn, P., Anderson, W. F., Zauber, A., Hawk, E., and Bertagnolli, M. 2005. Cardiovascular risk associated with celecoxib in a clinical trial for colorectal adenoma prevention. N Engl J Med 352:1071–80.

Suh, O., Mettlin, C., and Petrelli, N. J. 1993. Aspirin use, cancer, and polyps of the large bowel. Cancer 72:1171–7.

Takahashi, H., Kashima, T., Nomoto, S., Iwade, K., Tainaka, H., Shimizu, T., Nomizo, Y., Muramoto, N., Kimura, S., and Echizen, H. 1998. Comparisons between in-vitro and in-vivo metabolism of (*S*)-warfarin: catalytic activities of cDNA-expressed CYP2C9, its Leu359 variant and their mixture versus unbound clearance in patients with the corresponding CYP2C9 genotypes. Pharmacogenetics 8:365–73.

Taketo, M. M. 1998. Cyclooxygenase-2 inhibitors in tumorigenesis (part I). J Natl Cancer Inst 90:1529–36.

Thun, M. J., Namboodiri, M. M., and Heath, C. W., Jr. 1991. Aspirin use and reduced risk of fatal colon cancer. N Engl J Med 325:1593–6.

Turchanowa, L., Dauletbaev, N., Milovic, V., and Stein, J. 2001. Nonsteroidal anti-inflammatory drugs stimulate spermidine/spermine acetyltransferase and deplete polyamine content in colon cancer cells. Eur J Clin Invest 31:887–93.

Ulrich, C. M., Bigler, J., Sibert, J., Greene, E. A., Sparks, R., Carlson, C. S., and Potter, J. D. 2002. Cyclooxygenase 1 (COX1) polymorphisms in African-American and Caucasian populations. Hum Mutat 20:409–10.

Ulrich, C. M., Bigler, J., Sparks, R., Whitton, J., Sibert, J. G., Goode, E. L., Yasui, Y., and Potter, J. D. 2004. Polymorphisms in PTGS1 (=COX-1) and risk of colorectal polyps. Cancer Epidemiol Biomarkers Prev 13:889–93.

Ulrich, C. M., Whitton, J., Yu, J. H., Sibert, J., Sparks, R., Potter, J. D., and Bigler, J. 2005. PTGS2 (COX-2) −765G > C promoter variant reduces risk of colorectal adenoma among nonusers of nonsteroidal anti-inflammatory drugs. Cancer Epidemiol Biomarkers Prev 14:616–9.

Ulrich, C. M., Bigler, J., and Potter, J. D. 2006. Non-steroidal anti-inflammatory drugs for cancer prevention: promise, perils, and pharmacogenetics. Nat Rev Cancer 6:130–40.

UW-FHCRC Variation Discovery Resource (UW-FHCRC-VDR). http://pga.gs.washington.edu/.

UW-NIEHS Environmental Genome Project (NIEHS-EGP). http://egp.gs.washington.edu/.

Vane, J. R. 1971. Inhibition of prostaglandin synthesis as a mechanism of action for aspirin-like drugs. Nat New Biol 231:232–5.
Watanabe, K., Kawamori, T., Nakatsugi, S., Ohta, T., Ohuchida, S., Yamamoto, H., Maruyama, T., Kondo, K., Ushikubi, F., Narumiya, S., Sugimura, T., and Wakabayashi, K. 1999. Role of the prostaglandin E receptor subtype EP1 in colon carcinogenesis. Cancer Res 59:5093–6.
Watanabe, K., Kawamori, T., Nakatsugi, S., Ohta, T., Ohuchida, S., Yamamoto, H., Maruyama, T., Kondo, K., Narumiya, S., Sugimura, T., and Wakabayashi, K. 2000. Inhibitory effect of a prostaglandin E receptor subtype EP(1) selective antagonist, ONO-8713, on development of azoxymethane-induced aberrant crypt foci in mice. Cancer Lett 156:57–61.
Wolter, F., Ulrich, S., and Stein, J. 2004. Molecular mechanisms of the chemopreventive effects of resveratrol and its analogs in colorectal cancer: key role of polyamines? J Nutr 134:3219–22.
Yan, M., Rerko, R. M., Platzer, P., Dawson, D., Willis, J., Tong, M., Lawrence, E., Lutterbaugh, J., Lu, S., Willson, J. K. V., Luo, G., Hensold, J., Tai, H.-H., Wilson, K., and Markowitz, S. D. 2004. 15-Hydroxyprostaglandin dehydrogenase, a COX-2 oncogene antagonist, is a TGF-beta-induced suppressor of human gastrointestinal cancers. Proc Natl Acad Sci USA 101:17468–73.
Zhang, X., Miao, X., Tan, W., Ning, B., Liu, Z., Hong, Y., Song, W., Guo, Y., Shen, Y., Qiang, B., Kadlubar, F. F., and Lin, D. 2005. Identification of functional genetic variants in cyclooxygenase-2 and their association with risk of esophageal cancer. Gastroenterology 129:565–76.

Chapter 11
The Role of Chemical Carcinogens and Their Biotransformation in Colorectal Cancer

Loïc Le Marchand

As reviewed in Chap. 1, the epidemiology of colorectal cancer (CRC) suggests a predominant role for lifestyle factors in the etiology of this disease. A number of these risk factors, including smoking and consumption of well-done or processed meat, may lead to exposure to exogenous chemicals which are strongly suspected to cause cancer in humans. For the purposes of this chapter, in line with standard nutritional epidemiologic usage, red meat refers to meat from mammals, white meat to that from fowl and fish. Most of these chemicals require transformation by xenobiotic-metabolism enzymes in order to become active carcinogens that are capable of binding to DNA and inducing mutations. Specific lifestyle exposures, such as alcohol, certain phytochemicals, smoking, and exogenous estrogens, may also induce or inhibit many of these biotransformation enzymes. Thus, the large interindividual variation which is typically observed in the activity levels of these enzymes may be due to differences in lifestyle. That variation may also reflect genetic differences, because the genes that code for these enzymes often contain common inherited polymorphisms that affect activity. Consequently, exposures to chemical carcinogens through diet and smoking, along with these possible modifying factors, both environmental and genetic, have been investigated for their associations with CRC. It should be noted that because these exposures are particularly common in Western countries, it is possible that they may explain a sizable component of the excess CRC risk observed in the developed world. The purpose of this chapter is to review the available research on chemical carcinogens, biotransformation and modifying factors, as it relates to the risk of CRC in the general population.

L.Le Marchand
Epidemiology Program, Cancer Research Center of Hawaii, University of Hawaii
e-mail: loic@crch.hawaii.edu

J.D. Potter and N.M Lindor (eds.), *Genetics of Colorectal Cancer*,
DOI: 10.1007/978-0-387-09568-4_19,

Chemical Carcinogens

Heterocyclic Aromatic Amines

More than twenty known heterocyclic aromatic amines (HAAs) have been shown to form when meat or fish is cooked at high temperature (>250°C) to a "well-done" state. Many of these compounds are known to be carcinogenic in experimental animals, including nonhuman primates (Sugimura et al. 2004). Several epidemiologic studies have linked the consumption of well-done meat with an increased risk of cancer at several sites, including the large bowel (Cross and Sinha 2004), although the data have not been entirely consistent. The levels of HAAs formed in cooked meats are dependent upon the type of meat, the temperature and duration of cooking, the use of sauces and marinades, and a variety of other factors which, together, make exposure assessment in free-living individuals difficult. The most abundant HAAs in cooked meats are 2-amino-1-methyl-6-phenylimidazol[4,5-*b*] pyridine (PhIP), 2-amino-3,8-dimethylimidazol[4,5-*f*]quinoxaline (8-MeIQx), 2-amino-3,4,8-trimethylimidazol[4,5-*f*]quinoxaline (4,8-DiMeIQx), and 2-amino-9*H*-pyrido[2,3-*b*]indole (AαC). Studies that have estimated HAA intake have usually shown a weak main-effect association with colorectal adenoma or cancer (Sinha et al. 1999, 2001, 2005b; Nowell et al. 2002; Butler et al. 2003; Wu et al. 2006), but not always (Augustsson et al. 1999; Le Marchand et al. 2002a; Gunter et al. 2005; Shin et al. 2007). Ingestion of a realistic dose of PhIP has also been shown to result in significant PhIP–DNA adduct formation in the colon (Dingley et al. 1999).

Smoking is another common source of HAA exposure in humans. AαC and PhIP are the most abundant of the known HAAs formed in tobacco smoke, and these compounds are considered human carcinogens (Hecht 2003). These carcinogens can reach the large bowel through the circulation or, perhaps, through direct contact after oral ingestion with mucus or saliva.

Polycyclic Aromatic Hydrocarbons

polycyclic aromatic hydrocarbons (PAHs) are carcinogenic pyrolysis products that are present in tobacco smoke and in cured meats or smoked foods, or are formed when meat is cooked directly above an open heat source (e.g., by grilling or barbecuing) (Phillips 1999). Benzo[α]pyrene (B[α]P) is one of the most potent PAH carcinogens in animal studies (Goldstein et al. 1998). Several epidemiologic studies have reported an association of B[α]P exposure through grilled/barbecued meat intake with colorectal adenoma (Sinha et al. 2005a, b; Gunter et al. 2005). PAH–DNA adducts have been shown to be present in the colonic mucosa (Alexandrov et al. 1996), and levels rise in circulating leukocytes (a potential surrogate marker) as a result of smoking tobacco or eating charbroiled meat (Rothman et al. 1990; Kang et al. 1995). Finally, levels of PAH–DNA adducts in blood leukocytes have been found to be associated with risk of colorectal adenoma (Gunter et al. 2007).

Nitrosamines

Salted, smoked, and pickled foods, and meat processed with nitrate or nitrite are the main source of preformed N-nitroso compounds (NOCs) in the diet (Tricker 1997). Intake of processed meats has been more consistently and strongly associated with CRC than other red meats in recent meta-analyses of the literature (Sandhu et al. 2001; Norat et al. 2002). Estimated, dietary nitrate and nitrite intakes have also been associated with colorectal adenoma (Ward et al. 2007); similarly, intake of a common preformed dietary nitrosamine in the diet, *N*-nitrosomethylamine (NDMA), has been associated with CRC risk in a Finnish cohort study (Knekt et al. 1999).

Moreover, the feeding of cooked fresh red meat to humans has been shown to increase endogenous NOC formation in the large intestine, as measured by fecal NOC level; this was not seen with white meat or fish, or with amounts of red meat below the approximate average intake in Western countries (Bingham et al. 2002). The proposed mechanism underlying this relationship is that heme, present in red, but not white, meat stimulates the endogenous formation of NOCs (Bingham et al. 1996; Cross et al. 2003). These carcinogens are able to cause gastrointestinal tumors in animals (Bogovski and Bogovski 1981). O^6-methylguanine is a promutagenic adduct formed by many *N*-methyl-NOCs and is responsible for the mutagenicity and carcinogenicity of alkylating agents. A high red-meat intake has recently been shown to increase the proportion of exfoliated colonic cells staining positive for the NOC-specific DNA adduct O^6-carboxymethylguanine, demonstrating a link between red-meat intake and a promutagenic lesion in the colon (Lewin et al. 2006).

Tobacco smoke is a source of exposure to other nitrosamines that are potent carcinogens and may affect CRC risk. The tobacco-specific 4-(methylnitrosamino)-1-(3-pyridyl)-1-butanone (NNK) is not known to cause colorectal tumors in experimental animals. However, it has been shown to stimulate the growth of colon cancer cells in culture by activation of β-adrenoceptors (Wu et al. 2005).

Acrylamide

The general population is exposed to measurable amounts of acrylamide through smoking and consumption of heat-processed carbohydrate-rich foods (Tareke et al. 2002). Glycidamine, the metabolite of acrylamide, has been shown to induce mutations in bacteria (IARC 1994), as well as chromosomal alterations and cell transformation in mammalian cell lines (Dearfield et al. 1995). Glycidamine is also known to form DNA adducts in vivo (IARC 1994). To date, the epidemiologic data do not suggest an association between dietary acrylamide intake and CRC (Mucci et al. 2003; Pelucci et al. 2003; Dybing and Sanner 2003), although studies have been few and dietary exposure is difficult to assess because levels vary markedly with food processing conditions.

Biotransformation

Most chemical carcinogens require activation by biotransformation enzymes in order to become reactive and bind to DNA or other target proteins (see Table 11.1). These enzymes play an important role in the metabolism and elimination of a variety of xenobiotics, including drugs, toxins, and carcinogens. In general, Phase I enzymes catalyze reactions that increase the reactivity of hydrophobic compounds, preparing them for reactions catalyzed by Phase II enzymes. The latter generally increase water solubility and facilitate elimination of the compounds through the urine. Phase I enzymes are mostly cytochrome P450 (CYP) enzymes; Phase II enzymes include glutathione S-transferases (GSTs), sulfotransferases (SULTs), UDP-glucuronosyl transferases (UGTs), NADPH quinine oxidoreductase (NQO), N-acetyltransferases (NATs), and others.

Table 11.1 Xenobiotic-metabolizing enzymes

Enzymes	Reactions
Phase 1: "Oxygenases"	Oxidation, reduction, or hydrolytic reactions
Cytochrome P450s (CYPs)	N and S oxidation, dealkylation, aliphatic and aromatic hydroxylation, deamination, dehalogenation
Flavin-containing monooxygenases	Nitrogen, Sulfur, and P oxidation
Epoxide hydrolases	Hydrolysis of epoxides
Phase 2: "Transferases"	Conjugating with substrate
Sulfotransferases	Addition of sulfate
UDP*-glucoronosyltransferases (UGTs)	Addition of glucuruonic acid
Glutathione-*S*-transferases (GSTs)	Addition of glutathione
N-acetyltransferases (NATs)	Addition of acetyl group
Methyltransferases (MTs)	Addition of methyl group
Reducing enzymes	
Alcohol dehydrogenases	Reduction of alcohols
Aldehyde dehydrogenases	Reduction of aldehydes
NADPH**-quinone oxidoreductase (NQC)	Reduction of quinines

*UDP** uridine diphosphate; *NADPH*** reduced nicotinamide adenine dinucleotide phosphate. Information on specific genetic polymorphisms of these enzymes can be found at http://www.hgvbase.org/ (The Human Genome Variation Database), http://www.pharmgkb.org (Pharmacogenetics and Pharmacogenomics Knowledge Base (PharmGKB), http://www.ncbi.nlm.nih.gov/sites/entrez?db=omim (Online Mendelian Inheritance in Man)

Heterocyclic Amines

The major pathways in the metabolism of HAAs, such as those for 8-MeIQX and PhIP, have been well characterized (Turesky 2002). These compounds first undergo oxidation by CYP enzymes and, then, direct conjugation by UGTs or SULTs before being eliminated (Turesky et al. 1998; Stillwell et al. 1997). CYP1A2, which is principally expressed in the liver, is the major P450 involved in the oxidation of these HAAs (Turesky 2002). CYP 1A1 and 1B1 play a more minor role in the activation of these procarcinogens (Crofts et al. 1998; Shimada et al. 1999). N-hydroxylated metabolites can further undergo O-sulfonylation by SULT1A1 which leads to the formation of *N*-sulfonyloxy esters that can undergo heterocyclic cleavage generating nitrenium ions, the ultimate carcinogens. *N*-hydroxy-HCAs can also undergo O-acetylation by NAT2 and, to a lesser extent, NAT1, also contributing to the formation of nitrenium ions. Both NAT1 and NAT2 are expressed in intestinal cells (Hein 2002). In contrast, NAT1 and NAT2 are also involved in the detoxification of aromatic amines (to which tobacco smokers are exposed) by N-acetylation reaction.

Polycyclic Hydrocarbons

B[α]P, have been shown to be metabolized initially by CYP 1A1 or 1B1 to an epoxide [benzo(α)pyrene-7,8-epoxide] and, subsequently, hydrolyzed by microsomal epoxide hydrolase (EPHX1) to a dihydrodiol (benzo(α)pyrene-7,8-dihydrodiol) (Shimada et al. 1999; Rihs et al. 2005). CYP 1A1, 1B1, or 3A4 can then transform the dihydrodiol to a highly reactive diol-epoxide (benzo(α)pyrene-7,8-dihydrodiol-9,10-epoxide, BPDE) that can covalently bind to DNA, creating a PAH adduct which may, if the DNA is not repaired, induce mutation, predominantly in the form of G-to-T transversions (McCoull et al. 1999). PAH-diol-epoxide metabolites can be detoxified by GSTs, particularly GSTM1 and GSTP1, which exhibit substrate specificity and are expressed in the colon (Sundberg et al. 1997; Hoensch et al. 2006).

N-Nitroso Compounds

NDMA, a common preformed nitrosamine in the diet, undergoes hydroxylation and subsequent hydrolysis to an aldehyde and a monoalkylnitrosamine that rearranges and releases a carbocation that reacts with DNA bases (Loeppky 1999). The hydroxylation is catalyzed by CYP2E1 (Lin et al. 1999); other P450s, including CYP2A6, have also been implicated (Gonzalez and Gelboin 1993; Kamataki et al. 1999).

Modulators of Biotransformation

There are large interindividual differences in the rates of metabolism of drugs and carcinogens. These differences are not completely understood, but some environmental and genetic factors have been identified.

Environmental Modulators

Environmental factors that influence the metabolism of drugs and carcinogens in humans include diet, smoking, alcohol, drugs (e.g., phenobarbital, rifampicin, clotrimazole), herbal remedies (e.g., St. Johns wort), and exposure to environmental pollutants (e.g., PAHs, dioxin) (Conney 2003). Smoking is known to induce a number of CYP enzymes, such as 1A1, 1A2, and 1B1. Dietary factors, and the CYP enzymes that they induce, include caffeine (1A2), alcohol (2E1), well-done meat (1A1, 1A2). Increasing the ratio of protein to carbohydrate in the diet has also been shown to increase the oxidative metabolism of certain drugs (Conney 2003). In contrast, grapefruit is known to inhibit CYP3A4 and 1A2, whereas cruciferous vegetables (e.g., watercress and broccoli sprouts) inhibit CYP2E1 (Conney 2003; Cuthrell and Le Marchand 2006).

Cruciferous vegetables also induce Phase II enzymes. Consumption of watercress by smokers increased the excretion of glucuronidated metabolites of nicotine, suggesting that UGT activity is increased (Hecht et al. 1999). Consumption of Brussels sprouts for a week significantly increased plasma and intestinal GST levels in nonsmokers (Nijhoff et al. 1995). Butyrate, one of the major products of colonic microbial fermentation, has also been shown to induce GST in colon tumor cell lines and to protect against genotoxicity (Ebert et al. 2001). Recently, GST activity in the rectal mucosa has been shown to be affected by fruit and vegetable intake (Tijhuis et al. 2007). In future studies, GST activity in blood lymphocytes may serve as a convenient biomarker because it has been shown to correlate with GST activity in colon tissue (Szarka et al. 1995).

The mechanisms by which hydrolysis products of glucosinolates from cruciferous vegetables induce Phase II enzymes are relatively well understood. Isothiocyanates are known to increase the transcription of genes that contain an antioxidant response element (ARE), such as *GST*s and *NQO* (Higdon et al. 2007). Similarly, acid condensation products of indole-3-carbinol bind in the cytoplasm to the aryl hydrocarbon receptor (AhR) and complex with the AhR nucleus translocator (Arnt) protein to enter the nucleus (Safe 2001). This complex binds to specific DNA sequences, the xenobiotic response elements (XRE), and the transcription of the corresponding genes (e.g., *CYP1A1*, *1A2*, *1B1*) is enhanced.

It should be noted that the relationship between modulation of biotransformation enzymes and carcinogenesis is not straightforward. Although the induction of CYP enzymes that metabolize carcinogens usually inhibits carcinogenesis in experimental animals (presumably because detoxification pathways are enhanced to a greater extent than activation pathways), sometimes carcinogenesis is enhanced (Conney 2003). Furthermore, some enzyme inducers that inhibit carcinogenesis when given together with the carcinogen are tumor promoters when given after the carcinogen (Conney 2003).

Genetic Modulators

Inherited single-nucleotide polymorphisms (SNPs) or copy number variants (CNVs) in the genes coding for biotranformation enzymes often affect enzyme activity, either by affecting the expression of the gene or the transcription of the mRNA, or changing the amino acid sequence of the protein.

The human acetylator polymorphism was identified over 50 years ago when it was observed that the N-acetylation of isoniazid and sulfamethazine divided human populations into rapid, intermediate, and slow acetylator phenotypes. In recent years, over 25 human *NAT1* and *NAT2* alleles have been identified (Hein 2002), and the relationship of these alleles to phenotype has been relatively well characterized. Because *NAT2*4* confers high enzyme activity and is the most common allele in the originally studied population (Japanese), it is defined as the reference allele. Most epidemiologic studies of *NAT2* assessed three (M1, M2, M3) or four (+M4 in African Americans) alleles to individuals identified as having the slow phenotype. It has recently been shown that this approach results in some misclassification, and a more comprehensive genotyping method (with 12 variant alleles) has been proposed (Hein 2002). Similarly, for *NAT1*, *1*4* is defined as the reference allele (Hein 2002), and a comprehensive genotyping method has recently been proposed (Doll and Hein 2002). The *NAT1*10* allele has been associated with a rapid-acetylator phenotype, both in vitro and in vivo (Bell et al. 1995; Hein 2002), but this has not been confirmed by recombinant expression studies (Hein 2002). Thus, the relationship between *NAT1* genetic variants and phenotypic enzyme activity remains unclear.

Other biotransformation genes have been less comprehensively studied. Several genetic variants in *CYP1A2* have been investigated for association with enzyme activity assessed indirectly by caffeine-metabolism phenotyping. One polymorphism in exon 1 (*CYP1A2*1F*) has been associated with a lower inducibility in smokers (Sachse et al. 1999). A 2455 A-to-G substitution polymorphism in *CYP1A1*, resulting in a Ile462Val substitution in the heme binding region of exon 7, has been shown to be associated with an increased in vitro activity and/or inducibility (Landi et al. 1994; Crofts et al. 1994; Kiyohara et al. 1996). Moreover, human studies that have used urinary 1-hydroxypyrene as a marker of PAH activation have shown higher 1-OHP excretion in individuals with the polymorphism (Wu et al. 1998; Merlo et al. 1998; Nerurkar et al. 2000). A G1294C substitution in the *CYP1B1* gene is also thought to result in a more active enzyme variant (Shimada et al. 1999). As mentioned earlier, the regulation of *CYPs1A1*, *1A2*, and

1B1 expression is under the control of AhR, a ligand activated transcription factor (Swanson and Bradfield 1993). A polymorphism within the coding region of the *AHR* gene, which results in replacement of Arg by Lys at codon 554 (G1721A polymorphism), has been identified, in a Japanese population, with an allele frequency of 0.43 (Kawajiri et al. 1995) and shown to be associated with a threefold increase in induced CYP1A1 activity (Smart and Daly 2000).

The several-fold variation in EPHX1 activity in humans has partly been attributed to polymorphisms in exon 3 (Tyr113His) and exon 4 (His139Arg) of the *EPHX1* gene that result in amino acid substitutions. EPHX1 activity has been shown in vitro to be reduced (about 40%) with 113His and increased (about 25%) with 139Arg, possibly due to altered stability of the protein (Hassett et al. 1994; Laurenzana et al. 1998). The combined high activity alleles for these polymorphisms have been associated with increased BPDE DNA adducts (Pastorelli et al. 1998) and chromosomal aberrations (Cajas-Salazar et al. 2003).

The G638A polymorphism in *SULT1A1* results in an amino acid change (Arg to His) and decreased sulfotransferase activity, as measured in platelets (Ozawa et al. 1994). Functional polymorphisms have also been described in the *CYP2A6* gene by studying individuals who were deficient in their ability to metabolize the drug, coumarin, a known substrate (Fernandez-Salguero et al. 1995). The *CYP2A6*2* variant allele has a T-to-A substitution at codon 160 that leads to a leu-to-his change and reduced enzyme activity. The *CYP2A6*3* allelic variant may have resulted from a gene conversion between the wild-type allele and the neighboring *CYP2A7*. It has been suggested that this polymorphism confers reduced activity because of sequence similarity to *CYP2A7* which codes for an inactive enzyme (Fernandez-Salguero et al. 1995).

Sequence variations in *GST* genes are common and have been shown to result in changes in isoenzyme levels, either through deletion (*GSTM1* and *GSTT1*) or single nucleotide polymorphisms (e.g., *GSTP1* and *GSTA1*). The activity of GST isoenzymes in the rectal mucosa has been shown to be affected by these polymorphisms (Tijhuis et al. 2007).

Genetic Polymorphisms in Biotransformation Genes and CRC Risk

Using the considerable variation that exists in the prevalence of the rapid-acetylation phenotype across populations, a recent ecological study showed that, in combination with meat intake, some significant proportion of the international variability in CRC incidence can be attributed to *NAT2* genotype (Ognjanovic et al. 2006). Two recent reviews of past analytical studies of NAT phenotype or genotype and CRC or adenoma concluded that no consistent (main effect) association had been found (Brockton et al. 2000; Hein 2002). However, most studies that examined the combined effects of dietary exposure and the NAT2 phenotype or genotype reported a stronger effect on CRC or adenoma risk for meat (Roberts-Thomson et al. 1996), fried meat (Welfare et al. 1997), red meat (Chen et al. 1998), or a meat mutagen index (Kampman et al. 1999) in rapid/intermediate acetylators than in slow acetylators. Thus, these data provide some evidence for a joint effect of the NAT2 rapid phenotype and meat carcinogens on adenoma and CRC. Fewer studies have included *NAT1*10*, but some have also suggested a stronger association for this allele among subjects exposed to HAAs (Brockton et al. 2000; Hein 2002). Finally, two studies that examined the joint effects of the CYP1A2 and NAT2 rapid phenotypes found a marked increased risk for CRC in subjects who also were exposed to well-done meat (Lang et al. 1994; Le Marchand et al. 2001). Other studies have examined the interactions of *NAT2* and *NAT1* with smoking, but the results have been inconsistent (Brockton et al. 2000; Lilla et al. 2006). Unfortunately, all these studies were relatively small and lacked statistical power to test conclusively for interactions.

A smaller number of studies have examined other genes involved in meat carcinogen metabolism in relation to CRC. Individuals with the rapid CYP2A6 phenotype or low activity *GSTA1* genotype were found to be at increased risk, and those with the low activity *SULT1A1* genotype at lower risk, of CRC (Nowell et al. 2002). The *SULT1A1*2* (high activity) allele has also been associated with an increased CRC risk (Sun et al. 2005) and main-effect associations were also reported for SNPs in *UGT1A6* and *UGT1A7* (van der Logt et al. 2004; Hubner et al. 2006) for CRC and adenoma. These associations are consistent with the role of the corresponding enzymes in the biotransformation of HAAs or nitrosamines, although interactions with exposure to meat carcinogens were either not tested or not significant. Such measures are useful in strengthening the biological plausibility of these relationships. In this regard, the findings that two functional polymorphisms in *CYP2E1* modify the associations of red meat and processed meats with rectal cancer provide additional evidence for an association of nitrosamines with CRC (Le Marchand et al. 2002b).

CYP1A1 has also been investigated in relation to CRC. A main effect, as well as an interaction with smoking, was suggested in several studies (Sivaraman et al. 1994; Le Marchand 2002; Slattery et al. 2004). Moreover, smokers carrying both the *CYP1A1* Val462 and *NQO1* ser187 alleles have been reported to be at markedly increased risk of colorectal adenoma (Hou et al. 2005). The two *EPHX1* alleles in codons 113 and 139 associated with high predicted enzymatic activity have been reported to increase risk for colorectal adenoma and CRC, particularly among smokers or individuals who regularly eat well-done meat (Cortessis et al. 2001; Huang et al. 2005); however, not all studies found these associations (Robien et al. 2005; Mitroun et al. 2007).

A recent review of the literature on CRC and the *GSTM1* and *GSTT1* deletion polymorphisms concluded that no consistent main-effect association had been observed (Cotton et al. 2000). Similarly, studies that examined the combined effects of these polymorphisms and smoking have usually found no or weak interactions (Gertig et al. 1998; Cotton et al. 2000; Lüchtenborg et al. 2005; Huang et al. 2006). A cohort study in Singapore reported that intake of cruciferous vegetables [a source of isothiocyanates (ITC)] modified the risk of colorectal cancer in individuals with low GST activity (Seow et al. 2002): a 57% reduction in CRC risk was observed among high ITC consumers with both the *GSTM1* and *GSTT1* null genotypes. It remains to be seen whether such an effect can be detected in Western populations which typically have a markedly lower intake of cruciferous vegetables.

The most recent studies have examined the effects of a larger number of SNPs or genes. One study testing the associations of multiple SNPs in *CYP* genes and CRC reported a main-effect association for SNPs in *CYP1A2* and *CYP1B1* (Bethke et al. 2007). Another study of multiple combinations of *CYP* gene polymorphisms replicated the associations with *CYP1A2* and *CYP2E1* mentioned above and suggested that building a multigenic model might be a promising approach (Küry et al. 2007).

Conclusion and Research Needs

Many decades of laboratory research on chemical carcinogens have provided a rich foundation for the investigation of their biological effects and mechanisms of action in humans. Epidemiologic studies have confirmed that large segments of the population are exposed to significant doses of these compounds through diet and smoking, especially in developed countries where two-third of the world CRCs occur. An increase in red-meat and cigarette consumption, as seen among Japanese migrants to the US, and in Japan and Korea since the 1950s, has been followed by a rapid rise in CRC rates (Le Marchand 1999; Kono 2004). Case-control and prospective studies have provided suggestive evidence for the role of specific meat and tobacco carcinogens in the etiology of CRC. Evidence is also emerging for the additional role of environmental and genetic factors that enhance the biotransformation of these compounds into ultimate carcinogens. However, the available data are far from being conclusive. The challenges in measuring exposure to specific carcinogens in observational studies have been considerable. In addition, the studies conducted to date suffered from methodological limitations (insufficient sample size, lack of control for Type I error, confounding, etc.). The effects reported have been of low magnitude and, as a result, any potentially confirmatory studies need to be of much larger size. The complexity of the biological pathways involved is such that multiple biotransformation phenotypes, cofactors, and modifiers need to be considered, making these studies difficult to implement and expensive. Large existing prospective studies in which exposure information was collected prior to onset of disease should be particularly useful in minimizing recall and selection biases. New approaches using information on linkage disequilibrium need to be applied to scan comprehensively genetic variation at candidate loci for association with disease. Perhaps most importantly, biomarkers of long-term exposure need to be developed so that they can be related to cancer risk and/or used to validate questionnaire exposure information. Similarly, biomarkers of early biological effects (e.g., DNA adducts) that can be reliably measured on large numbers of samples are needed, since such measures have the advantage of integrating the effects of exposure, absorption, and individual biological response. These new tools and studies will ultimately improve our etiologic understanding of CRC, as well as that of other cancers, and may lead to new prevention and therapeutic approaches.

References

Alexandrov, K., Rojas, M., Kadlubar, F.F., Lang, N.P. and Bartsch, H. 1996. Evidence of *anti*-benzo[α]pyrene diolepoxide–DNA adduct formation in human colon mucosa. Carcinogenesis 17:2081–3.

Augustsson, K., Skog, K., Jagerstad, M., Dickman, P.W. and Steineck, G. 1999. Dietary heterocyclic amines and cancer of the colon, rectum, bladder, and kidney: a population-based study. Lancet 353:703–7.

Bell, D.A., Badawi, A.F., Lang, N.P., Ilett, K.F., Kadlubar, F.F. and Hirvonen, A. 1995. Polymorphism in the N-acetyltransferase 1 polyadenylation signal: association of NAT1*10 allele with higher N-acetylation activity in bladder and colon tissue. Cancer Res 55:5226–9.

Bethke, L., Webb, E., Sellick, G., Rudd, M., Penegar, S., Withey, L., Qureshi, M. and Houlston, R. 2007. Polymorphisms in the cytochrome P450 genes CYP1A2, CYP1B1, CYP3A4, CYP3A5, CYP11A1, CYP17A1, CYP19A1 and colorectal cancer risk. BMC Cancer 7:123.

Bingham, S.A., Pignatelli, B., Pollock, J.R., Ellul, A., Malaveille, C., Gross, G., Runswick, S., Cummings, J.H. and O'Neill, I.K. 1996. Does increased endogenous formation of N-nitroso compounds in the human colon explain the association between red meat and colon cancer? Carcinogenesis 17:515–23.

Bingham, S.A., Hughes, R. and Cross, A.J. 2002. Effect of white versus red meat on endogenous N-nitrosation in the human colon and further evidence of a dose response. J Nutr 132(Suppl): 3522S–5S.

Bogovski, P. and Bogovski, S. 1981. Animal species in which N-nitroso compounds induce cancer. Int J Cancer 27:471–4.

Brockton, N., Little, J., Sharp, L. and Cotton, S.C. 2000. N-acetyltransferase polymorphisms and colorectal cancer: A HuGE review. Am J Epidemiol 151: 846–61.

Butler, L.M., Sinha, R., Millikan, R.C., Martin, C.F., Newman, B., Gammon, M.D., Ammerman, A.S. and Sandler, R.S. 2003. Heterocyclic amines, meat intake, and association with colon cancer in a population-based study. Am J Epidemiol 157:434–45.

Cajas-Salazar, N., Au, W.W., Zwischenberger, J.B., Sierra-Torres, C.H., Salama, S.A., Alpard, S.K. and Tyring, S.K. 2003. Effect of epoxide hydrolase polymorphisms on chromosome aberrations and risk for lung cancer. Cancer Genet Cytogenet 145:97–102.

Chen, J., Stampfer, M.J., Hough, H.L., Garcia-Closas, M., Willett, C., Hennekens, C.H., Kelsey, K.T. and Hunter, D.J. 1998. A prospective study of N-acetyltransferase genotype, red meat intake, and risk of colorectal cancer. Cancer Res 58:3307–11.

Conney, A.H. 2003. Enzyme induction and dietary chemicals as approaches to cancer chemoprevention: the Seventh DeWitt S. Goodman Lecture. Cancer Res 63:7005–31.

Cortessis, V., Siegmund, K., Chen, Q., Zhou, N., Diep, A., Frankl, H., Lee, E., Zhu, Q.S., Haile, R. and Levy, D. 2001. A case-control study of microsomal epoxide hydrolase, smoking, meat consumption, glutathione S-transferase M3, and risk of colorectal adenomas. Cancer Res 61: 2381–5.

Cotton, S.C., Sharp, L., Little, J. and Brockton. N. 2000. Glutathione S-transferase polymorphisms and colorectal cancer: a HuGE review. Am J Epidemiol 1517–32.

Crofts, F., Taioli, E., Trachman, J., Cosma, G.N., Currie, D., Toniolo, P. and Garte, S.J. 1994. Functional significance of different human CYP1A1 genotypes. Carcinogenesis 15:2961–3.

Crofts, F.G., Sutter, T.R. and Strickland, P.T. 1998. Metabolism of 2-amino-1-methyl-6-phenylimidazo[4,5-b]pyridine by human cytochrome P4501A1, P4501A2 and P4501B1. Carcinogenesis 19:1969–73.

Cross, A.J. and Sinha, R. 2004. Meat-related mutagens/carcinogens in the etiology of colorectal cancer. Environ Mol Mutagen 44:44–55.

Cross, A.J., Pollock, J.R. and Bingham, S.A. 2003. Haem, not protein or inorganic iron, is responsible for endogenous intestinal N-nitrosation arising from red meat. Cancer Res 63:2358–60.

Cuthrell, K. and Le Marchand, L. 2006. Grapefruit and Cancer – A Review. In: B.S. Patil, J.S. Brodbelt, E.G. Miller and N.D. Turner (eds). Potential Health Benefits of Citrus. ACS Symposium Series 936, Washington, DC, pp 235–52.

Dearfield, K.L., Douglas, G.R., Ehling, U.H., Moore, M.M., Sega, G.A. and Brusick, D.J. 1995. Acrylamide – a review of its genotoxicity and assessment of heritable genetic risk. Mutat Res 330:71–99.

Dingley, K.H., Curtis, K.D., Nowell, S., Felton, J.S., Lang, N.P. and Turteltaub, K.W. 1999. DNA and protein adduct formation in the colon and blood of humans after exposure to a dietary-relevant dose of 2-amino-1-methyl-6-phenylimidazo[4,5-b]pyridine. Cancer Epidemiol Biomarkers Prev 8:507–12.

Doll, M.A. and Hein, D.W. 2002. Rapid method to distinguish frequent and/or functional polymorphisms in human *N-Acetyltransferase*-1. Anal Biochem 301:328–32.

Dybing, E. and Sanner, T. 2003. Risk assessment of acrylamide in foods. Toxicol Sci 75:7–15.

Ebert, M.N., Beyer-Sehlmeyer, G., Liegibel, U.M., Kautenburger, T., Becker, T.W. and Pool-Zobel, B.L. 2001. Butyrate induces glutathione S-transferase in human colon cells and protects from genetic damage by 4-hydroxy-2-nonenal. Nutr Cancer 41:156–64.

Fernandez-Salguero, P., Hoffman, S.M., Cholerton, S., Mohrenweiser, H., Raunio, H., Rautio, A., Pelkonen, O., Huang, J.D., Evans, W.E. and Idle, J.R. 1995. A genetic polymorphism in coumarin 7-hydroxylation: sequence of the human CYP2A genes and identification of variant CYP2A6 alleles. Am J Hum Genet 57:651–60.

Gertig, D.M., Stampfer, M., Haiman, C., Hennekens, C.H., Kelsey, K. Hunter, D.J. 1998. Glutathione S-transferase GSTM1 and GSTT1 polymorphisms and colorectal cancer risk: a prospective study. Cancer Epidemiol Biomarkers Prev 7:1001–5.

Goldstein, L.S., Weyand, E.H., Safe, S., Steinberg, M., Culp, S.J., Gaylor, D.W., Beland, F.A. and Rodriguez, L.V. 1998. Environ Health Perspect 106(Suppl 6):1325–30.

Gonzalez, F.J. and Gelboin, H.V. 1993. Role of human cytochrome P-450s in risk assessment and susceptibility to environmentally based disease. J Toxicol Environ Health 40:289–308.

Gunter, M.J., Probst-Hensch, N.M., Cortessis, V.K., Kulldorff, M., Haile, R.W. and Sinha, R. 2005. Meat intake, cooking-related mutagens and risk of colorectal adenoma in a sigmoidoscopy-based case-control study. Carcinogenesis 26:637–42.

Gunter, M.J., Divi, R.L., Kulldorff, M., Vermeulen, R., Haverkos, K.J., Kuo, M.M., Strickland, P., Poirier, M.C., Rothman, N. and Sinha, R. 2007. Leukocyte polycyclic aromatic hydrocarbon–DNA adduct formation and colorectal adenoma. Carcinogenesis 28:1426–9.

Hassett, C., Aicher, L., Sidhu, J.S. and Omiecinski, C.J. 1994. Human microsomal epoxide hydrolase: genetic polymorphism and functional expression in vitro of amino acid variants. Hum Mol Genet 3:421–8.

Hecht, S.S. 2003. Tobacco carcinogens, their biomarkers and tobacco-induced cancer. Nat Rev Cancer 3:733–44.

Hecht, S.S., Carmella, S.G. and Murphy, S.E. 1999. Effects of watercress consumption on urinary metabolites of nicotine in smokers. Cancer Epidemiol Biomarkers Prev 8:907–13.

Hein, D.W. 2002. Molecular genetics and function of NAT1 and NAT2: role in aromatic amine metabolism and carcinogenesis. Mutat Res 506–507:65–77.

Higdon, J.V., Delage, B., Williams, D.E. and Dashwood, R.H. 2007. Cruciferous vegetables and human cancer: epidemiologic evidence and mechanistic basis. Pharmacol Res 55:224–36.

Hoensch, H., Peters, W.H., Roelofs, H.M. and Kirch, W. 2006. Expression of the glutathione enzyme system of human colon mucosa by localisation, gender and age. Curr Med Res Opin 22:1075–83.

Hou, L., Chatterjee, N., Huang, W.Y., Baccarelli, A., Yadavalli, S., Yeager, M., Bresalier, R.S., Chanock, S.J., Caporaso, N..E, Ji, B.T., Weissfeld, J.L. and Hayes, R.B. 2005. CYP1A1 Val462 and NQO1 Ser187 polymorphisms, cigarette use, and risk for colorectal adenoma. Carcinogenesis 26:1122–8.

Huang, W.Y., Chatterjee, N., Chanock, S., Dean, M., Yeager, M., Schoen, R.E., Hou, L.F., Berndt, S.I., Yadavalli, S., Johnson, C.C. and Hayes, R.B. 2005. Microsomal epoxide hydrolase polymorphisms and risk for advanced colorectal adenoma. Cancer Epidemiol Biomarkers Prev 14:152–7.

Huang, K., Sandler, R.S., Millikan, R.C., Schroeder, J.C., North, K.E. and Hu, J. 2006. GSTM1 and GSTT1 polymorphisms, cigarette smoking, and risk of colon cancer: a population-based case-control study in North Carolina (United States). Cancer Causes Control 17:385–94.

Hubner, R.A., Muir, K.R., Liu, J.F., Logan, R.F., Grainge, M., Armitage, N., Shepherd, V., Popat, S., Houlston, R.S. and United Kingdom Colorectal Adenoma Prevention Consortium. 2006. Genetic variants of UGT1A6 influence risk of colorectal adenoma recurrence. Clin Cancer Res 12:6585–9.

IARC Monographs on the Evaluation of Carcinogenic Risks to Humans: Some Industrial Chemicals, No. 60. International Agency for Research on Cancer, Lyon, 1994, pp 389–433.

Kamataki, T., Nunoya, K., Sakai, Y., Kushida, H. and Fujita, K. 1999. Genetic polymorphism of CYP2A6 in relation to cancer. Mutat Res 428:125–30.

Kampman, E., Slattery, M.L., Bigler, J., Leppert, M., Samowitz, W., Caan, B.J. and Potter, J.D. 1999. Meat consumption, genetic susceptibility, and colon cancer risk: a United States Multicenter Case-control Study. Cancer Epidemiol Biomarkers Prev 8:15–24.

Kang, D.H., Rothman, N., Poirier, M.C., Greenberg, A., Hsu, C.H., Schwartz, B.S., Baser, M.E., Groopman, J.D., Weston, A. and Strickland, P.T. 1995. Interindividual differences in the concentration of 1-hydroxypyrene-glucuronide in urine and polycyclic aromatic hydrocarbon–DNA adducts in peripheral white blood cells after charbroiled beef consumption. Carcinogenesis 16:1079–85.

Kawajiri, K., Watanabe, J., Eguchi, H., Nakachi, K., Kiyohara, C. and Hayashi, S. 1995. Polymorphisms of the human Ah receptor gene are not involved in lung cancer. Pharmacogenetics 5:151–8.

Kiyohara, C., Hirohata, T. and Inutsuka, S. 1996. The relationship between aryl hydrocarbon hydroxylase and polymorphisms of the CYP1A1 gene. Jpn J Cancer Res 87:18–24.

Knekt, P., Jarvinen, R., Dich, J. and Hakulinen, T. 1999. Risk of colorectal and other gastro-intestinal cancers after exposure to nitrate, nitrite and N-nitroso compounds: a follow-up study. Int J Cancer 80:852–6.

Kono, S. 2004. Secular trend of colon cancer incidence and mortality in relation to fat and meat intake in Japan. Eur J Cancer Prev 13:127–32.

Küry, S., Buecher, B., Robiou-du-Pont, S., Scoul, C., Sebille, V., Colman, H., Le Houerou, C., Le Neel, T., Bourdon, J., Faroux, R., Ollivry, J., Lafraise, B., Chupin, L.D. and Bezieau, S. 2007. Combinations of cytochrome p450 gene polymorphisms enhancing the risk for sporadic colorectal cancer related to red meat consumption. Cancer Epidemiol Biomarkers Prev 16: 1460–7.

Landi, M.T., Bertazzi, P.A., Shields, P.G., Clark, G., Lucier, G.W., Garte, S.J., Cosma, G. and Caporaso, N.E. 1994. Association between CYP1A1 genotype, mRNA expression and enzymatic activity in humans. Pharmacogenetics 4:242–6.

Lang, N.P., Butler, M.A., Massengill, J., Lawson, M., Stotts, R.C., Hauer-Jensen, M. and Kadlubar F.F. 1994. Rapid metabolic phenotypes for acetyltransferase activity and cytochrome CYP1A2 and putative exposure to food-borne heterocyclic amine increase the risk of colorectal cancer or polyps. Cancer Epidemiol Biomarkers Prev 3:675–82.

Laurenzana, E.M., Hassett, C. and Omiecinski C.J. 1998. Post-transcriptional regulation of human microsomal epoxide hydrolase. Pharmacogenetics 8:157–67.

Le Marchand L. 1999. Combined influence of genetic and dietary factors on colorectal cancer incidence in Japanese Americans. Monograph Natl Cancer Inst 26:101–5.

Le Marchand L. Meat Intake, Metabolic Genes and Colorectal Cancer Risk. In: E. Riboli and R. Lambert (eds). Nutrition and Lifestyle: Opportunities for Cancer Prevention. IARC Scientific Publications No. 156, Lyon, 2002, pp 481–5.

Le Marchand, L., Hankin, J.H., Wilkens, L.R., Pierce, L.M., Franke, A.A., Kolonel, L.N., Seifried, A., Custer, L.J., Chang, W., Lum-Jones, A. and Donlon, T. 2001. Combined effect of well-done red meat, smoking and rapid NAT2 and CYP1A2 phenotypes in increasing colorectal cancer risk. Cancer Epidemiol Biomarkers Prev 10:1259–66.

Le Marchand, L., Hankin, J.H., Pierce, L.M., Sinha, R., Nerurkar, P.V., Franke, A., Wilkens, L.R., Kolonel, L.N., Donlon, T., Seifried, A., Custer, L.J., Lum-Jones, A. and Chang, W. 2002a. Well-done red meat, metabolic phenotypes and colorectal cancer in Hawaii. Mutat Res 506–507: 205–14.

Le Marchand, L., Donlon, T., Seifried, A. and Wilkens, L.R. 2002b. Red meat intake, CYP2E1 genetic polymorphisms and colorectal cancer risk. Cancer Epidemiol Biomarkers Prev 11: 1019–24.

Lewin, M.H., Bailey, N., Bandaletova, T., Bowman, R., Cross, A.J., Pollock, J., Shuker, D.E. and Bingham, S.A. 2006. Red meat enhances the colonic formation of the DNA adduct O_6-carboxymethyl guanine: implications for colorectal cancer risk. Cancer Res 66:1859–65.

Lilla, C., Verla-Tebit, E., Risch, A., Jager, B., Hoffmeister, M., Brenner, H. and Chang-Claude, J. 2006. Effect of *NAT1* and *NAT2* genetic polymorphisms on colorectal cancer risk associated with exposure to tobacco smoke and meat consumption. Cancer Epidemiol Prev 15:99–107.

Lin, H.L., Parsels, L.A., Maybaum, J. and Hollenberg, P.F. 1999. *N*-Nitrosodimethylamine-mediated cytotoxicity in a cell line expressing P450 2E1: evidence for apoptotic cell death. Toxicol Appl Pharmacol 157:117–24.

Loeppky, R.N. 1999. The mechanism of bioactivation of *N*-nitrosodiethanolamine. Drug Metab Rev 31:175–93.

Lüchtenborg, M., Weijenberg, M.P., Kampman, E., van Muijen, G.N., Roemen, G.M., Zeegers, M.P., Goldbohm, R.A., van 't Veer, P., de Goeij, A.F. and van den Brandt, P.A. 2005. Cigarette smoking and colorectal cancer: APC mutations, hMLH1 expression, and GSTM1 and GSTT1 polymorphisms. Am J Epidemiol 161:806–15.

McCoull, K.D., Rindgen, D., Blair, I.A. and Penning, T.M. 1999. Synthesis and characterization of polycyclic aromatic hydrocarbon o-quinone depurinating N7-guanine adducts. Chem Res Toxicol 12:237–46.

Merlo, F., Andreassen, A., Weston, A., Pan, C.F., Haugen, A., Valerio, F., Reggiardo, G., Fontana, V., Garte, S., Puntoni, R. and Abbondandolo, A. 1998. Urinary excretion of 1-hydroxypyrene as a marker for exposure to urban air levels of polycyclic aromatic hydrocarbons. Cancer Epidemiol Biomarkers Prev 7:147–55.

Mitroun P.N., Watson, M.A., Loktionov, A.S., Cardwell, C., Gunter, M.J., Atkin, W.S., Macklin, C.P., Cecil, T., Bishop, D.T., Primrose, J. and Bingham, S.A. 2007. Role of NQO1C609T and EPHX1 gene polymorphisms in the association of smoking and alcohol with sporadic distal colorectal adenomas: results from the UKFSS Study. Carcinogenesis 28:875–82.

Mucci, L.A., Dickman, P.W., Steineck, G., Adami, H.-O. and Augustsson, K. 2003. Dietary acrylamide and cancer of the large bowel, kidney, and bladder. Absence of an association in a population-based study in Sweden. Br J Cancer 88:84–89.

Nerurkar, P.V., Okinaka, L., Aoki, C., Seifried, A., Lum-Jones, A., Wilkens, L.R., and Le Marchand, L. 2000. CYP1A1, GSTM1, and GSTP1 genetic polymorphisms and urinary 1-hydroxypyrene excretion in non-occupationally exposed individuals. Cancer Epidemiol Biomarkers Prev 9:1119–22.

Nijhoff, W.A., Grubben, M.J., Nagengast, F.M., Jansen, J.B., Verhagen, H., Van Poppel, G., Peters, W.H. 1995. Effects of consumption of Brussels Sprouts on intestinal and lymphocytic glutathione S-transferases in humans. Carcinogenesis 16:2125–8.

Norat, T., Lukanova, A., Ferrari, P. and Riboli, E. 2002. Meat consumption and colorectal cancer risk: dose–response meta-analysis of epidemiological studies. Int J Cancer 98:241–56.

Nowell, S., Coles, B., Sinha, R., MacLeod, S., Luke Ratnasinghe, D., Stotts, C., Kadlubar, F.F., Ambrosone, C.B. and Lang, N.P. 2002. Analysis of total meat intake and exposure to individual heterocyclic amines in a case-control study of colorectal cancer: contribution of metabolic variation to risk. Mutat Res 506–507:175–85.

Ognjanovic, S., Yamamoto, J., Maskarinec, G. and Le Marchand, L. 2006. NAT2, meat consumption and colorectal cancer incidence: an ecological study among 27 countries. Cancer Causes Control 17:1175–82.

Ozawa, S., Chou, H.C., Kadlubar, F.F., Nagata, K., Yamazoe, Y. and Kato, R. 1994. Activation of 2-hydroxyamino-1-methyl-6-phenylimidazo[4,5-b] pyridine by cDNA-expressed human and rat arylsulfotransferases. Jpn J Cancer Res 85:1220–8.

Pastorelli, R., Guanci, M., Cerri, A., Negri, E., La Vecchia, C., Fumagalli, F., Mezzetti, M., Cappelli, R., Panigalli, T., Fanelli, R. and Airoldi, L. 1998. Impact of inherited polymorphisms in glutathione S-transferase M1, microsomal epoxide hydrolase, cytochrome P450 enzymes on DNA, and blood protein adducts of benzo(α)pyrene-diolepoxide. Cancer Epidemiol Biomarkers Prev 7:703–9.

Pelucci, C., Franschesci, S., Levi, F., Trichopoulos, D., Bosetti, C., Negri, E. and La Vacchia C. 2003. Fried potatoes and human cancer. Int J Cancer 105:558–60.

Phillips, D.H. 1999. Polycyclic aromatic hydrocarbons in the diet. Mutat Res 443:139–47.

Rihs, H.P., Pesch, B., Kappler, M., Rabstein, S., Rossbach, B., Angerer, J., Scherenberg, M., Adams, A., Wilhelm, M., Seidel, A. and Bruning, T. 2005. Occupational exposure to polycyclic aromatic hydrocarbons in German industries: association between exogenous exposure and urinary metabolites and its modulation by enzyme polymorphisms. Toxicol Lett 157: 241–55.

Roberts-Thomson, I.C., Ryan, P., Khoo, K.K., Hart, W.J., McMichael, A.J. and Butler, R.N. 1996. Diet, acetylator phenotype and risk of colorectal neoplasia. Lancet 347:1372–4.

Robien, K., Curtin, K., Ulrich, C.M., Bigler, J., Samowitz, W., Caan, B., Potter, J.D. and Slattery, M.L. 2005. Microsomal epoxide hydrolase polymorphisms are not associated with colon cancer risk. Cancer Epidemiol Biomarkers Prev 14:1350–2.

Rothman, N., Poirier, M.C., Baser, M.E., Hansen, J.A., Gentile, C., Bowman, E.D. and Strickland, P.T. 1990. Formation of polycyclic aromatic hydrocarbon–DNA adducts in peripheral white blood cells during consumption of charcoal-broiled beef. Carcinogenesis 11:1241–3.

Sachse, C., Brockmoller, J., Bauer, S. and Roots, I. 1999. Functional significance of a C → A polymorphism in intron 1 of the cytochrome P450 CYP1A2 gene tested with caffeine. Br J Clin Pharmacol 47:445–9.

Safe, S. 2001. Molecular biology of the Ahr receptor and its role in carcinogenesis. Toxicol Lett 120:1–7.

Sandhu, M.S., White, I.R. and McPherson, K. 2001. Systematic review of the prospective cohort studies on meat consumption and colorectal cancer risk: a meta-analytical approach. Cancer Epidemiol Biomarkers Prev 10:439–46.

Seow, A., Yuan, J.M., Sun, C.L., Van Den Berg, D., Lee, H.P. and Yu, M.C. 2002. Dietary isothiocyanates, glutathione S-transferase polymorphisms and colorectal cancer risk in the Singapore Chinese Health Study. Carcinogenesis 23:2055–61.

Shimada, T., Watanabe, J., Kawajiri, K., Sutter, T.R., Guengerich, F.P., Gillam, M.J. and Inoue, K. 1999. Catalytic properties of polymorphic human cytochrome P450 1B1 variants. Carcinogenesis 20:1607–13.

Shin, A., Shrubsole, M.J., Ness, R.M., Wu, H., Sinha, R., Smalley, W.E., Shyr, Y. and Zheng, W. 2007. Meat and meat-mutagen intake, doneness preference and the risk of colorectal polyps: the Tennessee Colorectal Polyp Study. Int J Cancer 121:136–42.

Sinha, R., Chow, W.H., Kulldorff, M., Denobile, J., Butler, J., Garcia-Closas, M., Weil, R., Hoover, R.N. and Rothman, N. 1999. Well-done, grilled red meat increases the risk of colorectal adenomas. Cancer Res 59:4320–4.

Sinha, R., Kulldorff, M., Chow, W.H., Denobile, J. and Rothman, N. 2001. Dietary intake of heterocyclic amines, meat-derived mutagenic activity, and risk of colorectal adenomas. Cancer Epidemiol Biomarkers Prev 10:559–62.

Sinha, R., Kulldorff, M., Gunter, M.J., Strickland, P. and Rothman, N. 2005a. Dietary benzo[a] pyrene intake and risk of colorectal adenoma. Cancer Epidemiol Biomarkers Prev 14:2030–4.

Sinha, R., Peters, U., Cross, A.J., Kulldorff, M., Weissfeld, J.L., Pinsky, P.F., Rothman, N. and Hayes, R.B. 2005b. Meat, meat cooking methods and preservation, and risk for colorectal adenoma. Cancer Res 65:8034–41.

Sivaraman, L., Leatham, M.P., Yee, J., Wilkens, L.R., Lau, A.F. and Le Marchand, L. 1994. CYP1A1 genetic polymorphisms and in situ colorectal cancer. Cancer Res 54:3692–5.

Slattery, M.L., Samowtiz, W., Ma, K., Murtaugh, M., Sweeney, C., Levin, T.R. and Neuhausen, S. 2004. CYP1A1, cigarette smoking, and colon and rectal cancer. Am J Epidemiol 160:842–52.

Smart, J. and Daly, A.K. 2000. Variation in induced CYP1A1 levels: relationship to CYP1A1, Ah receptor and GSTM1 polymorphisms. Pharmacogenetics 10:11–24.

Stillwell, W.G., Kidd, L.C, Wishnok, J.S., Tannenbaum, S.R. and Sinha, R. 1997. Urinary excretion of unmetabolized and phase II conjugates of 2-amino-1-methyl-6-phenylimidazo[4,5-*b*] pyridine and 2-amino-3,8-dimethylimidazo[4,5-*f*]quinoxaline in humans: Relationship to cytochrome P450 1A2 and *N*-acetyltransferase activity. Cancer Res 57:3457–64.

Sugimura, T., Wakabayashi, K., Nakagama, H. and Nagao, M. 2004. Heterocyclic amines: Mutagens/carcinogens produced during cooking of meat and fish. Cancer Sci 95:290–99.

Sun, X.F., Ahmadi, A., Arbman, G., Wallin, A., Asklid, D. and Zhang, H. 2005. Polymorphisms in sulfotransferase 1A1 and glutathione S-transferase P1 genes in relation to colorectal cancer risk and patients' survival. World J Gastroenterol 11:6875–9.

Sundberg, K., Widersten, M., Seidel, A., Mannervik, B. and Jernstrom, B. 1997. Glutathione conjugation of bay- and fjord-region diol epoxides of polycyclic aromatic hydrocarbons by glutathione transferases M1–1 and P1–1. Chem Res Toxicol 10:1221–7.

Swanson, H.I. and Bradfield, C.A. 1993. The AH-receptor: genetics, structure and function. Pharmacogenetics 3:213–30.

Szarka, C.E., Pfeiffer, G.R., Hum, S.T., Everley, L.C., Balshem, A.M., Moore, D.F., Litwin, S., Goosenberg, E.B., Frucht, H. and Engstrom, P.F. 1995. Glutathione S-transferase activity and glutathione S-transferase mu expression in subjects with risk for colorectal cancer. Cancer Res 55:2789–93.

Tareke, E., Rydberg, P., Karlsson, P., Eriksson, S. and Torqvist, M. 2002. Analysis of acrylamide, a carcinogen formed in heated food-stuffs. J Agric Food Chem 50:517–22.

Tijhuis, M.J., Visker, M.H., Aarts, J.M., Peters, W.H., Roelofs, H.M., den Camp, L.O., Rietjens, I.M., Boerboom, A.M., Nagengast, F.M., Kok, F.J. and Kampman, E. 2007. Glutathione S-transferase phenotypes in relation to genetic variation and fruit and vegetable consumption in an endoscopy-based population. Carcinogenesis 28:848–57.

Tricker, AR. 1997. N-nitroso compounds and man: sources of exposure, endogenous formation and occurrence in body fluids. Eur J Cancer Prev 6:226–68.

Turesky, R.J. 2002. Heterocyclic aromatic amine metabolism, DNA adduct formation, mutagenesis, and carcinogenesis. Drug Metab Rev 34:625–50.

Turesky, R.J., Constable, A., Richoz, J., Varga, N., Markovic, J., Martin, M.V., Guengerich, F.P. 1998. Activation of heterocyclic aromatic amines by rat and human liver microsomes and by purified rat and human cytochrome P450 1A2. Chem Res Toxicol 11:925–36.

van der Logt, E.M., Bergevoet, S.M., Roelofs, H.M., van Hooijdonk, Z., te Morsche, R.H., Wobbes, T., de Kok, J.B., Nagengast, F.M. and Peters, W.H. 2004. Genetic polymorphisms in UDP-glucuronosyltransferases and glutathione S-transferases and colorectal cancer risk. Carcinogenesis 25:2407–15.

Ward, M.H., Cross, A.J., Divan, H., Kulldorff, M., Nowell-Kadlubar, S., Kadlubar, F.F. and Sinha R. 2007. Processed meat intake, CYP2A6 activity and risk of colorectal adenoma. Carcinogenesis 28:1210–6.

Welfare, M.R., Cooper, J., Bassendine, M.F. and Daly, A.K. 1997. Relationship between acetylator status, smoking, diet and colorectal cancer risk in the north-east of England. Carcinogenesis 18: 1351–4.

Wu, M.T., Huang, S.L., Ho, C.K., Yeh, Y.F. and Christiani, D.C. 1998. Cytochrome P450 1A1 MspI polymorphism and urinary 1-hydroxypyrene concentrations in coke-oven workers. Cancer Epidemiol Biomarkers Prev 7:823–9.

Wu, W.K., Wong, H.P., Luo, S.W., Chan, K., Huang, F.Y., Hui, M.K., Lam, E.K., Shin, V.Y., Ye, Y.N., Yang, Y.H. and Cho, C.H. 2005. 4-(Methylnitrosamino)-1-(3-pyridyl)-1-butanone from cigarette smoke stimulates colon cancer growth via beta-adrenoceptors. Cancer Res 65:5272–7.

Wu, K., Giovannucci, E., Byrne, C., Platz, E.A., Fuchs, C., Willett, W.C. and Sinha, R. 2006. Meat mutagens and risk of distal colon adenoma in a cohort of U.S. men. Cancer Epidemiol Biomarkers Prev 15:1120–5.

Chapter 12
Calcium and Vitamin D

Roberd M. Bostick, Michael Goodman, and Eduard Sidelnikov

Introduction

Of the various agents tested in clinical trials against colorectal neoplasms, only two, calcium and nonsteroidal anti-inflammatory drugs (NSAIDs), have been found to have preventive efficacy, and calcium is the only one for which significant long-term adverse consequences have not been demonstrated. Vitamin D alone, and in combination with calcium, is now being tested in a large, long-term, randomized, placebo-controlled chemoprevention trial of sporadic colorectal adenoma recurrence, and is of current intense interest. This chapter summarizes calcium and vitamin D physiology and metabolism and the mechanistic, genetic, and epidemiologic evidence for these agents in preventing colorectal cancer.

Calcium and Vitamin D Physiology and Metabolism

Calcium

Calcium has a variety of functions in the body, including its "classical" functions in bone structure, current flow across excitable membranes, fusion and release of storage vesicles, and muscle contraction, and its "nonclassical functions" such as intracellular regulation of various enzymes, and regulation of cell proliferation and differentiation (Friedman 2006; Bringhurst et al. 2007; Bostick 2001; Chakrabarty et al. 2003). Tight regulation of calcium within narrow, low intracellular and high extracellular ranges is essential for human life, and is an intricate, homeostatic dance involving calcium intake, calcium absorption, and discharge from the intes-

R.M. Bostick (✉), M. Goodman, and E. Sidelnikov
Department of Epidemiology, Rollins School of Public Health, Emory University
e-mail: rmbosti@sph.emory.edu

J.D. Potter and N.M Lindor (eds.), *Genetics of Colorectal Cancer*,
DOI: 10.1007/978-0-387-09568-4_20,

tines, deposition and release from bone, and urinary filtering and excretion involving parathyroid hormone (PTH) and vitamin D intake, synthesis, and metabolism.

Calcium, present in a variety of foods, enters the body only through the intestine (Friedman 2006; Bringhurst et al. 2007). Active, vitamin D-dependent, transcellular transport occurs in the proximal duodenum and, to progressively lesser extent, in the jejunum and ileum; facilitated diffusion through the intercellular spaces throughout the small intestine accounts for the majority of the total calcium uptake. These processes also operate to a lesser extent in the large intestine. On average, about 30% of calcium consumed is absorbed, and the remainder is excreted in feces (Bostick 2001). Nonabsorbed fecal stream-calcium can bind with various compounds, such as bile acids, and/or be excreted as free calcium (Bostick 2001).

Calcium and the calcium-sensing receptor (CaSR), which transduces extracellular calcium binding into a variety of intracellular responses, appear to function together to regulate diverse cellular processes in a variety of cell types (Lamprecht and Lipkin 2001). Examples include roles in cell-cycle regulation and cell–cell and cell–matrix adhesion (Lamprecht and Lipkin 2001). Also, the CaSR, more than a calcium sensor, is a fairly broad spectrum sensor of small cationic molecules capable of transducing signals in response to heavy metals and cationic amino acids (Chattopadhyay and Brown 2006). As a probable amino acid sensor, the CaSR may also participate in the control of digestion, absorption, appetite, and somatic metabolism (Chattopadhyay and Brown 2006).

Human colonocytes proliferate only in a low calcium environment, and even modest increases in calcium concentration reduce proliferation and induce differentiation. Although plasma-free calcium levels are tightly regulated, calcium concentration generally increases from colonic crypt base to lumen, corresponding to crypt colonocyte proliferation and differentiation (Rodland 2004).

Vitamin D

Vitamin D has both "classical" endocrine functions (i.e., related to calcium homeostasis) and "nonclassical" autocrine/paracrine functions (i.e., not related to calcium homeostasis or functions) that operate through genomic (via the nuclear vitamin D receptor (VDR)) and nongenomic ("rapid responses" not involving the nuclear VDR) mechanisms (Lips 2006; Norman 2006; Holick 2007). The nonclassical autocrine/paracrine functions may be most relevant to colon carcinogenesis and prevention. The VDR is expressed in many human tissues, including the colon. These tissues also express CYP27B1 and CYP24 enzymes, which, respectively, synthesize and degrade the most potent activator of VDR, $1\alpha,25\text{-(OH)}_2$-vitamin D (collective term for $1\alpha,25\text{-(OH)}_2$-vitamins D_2 and D_3). Beyond calcium homeostasis, vitamin D has a role in cell-cycle regulation, promotes bile-acid degradation, and influences growth-factor signaling, inflammation, and immune function.

There are two precursors to active vitamin D hormones, provitamin D_3 and provitamin D_2 (Friedman 2006; Bringhurst et al. 2007; Holick 2007). Provitamin D_3 is synthesized in the skin, where, on exposure to ultraviolet radiation, it is converted to vitamin D_3, which may also be obtained from a few dietary sources. Provitamin D_2 is present only in plants, and vitamin D_2 exposure in humans is only from the diet and vitamin supplements. Because there is no practical difference between the antirachitic properties of vitamin D_2 and vitamin D_3 (the major circulating form) in humans, "vitamin D" is traditionally used as a collective term for vitamins D_2 and D_3. Vitamin D is transported through the circulation to the liver, where it is hydroxylated at the 25 position to 25-OH-vitamin D. This reaction is catalyzed by one or more enzymes with 25-hydroxylase activity, including CYP27A1, CYP2D6, CYP2R1, CYP2C11, CYP3A4, CYP2D25, and CYP2J3. 25-OH-vitamin D can be stored in the liver or be distributed widely via the circulation. Because 25-hydroxylation is not closely regulated, 25-OH-vitamin D levels reflect overall vitamin D status from combined dietary and sunlight sources. Over 95% of 25-OH-vitamin D in serum consists of 25-OH-vitamin D_3, which has a circulating half-life of about 20 days. Body fat is a large storage reservoir for 25-OH-vitamin D. In various tissues, 25-OH-vitamin D undergoes a second hydroxylation catalyzed by CYP27B1 at the 1α position to form 1α,25-$(OH)_2$-vitamin D_3, which is 100- to 1,000-fold more potent than 25-OH-vitamin D. 1α,25-$(OH)_2$-vitamin D produced in the kidneys is released into the circulation for its classical endocrine function in bone and calcium homeostasis, whereas 1α,25-$(OH)_2$-vitamin D, produced in other tissues, exerts its nonclassical autocrine/paracrine effects and is not released into the circulation because its synthesis is balanced with degradation (Fraser 1995). In contrast to 25-hydroxylation, endocrine and autocrine/paracrine 1α-hydroxylation is tightly regulated, and 1α,25-$(OH)_2$-vitamin D is short lived (hours to a few days); consequently, serum levels of 1α,25-$(OH)_2$-vitamin D do not reflect vitamin D status except during clear conditions of deficiency or excess (Holick 1990). CYP24, expressed in many vitamin D target tissues, initiates the degradation of 25-OH-vitamin D and 1α,25-$(OH)_2$-vitamin D to their excretory metabolites. Calcium can increase CYP24 gene transcription, but the major inducer is 1α,25-$(OH)_2$-vitamin D, thus promoting its own inactivation and limiting its biologic effects.

For its nonclassical, autocrine/paracrine functions, vitamin D modulates more than 200 responsive genes with a wide array of functions in a cell- and tissue-specific manner (Ebert et al. 2006; Yee et al. 2005). Functions identified to date include roles in regulating cell proliferation, differentiation, and apoptosis; growth-factor signaling; protection against oxidative stress; bile-acid and xenobiotic metabolism; immunomodulation; cell adhesion; DNA repair; and angiogenesis.

1α,25-$(OH)_2$-vitamin D is thought to act through genomic and nongenomic mechanisms (Lips 2006; Norman 2006; Holick 2007; Reichrath et al. 2007). Genomic effects are mediated via binding to the nuclear VDR. 1α,25-$(OH)_2$-vitamin D, being a relatively small, lipophilic molecule that easily penetrates the cell membrane, is taken up by the cell by simple diffusion and binds to the VDR, then the VDR binds to target DNA sequences as a heterodimer with the retinoid X receptor (RXR), recruiting a series of coactivators resulting in the induction of target gene expression. Nongenomic effects, or rapid responses (Lips 2006; Norman 2006; Holick 2007), may work through a plasma membrane receptor (apparently the VDR in a second location) and second messengers involved in regulation of voltage-gated calcium channels, opening of chloride channels, modulation of protein kinase activity, activation of mitogen-activated protein kinases, and a role in cell-cycle regulation. The nongenomic pathways lead to the onset of rapid biological responses (seconds to 1–2 min), including inhibition of cell proliferation and stimulation of cell differentiation.

Mechanisms of Calcium and Vitamin D in Colorectal Carcinogenesis

As described earlier, calcium and vitamin D are highly physiologically inter-related, and the growing list of putative mechanisms by which they may reduce risk of colorectal neoplasms reflects these inter-relationships. The importance of the intestinal tract in calcium homeostasis and the abundance of CaSRs and VDRs in the colon suggest that substantial and/or prolonged calcium and vitamin D exposures outside optimal ranges may adversely affect the colon. Modern exposures in industrialized countries to calcium and vitamin D are quite low by evolutionary and historical standards. The estimated average intake of calcium in Western diets is 740 mg daily (Bostick 2001), yet the calcium intake of all mammalian species (including chimpanzees) other than modern human is equivalent to a human intake of 1,500–2,000 mg daily, an amount that corresponds to the estimated intake of Paleolithic man (wild plant foods are high in calcium, whereas plants grown and marketed by modern industrial agricultural methods are low in calcium) (Bostick 2001). In contrast to the increasingly indoor lifestyles in industrialized countries, humans during the Paleolithic period were primarily outdoor gatherer-hunters exposed to sunlight most days, year round. Dark-skinned people who spend most of their time outdoors in sub-Saharan latitudes maintain 25-OH-vitamin D blood levels of about 150 nmol L^{-1}* (Hollis 2005), and various lines of evidence suggest that optimal levels may be 80–250 nmol L^{-1}, levels, achievable by total vitamin D exposures of 25–100 μg† daily (averaged over a year) (Holick 2007; Hollis 2005; Heaney 2005). In the United States (US) and Europe, half or more of the population maintains 25-OH-vitamin D blood levels below this range, and median dietary intakes of vitamin D in the US are around 2.5 μg daily. Both calcium and vitamin D influence bile-acid metabolism, affect genes/proteins in colon carcinogenic pathways, and modulate cell proliferation and differentiation, all thought to be important in colon carcinogenesis.

Calcium

The three most prominent hypotheses for protective effects of calcium against colorectal cancer involve (1) bile-acid scavenging, (2) direct effects on the cell cycle, and (3) modulation of E-cadherin and β-catenin expression via the CaSR. These potential

---

* nmol L^{-1} = μg/L × 2.5; μg/L = ng/ml.

† μg = IU/40.

mechanisms are probably complementary in reducing mutations and promoting safer patterns of cell-cycle events in colon crypt cells, as well as in the expression of various genes regulating the normal structure and function of the colon crypt.

Bile Acids

Bile acids, produced in response to fat intake and digestion, are mutagenic and otherwise damage cells, provoking compensatory hyperproliferation. Bile acids can be neutralized in the gut lumen by free calcium when calcium intake reaches 1,500–2,000 mg daily (Bostick 2001). In a corollary to this hypothesis, if calcium intake is high enough to bind bile acids and prevent cell injury, it may also prevent the consequent inflammatory response, thus suppressing COX-2 and promoting APC expression, which in turn suppresses proliferation (Boyapati et al. 2003).

Direct Effect on Cell Cycle

The hypothesis here, based on in vitro studies, is that free gut calcium has a direct effect on cell cycle, decreasing proliferation and increasing differentiation by as-yet-unclear mechanisms, perhaps involving interaction with E-cadherin (a calcium-dependent cell-adhesion molecule affected by the *wnt* pathway), cAMP, calmodulin, tyrosine kinase, ornithine decarboxylase, and/or the CaSR (Bostick 2001; Chakrabarty et al. 2003).

Calcium and the CaSR

The CaSR transduces extracellular calcium binding into a variety of intracellular responses, including pathways involved in proliferation, differentiation, and apoptosis control (Lamprecht and Lipkin 2001). This hypothesis, then, may also be the mechanism for the direct effect on cell cycle, but with added benefits for cell adhesion: the CaSR regulates colon epithelial cell proliferation and differentiation in vitro by upregulating E-cadherin expression and downregulating β-catenin binding to TCF4 (Chakrabarty et al. 2003). In a rat model, colon crypt epithelial cells acquire CaSR expression as they migrate and differentiate toward the apex of the crypt, and both calcium and $1\alpha,25\text{-}(OH)_2$-vitamin D_3 stimulate the promoter of the *CaSR* gene in an additive manner (Bhagavathula et al. 2005). Although free calcium levels in plasma are tightly regulated, there is a substantial calcium gradient in the colon crypt, with concentrations increasing from base to lumen (Rodland 2004). The gradients of calcium concentration and CaSR expression correlate with colonocyte proliferation and differentiation, and CaSR expression is inversely associated with differentiation in colorectal carcinomas. Thus, the CaSR and extracellular calcium may function together, at least in part, by suppressing β-catenin/TCF4 activation to tightly regulate colon epithelial cell growth and differentiation programs; disruption of CaSR expression or function may circumvent normal proliferation, differentiation, and cell–cell and cell–matrix adhesion, thus promoting carcinogenesis (Lamprecht and Lipkin 2001; Peters et al. 2004a; Kallay et al. 2000). The CaSR may also serve a broader role as a "nutrient receptor," recognizing nutrients (e.g., amino acids) other than divalent cations (Brown 2005), thus playing additional roles in reducing risk of colorectal neoplasms.

Vitamin D

The four most prominent hypotheses for the role of vitamin D in protecting against colorectal cancer involve (1) bile-acid catabolism, (2) direct effects on the cell cycle, (3) growth-factor signaling, and (4) immunomodulation. As with calcium, these potential mechanisms are probably complementary.

Bile Acids

Vitamin D, as well as the secondary bile acid, lithocholic acid (LCA), can activate the VDR, which induces expression in vivo of CYP3A4, which, in turn, detoxifies LCA in the intestine and liver (Makishima et al. 2002). Such increased bile-acid detoxification may yield the same result as bile-acid neutralization by calcium: reduced DNA mutation, cell damage, compensatory hyperproliferation, and inflammation.

Direct Effect on Cell Cycle

Vitamin D is thought to protect against colorectal neoplasia by reducing epithelial cell proliferation, inducing differentiation, and promoting apoptosis (Bostick 2001), activities that are mediated in part by the VDR (Lips 2006; Norman 2006). These activities probably occur through colon tissue autocrine/paracrine synthesis of $1\alpha,25\text{-(OH)}_2$-vitamin D. Ligand-bound VDR can arrest cells in G1, probably through modulating cell-cycle proteins such as cyclin D1 (van den Bemd et al. 2000; Haussler et al. 1998; Moffatt et al. 2001). The apoptotic effects of vitamin D may occur through inducing *bak* and *TGFβ* or inhibiting bcl-2 expression (van den Bemd et al. 2000; Diaz et al. 2000). Vitamin D also can reduce expression of *c-myc*, *c-fos*, and *c-jun* oncogenes, and suppress telomerase and angiogenesis (van den Bemd et al. 2000; Haussler et al. 1998; Tong et al. 1998, 1999).

Growth-Factor Signaling

Vitamin D signaling and several growth-factor pathways (insulin, insulin-like growth factor, growth hormone, epidermal growth factor, vascular epithelial growth factor, and transforming growth factor and their relevant binding proteins and receptors) interact in ways that result in growth suppression. Vitamin D: interferes with epidermal growth-factor (EGF) signaling (perhaps by reducing expression of the EGF receptor) (van den Bemd et al. 2000; Tong et al. 1998; Tong et al. 1999); reduces expression of the insulin-like growth factor-1 (IGF-1) receptor (van den Bemd et al. 2000); and inhibits IGF-1 signaling generally (Xie et al. 1999). In vitro, although vitamin D has been shown not to affect total secreted TGFβ, it increased the amount of the active form (Chen et al. 2002). Further, 1α,25-$(OH)_2$-vitamin D_3: sensitized colon-cancer cell lines to TGFβ growth inhibition; increased IGF-IIR expression, increasing activation of latent TGFβ; and, in combination with TGFβ, reduced cell proliferation. SMAD3, a downstream protein in the TGFβ signaling pathway, is a coactivator of the VDR and positively regulates the vitamin D signaling pathway (Harris and Go 2004).

Immunomodulation

Vitamin D appears to have important effects on immunity and control of inflammation (Yee et al. 2005; Reichrath et al. 2007; Cannell et al. 2006; Cantorna 2006) that may be relevant to colon carcinogenesis and prevention. The colon is a reservoir for microbes, and inflammation is an established risk factor for colorectal cancer. Various cell types involved in immunologic reactions express VDR and CYP27B1. Local 1α,25-$(OH)_2$-vitamin D synthesis in immune cells is considered critically important for regulating and controlling immune responses. The role of vitamin D in immunomodulation is reviewed in detail elsewhere (Yee et al. 2005; Reichrath et al. 2007; Cannell et al. 2006; Cantorna 2006). The growing list of vitamin-D-responsive inflammation-control genes includes those for IL-2, IL-4, IL-5, IL-6, IL-10, IL-10R, IL-12, IFN-γ, lymphotoxin, TNFα, and GM-CSF. In prostate-cancer cell cultures, 1α,25-$(OH)_2$-vitamin D_3 decreased COX-2 expression, while increasing 15-PGDH and inhibiting EP2 and FP prostaglandin receptor expression. This area of research is new and it has not been directly linked to colorectal carcinogenesis and prevention; thus, the relative contribution, to colon carcinogenesis, of vitamin D effects on immunomodulation is currently unclear.

Epidemiology of Calcium, Vitamin D, and Colorectal Neoplasms

Human epidemiologic studies are motivated to a large extent by the mechanistic evidence discussed in previous sections and by the results of animal experiments, which have been almost entirely consistent in finding that calcium and vitamin D reduce colorectal tumorigenesis (Wargovich and Baer 1989). There have been numerous observational studies of calcium and vitamin D and risk of colorectal neoplasms, but few addressed interactions of these agents with interindividual genetic differences, and there have been few clinical trials. In this section, the epidemiologic evidence for modulation of colorectal cancer risk is summarized, first for calcium and then for vitamin D. For each agent, we first review their main-effect associations (i.e., associations not involving interactions with other agents or genetic polymorphisms) with colorectal cancer and then with colorectal adenoma, organized by type of epidemiologic study, followed by a review of studies of associations of genotypes of relevant genes in calcium or vitamin D metabolism and physiology (alone and in combination with calcium or vitamin D) with colorectal neoplasms. This is followed by a review of studies that investigated potential calcium-vitamin D interactions.

Calcium

Observational Studies of Colorectal Cancer and Calcium

Data from the numerous observational studies – especially from the prospective cohort studies – are consistent with the hypothesis that higher intakes of calcium reduce risk of colorectal cancer. Of 42 analyses of data from analytic observational studies of calcium and colorectal cancer (22 case-control studies and 20 prospective cohort studies), 30 (71%) found inverse associations, of which 16 were statistically significant, three found null associations, and nine found increased risk with higher intake, none of which was statistically significant. There is more consistency among the prospective cohort than the case-control studies: of the 20 cohort studies, 18 (90%) found inverse associations, of which eight were statistically significant. A pooled analysis of 10 cohort studies from five countries reported a statistically significant 22% reduction in risk for incident colorectal cancer among those consuming the highest versus the lowest levels of calcium (Cho et al. 2004). Since that pooled analysis, five new prospective studies of calcium and colorectal cancer have been reported, four of which reported relative risks (RRs) between 0.67 and 0.74 (of which three were statistically significant) (Flood et al. 2005; Kesse et al. 2005; Larsson et al. 2006; Park et al. 2007) and one reported a statistically nonsignificant RR of 1.2 (Lin et al. 2005). The proportions of inverse associations are comparable for men and women and for colon and rectal cancers.

Observational Studies of Colorectal Adenoma and Calcium

The results of the relatively fewer calcium and colorectal adenoma studies support those of the colorectal cancer studies. Of 11 observational studies of calcium and colorectal adenoma (eight primary case-control studies, two case-control studies nested in cohort studies, and one prospective study in a clinical-trial cohort), nine (82%) found inverse associations, of which one was statistically significant, and two found statistically nonsignificant increases in risk with higher intake.

Calcium and Clinical Trials of Biomarkers

Of at least 17 trials of calcium and colorectal epithelial cell proliferation, most were pilot studies with significant methodological limitations that largely reported beneficial responses (Bostick 1997). There are only two full-scale clinical trials (Baron et al. 1995; Bostick et al. 1995). One found no evidence for a reduction in the overall proliferative rate, but a marked, statistically significant downward shift (normalization) of the colon-crypt proliferative zone (Bostick et al. 1995). The second, an adjunct study to a calcium and adenoma recurrence trial (the Calcium Polyp Prevention Study, see later), found no effect on proliferation (Baron et al. 1995); however, in contrast to the first study, this study had poor reader reliability for proliferation markers ($r=0.41$ vs. 0.94 in the first study). Three small, pilot trials of calcium and biomarkers of apoptosis and cell differentiation reported inconclusive results (Holt et al. 1998). Several small trials investigating the calcium/bile-acid hypothesis found that calcium decreased the concentration and excretion of free bile acids and the cytotoxicity of fecal water (Alberts et al. 1996; Glinghammar et al. 1997; Govers et al. 1996; Lupton et al. 1996).

Calcium and Clinical Trials of Colorectal Neoplasms

There have been five preliminary and two major clinical trials of calcium and adenoma recurrence, and one major trial of prevention of incident colorectal cancer. In a US multicenter, randomized, double-blind, placebo-controlled clinical trial ($n=913$) of calcium supplementation and adenoma recurrence (the Calcium Polyp Prevention Study) (Baron et al. 1999), persons with at least one adenoma at baseline colonoscopy were randomized to either placebo or 1,200 mg of elemental calcium daily. Adenomas detected after a 1-year follow-up colonoscopy up to and including a 4-year follow-up colonoscopy were considered recurrent. The RR for any metachronous adenoma was 0.85 (95% confidence interval (95% CI: 0.74–0.98)); for the average number of adenomas, 0.76 (95% CI: 0.60–0.96); and for advanced adenomas, 0.46 (95% CI: 0.26–0.83) (Wallace et al. 2004). After 5 years of post-trial follow-up of 597 participants, the decreased risk for metachronous adenomas persisted (RR: 0.63, 95% CI: 0.46–0.87) (Grau et al. 2007). A smaller European trial ($n=665$) tested the effect of 2,000 mg of elemental calcium daily on metachronous adenoma, and found a statistically nonsignificant reduction RR: 0.66, 95% CI: 0.38–1.17) (Bonithon-Kopp et al. 2000). A meta-analysis of all seven adenoma recurrence trials yielded a summary RR of 0.80 (95% CI: 0.68–0.93) (Shaukat et al. 2005). Finally, in the Women's Health Initiative randomized, double-blind, placebo-controlled clinical trial ($n=36{,}282$ postmenopausal women) of 1,000 mg of elemental calcium plus 10 μg of vitamin D versus placebo over an average of 7 years, no prevention of incident, invasive colorectal cancers was found (RR: 1.08, 95% CI: 0.86–1.34) (Wactawski-Wende et al. 2006). However, these results are difficult to interpret because of the low adherence in the active treatment group (only 60% took 80% or more of their pills) and the high rate of drop-in in the placebo group (69% took calcium and vitamin D supplements on their own, resulting in intakes twice those of the national averages), the low doses administered, and the short length of follow-up for such a downstream endpoint. In summary, higher calcium intakes reduce metachronous colorectal adenomas, but there has been no adequate test of whether it can reduce incidence of colorectal carcinoma.

Observational Studies of CaSR Gene Polymorphisms and Colorectal Neoplasms

The human *CaSR* gene, located on chromosome 3q13.3-q21, has eight exons, of which exons 2–7 encode the 1,078 amino acid CaSR protein (Harding et al. 2006; Heath et al. 1996). Two promoter regions, which contain vitamin D response elements (VDREs), and more than 600 genetic variants have been found. Some rare genetic variants that lead to amino acid substitutions that alter the CaSR function, cause some familial calcium homeostasis-related disorders. Three single nucleotide polymorphisms (SNPs) in exon 7 (A986S, G990R, and Q1011E) that cause amino acid changes but do not appear to cause overt calcium homeostasis disturbances, were investigated in relation to risk for colorectal neoplasms in two studies (Speer et al. 2002; Peters et al. 2001). In a small ($n=56$ cases, 112 controls), hospital-based case-control study in Hungary, no association was found between *CaSR* A986S genotypes and rectal cancer (Speer et al. 2002). In a large ($n=716$ cases and 729 controls), sigmoidoscopy-based case-control study of distal colorectal adenomas nested in the Prostate, Lung, Colorectal, and Ovarian Cancer Screening Trial (PLCO trial), no statistically significant associations between these three *CaSR* genotypes and distal adenomas were found (Peters et al. 2001); however, the statistically nonsignificant results suggested that there may be lower risk for advanced adenomas with one diplotype (i.e., two haplotypes in combination) alone and in combination with higher calcium intake.

Studies of Calcium in Interaction with Other Agents, Risk Factors, and Genotypes in Relation to Risk of Colorectal Neoplasms

Possible interactions of calcium with various other agents, risk factors, and polymorphisms of genes other than *CaSR* have been reported. Studies of potential interactions of calcium with vitamin D are reviewed later. With the bile-acid hypothesis in mind, a few studies investigated possible interactions of calcium with fat intake; however, their sample sizes were too small to investigate interactions adequately and the results have been inconsistent, thus providing no solid answers. Based on the bile-acid/inflammation hypothesis, it was hypothesized that the calcium-colorectal neoplasm association may be modified by NSAIDs (Boyapati et al. 2003). In support of this, a colonoscopy-based case-control study of incident, nonfamilial colorectal adenomas ($n=177$ cases, 228 controls) in North Carolina reported a marked reduction of risk with higher calcium intakes among those not using NSAIDs (odds ratio (OR): 0.36, 95% CI: 0.15–0.85), but not among NSAID users (Boyapati et al. 2003). Similar results were seen in secondary analyses of the Calcium Polyp Prevention Study: among 832 participants, the RR for metachronous adenoma with calcium was 0.64 (CI: 0.46–0.90) among those who did not take aspirin and/or NSAIDs, but 0.93 (CI: 0.73–1.19) among those who did (Grau et al. 2005).

Because of the inter-relationships of calcium and vitamin D, potential interactions of calcium with *VDR* polymorphisms have been investigated in a few studies. In a case-control study of colorectal cancer ($n=217$ cases, 890 controls) nested in a large cohort of Singapore Chinese (Wong et al. 2003), associations of *VDR FokI* genotypes with colorectal cancer differed, depending on levels of calcium intakes, with the lowest risk seen among persons with the FF genotype who had high calcium intakes. In a health maintenance organization population-based case-control study of colorectal adenoma in Los Angeles, California ($n=373$ cases, 394 controls), a *VDR FokI* genotype-calcium interaction was also suggested; however, the lowest risk for large adenomas was seen among persons with the FF genotype who also had low calcium intakes (Ingles et al. 2001). In a population-based case-control study of colorectal adenoma in Maryland ($n=239$ cases, 228 controls), a modest inverse association between total calcium intake and adenomas did not differ by *FokI VDR* genotype (Peters et al. 2001). Among 803 participants in the Calcium Polyp Prevention Study, there was no evidence that the effect of calcium supplementation on adenoma recurrence was modified by *FokI VDR* polymorphisms (Grau et al. 2003).

In a population-based case-control study of incident colon and rectal cancer ($n=2,306$ cases, 2,749 controls) there was evidence that risk for colon, but not rectal, cancer was lowest in persons homozygous for both the *VDR* short *Poly(A)* polymorphism (SS) and the *BsmI* B polymorphism (BB) who also had high calcium intakes (Slattery et al. 2004b). In a colonoscopy-based case-control study of colorectal adenoma in Minneapolis, Minnesota ($n=393$ cases, 406 controls) (Kim et al. 2001), risk tended to be lowest among those with the *BsmI BB* genotype who had low calcium intakes. However, a colonoscopy-based case-control study of colorectal adenoma in North Carolina ($n=177$ cases, 228 controls) found evidence for lowest risk among those with a *VDR BsmI b* allele who had higher calcium intakes (Boyapati et al. 2003). In the same study population, associations of calcium intake with adenomas did not differ according to *VDR Tru9I* genotypes (Gong et al. 2005a). Among 803 participants in the Calcium Polyp Prevention Study, there was no evidence that the effect of calcium supplementation on adenoma recurrence was modified by *VDR TaqI* polymorphisms (Grau et al. 2003).

Based on knowledge of the two major molecular pathways to colorectal cancer, the APC-β-catenin-Tcf pathway (wingless pathway) and the mismatch repair pathway, a calcium association with colorectal neoplasms has been investigated in conjunction with *APC*, *CCND1* (the cyclin D1 gene, a downstream transcription target of the APC pathway), *K-ras*, and microsatellite instability (MSI). The only study to report a calcium-*APC* genotype interaction, a case-control study of colorectal cancer in Portugal ($n=196$ cases, 200 controls (a convenience sample of "healthy blood donors and health care workers")) reported evidence for an interaction of calcium with *APC* D1822V genotypes in which risk was lowest in persons with a high calcium intake and at least one V allele (Guerreiro et al. 2007). A colonoscopy-based case-control study of colorectal adenoma in North Carolina (n=161 cases, 213 controls) reported evidence for a calcium-*CCND1* A870G genotype interaction in which risk was lowest for persons with high calcium intakes and the GG genotype (Lewis et al. 2003). In a population-based case-control study of invasive colorectal cancer in North Carolina (n =486 cases, 1,048 controls), an inverse association between calcium and colorectal cancer did not differ by MSI status (Satia et al. 2005). In a case-case (n = 108) analysis of colorectal cancer in a Mediterranean population, relative to persons with tumors without *K-ras* mutations, there was a reduced risk for tumors *with K-ras mutations* in association with higher calcium intake (Bautista et al. 1997); however, in a colonoscopy-based case-control study of colorectal adenoma in the Netherlands (n=534 cases, 704 controls), higher calcium intake was associated with reduced risk for adenomas *without K-ras mutations*, but not for adenomas with *K-ras* mutations (Wark et al. 2006). There have been too few studies of calcium in relation to colon carcinogenesis pathway genes to draw any conclusions about differential associations by colon-carcinogenesis-pathway genotypes or acquired mutations.

Vitamin D

Based on clinical and ecologic observations of a correlation between sun exposure and colorectal cancer incidence, Garland and Garland proposed, in 1980, that vitamin D insufficiency may increase colon-cancer incidence and mortality (Garland and Garland 1980). Several ecologic studies have confirmed a correlation between sunlight exposure and colorectal cancer occurrence. Beyond these early hypothesis-generating studies, there is now a considerable and evolving literature on a vitamin D-colorectal neoplasms association; it includes studies of associations of colorectal neoplasms with questionnaire-based measures of dietary (food and supplements) vitamin D intake, circulating levels of vitamin D metabolites, and *VDR* gene polymorphisms.

Observational Studies of Colorectal Cancer and Vitamin D

Of 30 reported analytic observational studies of vitamin D and colorectal cancer (17 case-control studies and 13 prospective cohort studies), 20 (67%) found inverse associations, of which six were statistically significant, six found null associations, and four found statistically nonsignificant evidence of higher risks with higher intake. A pooling project of 5 (of 10) cohort studies with total vitamin D intake (diet plus supplements) assessment, reported, in 2004, a statistically nonsignificant

7% reduction in risk for incident colorectal cancer among those consuming the highest levels of vitamin D (blood levels were not assessed) versus those consuming the lowest levels (Cho et al. 2004). The risk estimates from three subsequent prospective cohort studies (Kesse et al. 2005; Park et al. 2007; Lin et al. 2005) were similar to that pooled estimate. The proportions of inverse associations are comparable for men and women and for colon and rectal cancers.

Because studies that investigated dietary vitamin D did not account for vitamin D exposure from sunlight (which provides 90–95% of vitamin D for most people (Holick 2007), and because vitamin D fortification of milk products (the primary source of dietary vitamin D in the US) may be inconsistent (Chen et al. 2007), there is probably serious error and misclassification of total vitamin D exposures based on diet alone which almost certainly biases findings toward the null. Whereas only 15 of 25 (60%) studies that assessed dietary vitamin D found inverse associations (of which five were statistically significant), all five studies that assessed vitamin D exposure with 25-OH-vitamin D blood levels found inverse associations (Wactawski-Wende et al. 2006; Braun et al. 1995; Feskanich et al. 2004; Garland et al. 1989; Tangrea et al. 1997) (of which one was statistically significant (Wactawski-Wende et al. 2006). This greater consistency for the 25-OH-vitamin-D-blood level studies may be remarkable given the low blood levels in the studies (mean levels in controls were generally less than 82 nmol L^{-1} – now considered by many to be the lower limit for vitamin D sufficiency (Holick 2007; Hollis 2005; Heaney 2005). However, given that there have been only five studies that assessed 25-OH-vitamin D blood levels, this greater proportional "consistency" should be viewed as only suggestive. There have been only six studies that investigated blood levels of 1α,25-$(OH)_2$-vitamin D with colorectal neoplasms; however, as might be expected from this tightly regulated, poor indicator of vitamin D exposure, the results have been null.

Observational Studies of Colorectal Adenoma and Vitamin D as a Main Effect

Of 21 reported analyses of data from analytic observational studies of vitamin D and colorectal adenoma (12 primary case-control studies, four case-control studies nested in prospective cohort studies, and five prospective studies in clinical-trial cohorts), 12 (57%) found inverse associations of which three were statistically significant, seven found null associations, and two found statistically nonsignificant increased risk. Eight of 15 (53%) that assessed dietary vitamin D intakes found inverse associations, of which one was statistically significant, whereas four of six (67%) studies that assessed 25-OH-vitamin D blood levels found inverse associations, of which two were statistically significant (Peters et al. 2001; Grau et al. 2003; Jacobs et al. 2007; Levine et al. 2001; Peters et al. 2004b; Platz et al. 2000). Although adenomas have received less study, at this time, the results of vitamin D and colorectal adenoma are fairly consistent with those for colorectal cancer.

Observational Studies of Colorectal Neoplasms and VDR and Vitamin D-Metabolizing Enzyme Gene Polymorphisms

As described earlier, receptors and enzymes important in the activity and metabolism of vitamin D include the VDR, vitamin D 25-hydroxylase(s), CYP27B1, CYP24, and CYP3A4. There have been numerous reported SNPs in the genes encoding for these receptors and enzymes; however, only a few polymorphisms of the *VDR* gene have been investigated in relation to risk of colorectal neoplasms. Four SNPs located near the 3′ region of the *VDR* gene are identified by their restriction endonuclease cleavage sites and include three intronic mutations, G>A (*BsmI*), C>A (*ApaI*), G>A (*Tru9I*), and a T>C mutation in exon 9 (*TaqI*) (McCullough et al. 2007). Although not known to have functional consequences, these SNPs are in strong linkage disequilibrium with a *Poly(A)* microsatellite repeat polymorphism in the 3′ untranslated region that may serve as a marker for functionally different alleles (Sweeney et al. 2006). At the 5′ end of the *VDR* gene (intron 2), there is a thymine/cytosine (T/C) polymorphism in the first two potential start (ATG) codons that are separated by three codons. This polymorphism, not in linkage disequilibrium with the other variants described earlier (Slattery et al. 2004a), results in two alleles that can be distinguished by a restriction fragment length polymorphism using the endonuclease *FokI* (Huang et al. 2006).

There have been a few reports from case-control studies of VDR variation and colorectal cancer. Several publications described analyses from a combined case-control dataset that included information on 2,450 colorectal cancer cases and 2,821 controls from three sites: the state of Utah, the Kaiser Permanente Medical Care Program of northern California, and the Twin Cities metropolitan area, Minnesota (colon study only). An analysis limited to a subset of participants from Utah investigated associations with three of the five linked *VDR* 3′ region polymorphisms (*Poly(A)*, *BsmI*, *TaqI*) and the 5′ region *FokI* polymorphism, individually and in combination (Slattery et al. 2001). On the one hand, in analyses involving individual polymorphic genotypes, compared to persons who were homozygous for common alleles, those who were homozygous for variant alleles tended to be at slightly lower risk; however, none of these associations was statistically significant. On the other hand, the combined SSBBtt genotype was associated with a statistically significant halving of risk. In an analysis that included the Utah and California sites, the results for *BsmI* were null, whereas those for the *Poly(A)* SS and the *BsmI* + *Poly(A)* (SSBB) genotypes were moderately inverse (ORs: 0.79 (95% CI: 0.56–0.96) and 0.82 (95% CI: 0.69–0.98), respectively) (Slattery et al. 2004b). There was no interaction between *VDR* polymorphisms and vitamin D intake but some evidence of an interaction between calcium intake and *BsmI*/*Poly(A)* diplotypes for rectal, but not colon cancer. In two analyses of data from all three participating states, the *FokI* Ff genotype was associated with a 10% statistically nonsignificant reduction in colorectal cancer risk (OR: 0.90, 95% CI: 0.80–1.02), whereas the ff genotype was associated with a somewhat more pronounced decreased risk (OR: 0.81, 95% CI: 0.68–0.96) (Murtaugh et al. 2006). The common haplotype bLF, containing the *BsmI* b, *Poly(A)* long (L) and *FokI* F alleles, was associated with a modestly increased risk of colon cancer (OR: 1.15, 95% CI: 1.03–1.28) as was, somewhat more strangely, the rare BLF haplotype (OR: 2.40, 95% CI: 1.43–4.02) (Sweeney et al. 2006). No case-control differences were detected for rectal cancer.

The only other population-based study of *VDR* genotypes and colorectal cancer was conducted in South Korea (Park et al. 2006). Colorectal cancer patients ($n = 190$) who underwent surgical treatment for colorectal cancer at a large medical center in Seoul were compared to 318 healthy controls with no history of colorectal cancer. Colorectal cancer risk was statistically significantly decreased for those with the *FokI* ff (compared to FF) (OR: 0.35, 95% CI: 0.19–0.65) and increased for those with the *ApaI* aa (compared to AA) (OR: 2.22, 95% CI: 1.12–4.40) genotypes.

There have been five VDR-colorectal adenoma case-control studies, all endoscopy-based. In the previously mentioned colonoscopy-based case-control study of incident, colorectal adenoma in North Carolina, the *VDR Tru9I* polymorphism variant u allele was associated with a modest, and not statistically significant, decreased risk for adenoma: ORs 0.88 (95% CI: 0.17–4.55) and 0.69 (95% CI: 0.40–1.25) for the Uu and uu genotypes, respectively (Gong et al. 2005a), with no evidence for an interaction with vitamin D intake. In the same population, there was no evidence for an association with the *VDR BsmI* polymorphism or for a *BsmI*-vitamin D intake interaction (Boyapati et al. 2003). Two similarly designed case-control studies found weak, statistically nonsignificant lower colorectal adenoma risk among those with *VDR FokI* Ff or ff genotypes (Peters et al. 2001; Ingles et al. 2001). In another similarly designed case-control study (Kim et al. 2001), relative to the *VDR BsmI bb* genotype, neither the *Bb* nor the *BB* genotype was strongly associated with risk of colorectal adenomas; however, participants in the lowest tertile of vitamin D intake who had the *BB* genotype were at lower risk (OR: 0.24, 95% CI: 0.08–0.76) than those in the highest tertile of vitamin D intake who had the *bb* genotype ($p_{\text{interaction}} = 0.07$).

Finally, three studies reported on associations between colorectal neoplasms and *VDR TaqI* and *FokI* polymorphisms. A case-control study of colorectal adenoma nested within the PLCO trial ($n = 239$ cases, 228 controls) found no evidence for associations with *VDR TaqI* genotypes (Peters et al. 2004a). In the other nested case-control study ($n = 217$ cases, 890 controls), in a cohort of Singapore Chinese, *VDR FokI* polymorphisms were associated with moderately higher colorectal cancer risk: ORs 1.51 (95% CI: 1.00–2.29) and 1.84 (95% CI: 1.15–2.94) for the Ff and ff genotypes, respectively (Wong et al. 2003). However, in a cohort analysis of the Calcium Polyp Prevention Study, there was no evidence for an association of *VDR TaqI* or *FokI* genotypes with metachronous adenomas (Grau et al. 2003).

Studies of Colorectal Neoplasms and Vitamin D and VDR Genotypes and Their Interaction with Other Genotypes

In the only study to investigate an interaction between vitamin D intake and polymorphisms of the *PPARγ* gene, a colonoscopy-based case-control study of incident, nonfamilial colorectal adenomas in North Carolina ($n = 163$ cases, 212 controls), there was no evidence for an interaction between vitamin D intake and *PPARγ Pro12Ala* genotypes (Gong et al. 2005b). Because there is evidence for transcriptional crosstalk between the *VDR* and the androgen receptor (*AR*) genes, the previously described Utah-California-Minnesota case-control study of colorectal cancer investigated interactions among vitamin D exposure, *AR* CAG repeats, and various *VDR* polymorphisms (Slattery et al. 2006). In this study (the only reported study to have investigated these potential interactions) there was evidence for: (1) an *AR* CAG repeat polymorphism-vitamin D exposure (diet and sunlight) interaction limited to men, such that risk for colon cancer was highest, but risk for rectal cancer was lowest, in those with 23 or more *AR* CAG repeats who also were in the lowest tertile of vitamin D intake or sunlight exposure; (2) an interaction of the *AR* CAG repeat polymorphism with *VDR Poly(A)/BsmI* LL/bb genotypes in relation to rectal cancer, again limited to men; and (3) an interaction of the *AR* CAG repeat polymorphism with *VDR Fok1* genotypes limited to women.

In the only study to investigate whether a vitamin D intake-colorectal cancer association differs according to whether the tumors were *K-ras* mutation positive or negative, a case–case ($n = 108$) analysis of colorectal cancer in a Mediterranean population, there was a tendency toward reduced risk, primarily for tumors with *K-ras* mutations (Bautista et al. 1997).

There have been too few studies such as these to draw any conclusions about whether vitamin D interacts with various genotypes or whether VDR genotypes interact with other genotypes to modify risk for colorectal neoplasms generally (or for mutation-defined subsets) in humans.

Calcium Plus Vitamin D

Of the numerous observational epidemiologic studies of calcium and vitamin D and colorectal neoplasms, only 13 have reported investigating whether they may synergistically modify risk of colorectal neoplasms and, of these, only four presented complete data for assessing interactions (Grau et al. 2003; Levine et al. 2001; Oh et al. 2007; Zheng et al. 1998). Only two of these four studies, both of adenomas, measured 25-OH-vitamin D blood levels (Grau et al. 2003; Levine et al. 2001). Some evidence of interaction between calcium and vitamin D was suggested in a small cohort study of colorectal cancer in men, a cohort study of rectal cancer in women, a multinational pooled analysis of 10 cohort studies, two sigmoidoscopy-based case-control studies of adenomas, a nested case-control analysis of distal colorectal adenoma in a cohort of female nurses, and a cohort analysis of a calcium supplementation-adenoma recurrence trial. Of these, the cohort analysis of the calcium supplementation trial (Grau et al. 2003) deserves the most attention. First, calcium exposure was from a randomized intervention of 1,200 mg of elemental calcium daily versus placebo and, second, although vitamin D was not an intervention exposure, serum 25-OH-vitamin D was measured. As previously described, this trial found a statistically significant reduction in metachronous adenoma with calcium supplementation (Baron et al. 1999). In a separate cohort analysis, adenoma recurrence in the calcium-intervention group was found only among those with 25-OH-vitamin D levels greater than the cohort median (72.8 nmol L^{-1}) (RR: 0.71, 95% CI: 0.57–0.89; $p_{interaction} = 0.01$) (Grau et al. 2003). Although these results suggest that calcium and vitamin D may interact in humans to reduce risk of colorectal neoplasms, they are far from conclusive. Assessment of vitamin D exposure, using 25-OH-vitamin D blood levels, may be particularly important for investigating calcium-vitamin D interactions.

Overall Summary and Conclusions

Calcium and vitamin D have become prominent in understanding the etiology and prevention of colorectal cancer. Recent discoveries of the CaSR, the VDR, and CYP27B1 and CYP24 in colon tissue; multiple actions of calcium and vitamin D not related to calcium homeostasis; and specific roles of calcium and vitamin D in maintaining normal colon crypt structure, function, and protection collectively provide the basis for expecting protective effects of calcium and vitamin D against colorectal carcinogenesis. The strong rationale for this expectation includes protection against mutagenic/mitogenic bile acids, modulation of the cell-cycle and cell adhesion in colon crypt colonocytes, growth-factor modulation, and immunomodulation. Evidence from the observational epidemiologic studies for protection against colorectal neoplasms by higher intakes of calcium is very strong, especially that from the prospective cohort studies. This observational evidence is strongly supported by clinical-trial findings of reduced adenoma recurrence with calcium supplementation. The human evidence for a protective effect of vitamin D against colorectal neoplasms is not as strong as that for calcium, probably because most vitamin D exposure is from sunlight and most of the observational epidemiologic studies of vitamin D and colorectal neoplasms investigated only poorly measured dietary exposures. Serum 25-OH-vitamin D, the best indicator of total vitamin D exposure, was inversely associated with colorectal cancer in all five studies that measured it. There are no reported clinical trials of vitamin D and metachronous colorectal adenoma, but one is underway. Overall, human studies investigating calcium-vitamin D interactions have been inconclusive, probably because almost all of them relied on assessing vitamin D exposure strictly through dietary intakes. The genes for the CaSR, the VDR, vitamin D-metabolizing enzymes, and other relevant proteins are polymorphic, but few studies have investigated their associations with colorectal neoplasms, alone or in combination with one another, with calcium, or with vitamin D. To date, there is neither strong nor consistent evidence that such genetic variation plays a role in colorectal carcinogenesis; however, this line of investigation is in its infancy. The inverse calcium-colorectal cancer association is probably causal, and the inverse vitamin D-colorectal cancer association holds a similar promise.

References

Alberts DS, Ritenbaugh C, Story JA, Aickin M, Rees-McGee S, Buller MK, Atwood J, Phelps J, Ramanujam PS, Bellapravalu S, Patel J, Bextinger L, Clark L (1996) Randomized, double-blinded, placebo-controlled study of effect of wheat bran fiber and calcium on fecal bile acids in patients with resected adenomatous colon polyps. J Natl Cancer Inst 88:81–92

Baron JA, Tosteson TD, Wargovich MJ, Sandler R, Mandel J, Bond J, Haile R, Summers R, van Stolk R, Rothstein R (1995) Calcium supplementation and rectal mucosal proliferation: a randomized controlled trial. J Natl Cancer Inst 87:1303–1307

Baron JA, Beach M, Mandel JS, van Stolk RU, Haile RW, Sandler RS, Rothstein R, Summers RW, Snover DC, Beck GJ, Bond JH, Greenberg ER (1999) Calcium supplements for the prevention of colorectal adenomas. Calcium Polyp Prevention Study Group. N Engl J Med 340:101–107

Bautista D, Obrador A, Moreno V, Cabeza E, Canet R, Benito E, Bosch X, Costa J (1997) Ki-ras mutation modifies the protective effect of dietary monounsaturated fat and calcium on sporadic colorectal cancer. Cancer Epidemiol Biomarkers Prev 6:57–61

Bhagavathula N, Kelley EA, Reddy M, Nerusu KC, Leonard C, Fay K, Chakrabarty S, Varani J (2005) Upregulation of calcium-sensing receptor and mitogen-activated protein kinase signalling in the regulation of growth and differentiation in colon carcinoma. Br J Cancer 93:1364–1371

Bonithon-Kopp C, Kronborg O, Giacosa A, Rath U, Faivre J (2000) Calcium and fibre supplementation in prevention of colorectal adenoma recurrence: a randomised intervention trial. European Cancer Prevention Organisation Study Group. Lancet 356:1300–1306

Bostick RM (1997) Human studies of calcium supplementation and colorectal epithelial cell proliferation. Cancer Epidemiol Biomarkers Prev 6:971–980

Bostick RM (2001) Diet and nutrition in the etiology and primary prevention of colon cancer. In: Bendich A, Deckelbaum R (eds) Preventive Nutrition: The Comprehensive Guide for Health Professionals. Humana, Totowa, NJ, pp 47–96

Bostick RM, Fosdick L, Wood JR, Grambsch P, Grandits GA, Lillemoe TJ, Louis TA, Potter JD (1995) Calcium and colorectal epithelial cell proliferation in sporadic adenoma patients: a randomized, double-blinded, placebo-controlled clinical trial. J Natl Cancer Inst 87:1307–1315

Boyapati SM, Bostick RM, McGlynn KA, Fina MF, Roufail WM, Geisinger KR, Wargovich M, Coker A, Hebert JR (2003) Calcium, vitamin D, and risk for colorectal adenoma: dependency on vitamin D receptor BsmI polymorphism and nonsteroidal anti-inflammatory drug use? Cancer Epidemiol Biomarkers Prev 12:631–637

Braun MM, Helzlsouer KJ, Hollis BW, Comstock GW (1995) Colon cancer and serum vitamin D metabolite levels 10–17 years prior to diagnosis. Am J Epidemiol 142:608–611

Bringhurst FR, Demay MB, Krane SM, Kronenberg HM (2007) Bone and mineral metabolism in health and disease. In: Kasper DL, Braunwald E, Fauci AS, Hauser SL, Longo DL, Jameson L, Isselbacher KJ (eds) Harrison's Principles of Internal Medicine. McGraw-Hill, New York

Brown EM (2005) Calcium-sensing receptor. In: Feldman D, Pike JW, Glorieux FH (eds) Vitamin D2. Elsevier Academic, Boston, pp 551–562

Cannell JJ, Vieth R, Umhau JC, Holick MF, Grant WB, Madronich S, Garland CF, Giovannucci E (2006) Epidemic influenza and vitamin D. Epidemiol Infect 134:1129–1140

Cantorna MT (2006) Vitamin D and its role in immunology: multiple sclerosis, and inflammatory bowel disease. Prog Biophys Mol Biol 92:60–64

Chakrabarty S, Radjendirane V, Appelman H, Varani J (2003) Extracellular calcium and calcium sensing receptor function in human colon carcinomas: promotion of E-cadherin expression and suppression of beta-catenin/TCF activation. Cancer Res 63:67–71

Chattopadhyay N, Brown EM (2006) Role of calcium-sensing receptor in mineral ion metabolism and inherited disorders of calcium-sensing. Mol Genet Metab 89:189–202

Chen A, Davis BH, Sitrin MD, Brasitus TA, Bissonnette M (2002) Transforming growth factor-beta 1 signaling contributes to Caco-2 cell growth inhibition induced by 1,25(OH)(2)D(3). Am J Physiol Gastrointest Liver Physiol 283:G864–874

Chen TC, Chimeh F, Lu Z, Mathieu J, Person KS, Zhang A, Kohn N, Martinello S, Berkowitz R, Holick MF (2007) Factors that influence the cutaneous synthesis and dietary sources of vitamin D. Arch Biochem Biophys 460:213–217

Cho E, Smith-Warner SA, Spiegelman D, Beeson WL, van den Brandt PA, Colditz GA, Folsom AR, Fraser GE, Freudenheim JL, Giovannucci E, Goldbohm RA, Graham S, Miller AB, Pietinen P, Potter JD, Rohan TE, Terry P, Toniolo P, Virtanen MJ, Willett WC, Wolk A, Wu K, Yaun SS, Zeleniuch-Jacquotte A, Hunter DJ (2004) Dairy foods, calcium, and colorectal cancer: a pooled analysis of 10 cohort studies. J Natl Cancer Inst 96:1015–1022

Diaz GD, Paraskeva C, Thomas MG, Binderup L, Hague A (2000) Apoptosis is induced by the active metabolite of vitamin D3 and its analogue EB1089 in colorectal adenoma and carcinoma cells: possible implications for prevention and therapy. Cancer Res 60:304–312

Ebert R, Schutze N, Adamski J, Jakob F (2006) Vitamin D signaling is modulated on multiple levels in health and disease. Mol Cell Endocrinol 248:149–159

Feskanich D, Ma J, Fuchs CS, Kirkner GJ, Hankinson SE, Hollis BW, Giovannucci EL (2004) Plasma vitamin D metabolites and risk of colorectal cancer in women. Cancer Epidemiol Biomarkers Prev 13:1502–1508

Flood A, Peters U, Chatterjee N, Lacey JV, Schairer C, Schatzkin A (2005) Calcium from diet and supplements is associated with reduced risk of colorectal cancer in a prospective cohort of women. Cancer Epidemiol Biomarkers Prev 14:126–132

Fraser DR (1995) Vitamin D. Lancet 345:104–107

Friedman PA (2006) Agents affecting mineral ion homeostasis and bone turnover. In: Brunton LL, Lazo JS, Parker KL (eds) Goodman and Gilman's The Pharmacologic Basis of Therapeutics. McGraw-Hill, New York, pp 1647–1678

Garland CF, Garland FC (1980) Do sunlight and vitamin D reduce the likelihood of colon cancer? Int J Epidemiol 9:227–231

Garland CF, Comstock GW, Garland FC, Helsing KJ, Shaw EK, Gorham ED (1989) Serum 25-hydroxyvitamin D and colon cancer: eight-year prospective study. Lancet 2:1176–1178

Glinghammar B, Venturi M, Rowland IR, Rafter JJ (1997) Shift from a dairy product-rich to a dairy product-free diet: influence on cytotoxicity and genotoxicity of fecal water – potential risk factors for colon cancer. Am J Clin Nutr 66:1277–1282

Gong YL, Xie DW, Deng ZL, Bostick RM, Miao XJ, Zhang JH, Gong ZH (2005a) Vitamin D receptor gene Tru9I polymorphism and risk for incidental sporadic colorectal adenomas. World J Gastroenterol 11:4794–4799

Gong Z, Xie D, Deng Z, Bostick RM, Muga S, Hurley TG, Hebert JR (2005b) PPARγ Pro12Ala polymorphism and risk for incident sporadic colorectal adenomas. Carcinogenesis 26:579–585

Govers MJ, Termont DS, Lapre JA, Kleibeuker JH, Vonk RJ, Van der Meer R (1996) Calcium in milk products precipitates intestinal fatty acids and secondary bile acids and thus inhibits colonic cytotoxicity in humans. Cancer Res 56:3270–3275

Grau MV, Baron JA, Sandler RS, Haile RW, Beach ML, Church TR, Heber D (2003) Vitamin D, calcium supplementation, and colorectal adenomas: results of a randomized trial. J Natl Cancer Inst 95:1765–1771

Grau MV, Baron JA, Barry EL, Sandler RS, Haile RW, Mandel JS, Cole BF (2005) Interaction of calcium supplementation and nonsteroidal anti-inflammatory drugs and the risk of colorectal adenomas. Cancer Epidemiol Biomarkers Prev 14:2353–2358

Grau MV, Baron JA, Sandler RS, Wallace K, Haile RW, Church TR, Beck GJ, Summers RW, Barry EL, Cole BF, Snover DC, Rothstein R, Mandel JS (2007) Prolonged effect of calcium supplementation on risk of colorectal adenomas in a randomized trial. J Natl Cancer Inst 99:129–136

Guerreiro CS, Cravo ML, Brito M, Vidal PM, Fidalgo PO, Leitao CN (2007) The D1822V APC polymorphism interacts with fat, calcium, and fiber intakes in modulating the risk of colorectal cancer in Portuguese persons. Am J Clin Nutr 85:1592–1597

Harding B, Curley AJ, Hannan FM, Christie PT, Bowl MR, Turner JJ, Barber M, Gillham-Nasenya I, Hampson G, Spector TD, Thakker RV (2006) Functional characterization of calcium sensing

receptor polymorphisms and absence of association with indices of calcium homeostasis and bone mineral density. Clin Endocrinol 65:598–605
Harris DM, Go VL (2004) Vitamin D and colon carcinogenesis. J Nutr 134:3463S–3471S
Haussler M, Whitfield G, Haussler C, Hsieh J, Thompson P, Selznick S, Dominguez C, Jurutka P (1998) The nuclear vitamin D receptor: biological and molecular regulatory properties revealed. J Bone Miner Res 13:325–349
Heaney RP (2005) The vitamin D requirement in health and disease. J Steroid Biochem Mol Biol 97:13–19
Heath H, Odelberg S, Jackson CE, Teh BT, Hayward N, Larsson C, Buist NR, Krapcho KJ, Hung BC, Capuano IV, Garrett JE, Leppert MF (1996) Clustered inactivating mutations and benign polymorphisms of the calcium receptor gene in familial benign hypocalciuric hypercalcemia suggest receptor functional domains. J Clin Endocrinol Metab 81:1312–1317
Holick M (1990) The use and interpretation of assays for vitamin D and its metabolites. J Nutr 120:1464–1469
Holick MF (2007) Vitamin D deficiency. N Engl J Med 357:266–281
Hollis BW (2005) Circulating 25-hydroxyvitamin D levels indicative of vitamin D sufficiency: implications for establishing a new effective dietary intake recommendation for vitamin D. J Nutr 135:317–322
Holt PR, Atillasoy EO, Gilman J, Guss J, Moss SF, Newmark H, Fan K, Yang K, Lipkin M (1998) Modulation of abnormal colonic epithelial cell proliferation and differentiation by low-fat dairy foods: a randomized controlled trial. J Am Med Assoc 280:1074–1079
Huang SP, Huang CY, Wu WJ, Pu YS, Chen J, Chen YY, Yu CC, Wu TT, Wang JS, Lee YH, Huang JK, Huang CH, Wu MT (2006) Association of vitamin D receptor FokI polymorphism with prostate cancer risk, clinicopathological features and recurrence of prostate specific antigen after radical prostatectomy. Int J Cancer 119:1902–1907
Ingles SA, Wang J, Coetzee GA, Lee ER, Frankl HD, Haile RW (2001) Vitamin D receptor polymorphisms and risk of colorectal adenomas (United States). Cancer Causes Control 12:607–614
Jacobs ET, Alberts DS, Benuzillo J, Hollis BW, Thompson PA, Martinez ME (2007) Serum 25(OH)D levels, dietary intake of vitamin D, and colorectal adenoma recurrence. J Steroid Biochem Mol Biol 103:752–756
Kallay E, Bajna E, Wrba F, Kriwanek S, Peterlik M, Cross HS (2000) Dietary calcium and growth modulation of human colon cancer cells: role of the extracellular calcium-sensing receptor. Cancer Detect Prev 24:127–136
Kesse E, Boutron-Ruault MC, Norat T, Riboli E, Clavel-Chapelon F (2005) Dietary calcium, phosphorus, vitamin D, dairy products and the risk of colorectal adenoma and cancer among French women of the E3N-EPIC prospective study. Int J Cancer 117:137–144
Kim HS, Newcomb PA, Ulrich CM, Keener CL, Bigler J, Farin FM, Bostick RM, Potter JD (2001) Vitamin D receptor polymorphism and the risk of colorectal adenomas: evidence of interaction with dietary vitamin D and calcium. Cancer Epidemiol Biomarkers Prev 10:869–874
Lamprecht SA, Lipkin M (2001) Cellular mechanisms of calcium and vitamin D in the inhibition of colorectal carcinogenesis. Ann N Y Acad Sci 952:73–87
Larsson SC, Bergkvist L, Rutegard J, Giovannucci E, Wolk A (2006) Calcium and dairy food intakes are inversely associated with colorectal cancer risk in the Cohort of Swe dish Men. Am J Clin Nutr 83:667–673
Levine AJ, Harper JM, Ervin CM, Chen YH, Harmon E, Xue S, Lee ER, Frankel HD, Haile RW (2001) Serum 25-hydroxyvitamin D, dietary calcium intake, and distal colorectal adenoma risk. Nutr Cancer 39:35–41
Lewis RC, Bostick RM, Xie D, Deng Z, Wargovich MJ, Fina MF, Roufail WM, Geisinger KR (2003) Polymorphism of the cyclin D1 gene, CCND1, and risk for incident sporadic colorectal adenomas. Cancer Res 63:8549–8553
Lin J, Zhang SM, Cook NR, Manson JE, Lee IM, Buring JE (2005) Intakes of calcium and vitamin D and risk of colorectal cancer in women. Am J Epidemiol 161:755–764
Lips P (2006) Vitamin D physiology. Prog Biophys Mol Biol 92:4–8

Lupton JR, Steinbach G, Chang WC, O'Brien BC, Wiese S, Stoltzfus CL, Glober GA, Wargovich MJ, McPherson RS, Winn RJ (1996) Calcium supplementation modifies the relative amounts of bile acids in bile and affects key aspects of human colon physiology. J Nutr 126:1421–1428

Makishima M, Lu TT, Xie W, Whitfield GK, Domoto H, Evans RM, Haussler MR, Mangelsdorf DJ (2002) Vitamin D receptor as an intestinal bile acid sensor. Science 1296:1313–1316

McCullough ML, Stevens VL, Diver WR, Feigelson HS, Rodriguez C, Bostick RM, Thun MJ, Calle EE (2007) Vitamin D pathway gene polymorphisms, diet, and risk of postmenopausal breast cancer: a nested case-control study. Breast Cancer Res 9:R9

Moffatt KA, Johannes WU, Hedlund TE, Miller GJ (2001) Growth inhibitory effects of 1alpha, 25-dihydroxyvitamin D(3) are mediated by increased levels of p21 in the prostatic carcinoma cell line ALVA-31. Cancer Res 61:7122–7129

Murtaugh MA, Sweeney C, Ma KN, Potter JD, Caan BJ, Wolff RK, Slattery ML (2006) Vitamin D receptor gene polymorphisms, dietary promotion of insulin resistance, and colon and rectal cancer. Nutr Cancer 55:35–43

Norman AW (2006) Minireview: vitamin D receptor: new assignments for an already busy receptor. Endocrinology 147:5542–5548

Oh K, Willett WC, Wu K, Fuchs CS, Giovannucci EL (2007) Calcium and vitamin D intakes in relation to risk of distal colorectal adenoma in women. Am J Epidemiol 165:1178–1186

Park K, Woo M, Nam J, Kim JC (2006) Start codon polymorphisms in the vitamin D receptor and colorectal cancer risk. Cancer Lett 237:199–206

Park SY, Murphy SP, Wilkens LR, Nomura AM, Henderson BE, Kolonel LN (2007) Calcium and vitamin D intake and risk of colorectal cancer: the Multiethnic Cohort Study. Am J Epidemiol 165:784–793

Peters U, McGlynn KA, Chatterjee N, Gunter E, Garcia-Closas M, Rothman N, Sinha R (2001) Vitamin D, calcium, and vitamin D receptor polymorphism in colorectal adenomas. Cancer Epidemiol Biomarkers Prev 10:1267–1274

Peters U, Chatterjee N, McGlynn KA, Schoen RE, Church TR, Bresalier RS, Gaudet MM, Flood A, Schatzkin A, Hayes RB (2004a) Calcium intake and colorectal adenoma in a US colorectal cancer early detection program. Am J Clin Nutr 80:1358–1365

Peters U, Hayes RB, Chatterjee N, Shao W, Schoen RE, Pinsky P, Hollis BW, McGlynn KA (2004b) Circulating vitamin D metabolites, polymorphism in vitamin D receptor, and colorectal adenoma risk. Cancer Epidemiol Biomarkers Prev 13:546–552

Platz EA, Hankinson SE, Hollis BW, Colditz GA, Hunter DJ, Speizer FE, Giovannucci E (2000) Plasma 1,25-dihydroxy- and 25-hydroxyvitamin D and adenomatous polyps of the distal colorectum. Cancer Epidemiol Biomarkers Prev 9:1059–1065

Reichrath J, Lehmann B, Carlberg C, Varani J, Zouboulis CC (2007) Vitamins as hormones. Horm Metab Res 39:71–84

Rodland KD (2004) The role of the calcium-sensing receptor in cancer. Cell Calcium 35:291–295

Satia JA, Keku T, Galanko JA, Martin C, Doctolero RT, Tajima A, Sandler RS, Carethers JM (2005) Diet, lifestyle, and genomic instability in the North Carolina Colon Cancer Study. Cancer Epidemiol Biomarkers Prev 14:429–436

Shaukat A, Scouras N, Schunemann HJ (2005) Role of supplemental calcium in the recurrence of colorectal adenomas: a metaanalysis of randomized controlled trials. Am J Gastroenterol 100:390–394

Slattery ML, Yakumo K, Hoffman M, Neuhausen S (2001) Variants of the VDR gene and risk of colon cancer (United States). Cancer Causes Control 12:359–364

Slattery ML, Murtaugh M, Caan B, Ma KN, Wolff R, Samowitz W (2004a) Associations between BMI, energy intake, energy expenditure, VDR genotype and colon and rectal cancers (United States). Cancer Causes Control 15:863–872

Slattery ML, Neuhausen SL, Hoffman M, Caan B, Curtin K, Ma KN, Samowitz W (2004b) Dietary calcium, vitamin D, VDR genotypes and colorectal cancer. Intl J Cancer 111:750–756

Slattery ML, Sweeney C, Murtaugh M, Ma KN, Caan BJ, Potter JD, Wolff R (2006) Associations between vitamin D, vitamin D receptor gene and the androgen receptor gene with colon and rectal cancer. Int J Cancer 118:3140–3146

Speer G, Cseh K, Mucsi K, Takacs I, Dworak O, Winkler G, Szody R, Tisler A, Lakatos P (2002) Calcium-sensing receptor A986S polymorphism in human rectal cancer. Int J Colorectal Dis 17:20–24

Sweeney C, Curtin K, Murtaugh MA, Caan BJ, Potter JD, Slattery ML (2006) Haplotype analysis of common vitamin D receptor variants and colon and rectal cancers. Cancer Epidemiol Biomarkers Prev 15:744–749

Tangrea J, Helzlsouer K, Pietinen P, Taylor P, Hollis B, Virtamo J, Albanes D (1997) Serum levels of vitamin D metabolites and the subsequent risk of colon and rectal cancer in Finnish men. Cancer Causes Control 8:615–625

Tong WM, Kallay E, Hofer H, Hulla W, Manhardt T, Peterlik M, Cross HS (1998) Growth regulation of human colon cancer cells by epidermal growth factor and 1,25-dihydroxyvitamin D3 is mediated by mutual modulation of receptor expression. Eur J Cancer 34:2119–2125

Tong WM, Hofer H, Ellinger A, Peterlik M, Cross HS (1999) Mechanism of antimitogenic action of vitamin D in human colon carcinoma cells: relevance for suppression of epidermal growth factor-stimulated cell growth. Oncol Res 11:77–84

van den Bemd GJ, Pols HA, van Leeuwen JP (2000) Anti-tumor effects of 1,25-dihydroxyvitamin D3 and vitamin D analogs. Curr Pharm Des 6:717–732

Wactawski-Wende J, Kotchen JM, Anderson GL, Assaf AR, Brunner RL, O'Sullivan MJ, Margolis KL, Ockene JK, Phillips L, Pottern L, Prentice RL, Robbins J, Rohan TE, Sarto GE, Sharma S, Stefanick ML, Van Horn L, Wallace RB, Whitlock E, Bassford T, Beresford SA, Black HR, Bonds DE, Brzyski RG, Caan B, Chlebowski RT, Cochrane B, Garland C, Gass M, Hays J, Heiss G, Hendrix SL, Howard BV, Hsia J, Hubbell FA, Jackson RD, Johnson KC, Judd H, Kooperberg CL, Kuller LH, LaCroix AZ, Lane DS, Langer RD, Lasser NL, Lewis CE, Limacher MC, Manson JE (2006) Calcium plus vitamin D supplementation and the risk of colorectal cancer. N Engl J Med 354:684–696

Wallace K, Baron JA, Cole BF, Sandler RS, Karagas MR, Beach MA, Haile RW, Burke CA, Pearson LH, Mandel JS, Rothstein R, Snover DC (2004) Effect of calcium supplementation on the risk of large bowel polyps. J Natl Cancer Inst 96:921–925

Wargovich MJ, Baer AR (1989) Basic and clinical investigations of dietary calcium in the prevention of colorectal cancer. Prev Med 18:672–679

Wark PA, Van der Kuil W, Ploemacher J, Van Muijen GN, Mulder CJ, Weijenberg MP, Kok FJ, Kampman E (2006) Diet, lifestyle and risk of K-ras mutation-positive and -negative colorectal adenomas. Int J Cancer 119:398–405

Wong HL, Seow A, Arakawa K, Lee HP, Yu MC, Ingles SA (2003) Vitamin D receptor start codon polymorphism and colorectal cancer risk: effect modification by dietary calcium and fat in Singapore Chinese. Carcinogenesis 24:1091–1095

Xie S, Pirianov G, Colston K (1999) Vitamin D analogues suppress IGF-I signalling and promote apoptosis in breast cancer cells. Euro J Cancer 35:1717–1723

Yee YK, Chintalacharuvu SR, Lu J, Nagpal S (2005) Vitamin D receptor modulators for inflammation and cancer. Mini Rev Med Chem 5:761–778

Zheng W, Anderson KE, Kushi LH, Sellers TA, Greenstein J, Hong CP, Cerhan JR, Bostick RM, Folsom AR (1998) A prospective cohort study of intake of calcium, vitamin D, and other micronutrients in relation to incidence of rectal cancer among postmenopausal women. Cancer Epidemiol Biomarkers Prev 7:221–225

Index

A

A66G, 229
A67G, 225
A459D, 170
A636P missense alteration, 148, 150
A986S, 282
A1298C, 226
A2756G, 228
Aberrant crypt foci (ACF), 53, 99
AC. *See* Amsterdam Criteria
ACF. *See* Aberrant crypt foci
Acrylamide, 259–260
ACVR1, 199
Adenoma-carcinoma sequence, 95–96
Adenomas, 5
  initiation of, 98–100
  malignant potential of, 96–97
  metachronous, 8–9
  multiplicity, 97–98
  serrated, 14, 104–105, 108
  sessile serrated, 82, 108
  tubular, 96
  tubulovillous, 96
  villous, 96, 103–104
  vitamin D and, 285
ADH. *See* Alcohol dehydrogenase
Adrenal tumors, 127
AFM. *See* American Founder Mutation
Alcohol, 13
Alcohol dehydrogenase (ADH), 230
ALK-1, 199
Allied lesions, 106–109
ALOXs. *See* Arachidonate lipoxygenases
Alu, 76, 150
Alu repeats, 147
American Founder Mutation (AFM), 150
AMPK, 190
Amsterdam Criteria (AC), 143, 144, 154, 179, 180
  families fulfilling, 181
Androgen receptor (AR), 288
Animal fat, 11
Anthropometry, 12–13
Antioxidant response element (ARE), 262
AOM. *See* Azoxymethane
Apal, 287
APC gene, 1, 6, 14, 27, 28, 35, 36, 51, 54, 61, 98, 99, 105, 110, 123, 171, 284
  chromosomal instability and, 58–65
  deficiency, 64
  distal germline mutations, 126
  exon structure, 130
  FAP and, 128–129
  functional domains of, 59–60
  germline mutations, 129–130
  inactivation of, 111
  mutation, 100–101, 131, 205
  N-terminus fragment of, 63
  polymorphisms, 132
  protein motifs, 130
AR. *See* Androgen receptor
Arachidonate lipoxygenases (ALOXs), 243, 247
ARE. *See* Antioxidant response element
Arg27His, 224
Arg(239)Gin, 229
Art(415)Cys, 229
Ascorbate, 7
Ashkenazi Jewish population, 148–149, 203, 205
Aspirin, 16, 239
Attenuated FAP, 127, 171
Aurora-A, 57, 66
Azoxymethane (AOM), 38

B

Bannayan-Riley-Ruvalcaba syndrome, 195, 196, 197

Basal Cell Nevus syndrome, 195, 196
Base-excision repair (BER), 169–170, 173
BAT25, 143
BAT26, 143
Bax, 210
Beckwith-Wiedemann syndrome, 72
BER. *See* Base-excision repair
Bethesda Guidelines, 144, 156
Bile acids, 277, 278
Biomarkers, 281
Biotransformation of carcinogens, 257, 260–261
- CRC risk and, 264–266
- environmental modulators, 262–263
- modulators of, 262–264

BLM gene, 55, 150, 203
- biomechanical properties, 204–205
- haploinsufficiency, 206
- mutation, 205–206

Bloom, David, 203
Bloom's Syndrome Registry, 205
Bloom syndrome (BS), 150, 203
BMPR1A, 198–199, 200
BMPR2, 199
BMP-signaling pathways, 198
Bottom-up model, 99–100
BRAF, 74, 80–83, 98, 105–108, 110, 184
- hyperplastic polyps with mutations, 107
- mutation, 112, 128–129, 185
- V600E mutation, 145

BRCA1, 55, 150
- germline mutations, 210

BRCA2, 55, 150
- germline mutations, 210

BS. *See* Bloom syndrome
Bsml, 286, 287
BUB1, 56, 62, 66
BUB1B, 64
BUB3, 56, 61, 62
BubR1, 62

C

C57BL/6, 28
C-149T, 230
C-401T, 225
C452CT, 225
C452T, 225
C677T, 222
C699T, 224
C1043T, 225
C1080T, 224
C1985T, 224
CACC. *See* Colitis-associated colon cancer
CACNA1G, 81
Caco, 62
Calbindin-D9K promoter, 30
Calcium, 1, 9, 11–12
- CaSR and, 277–278
- cell cycle and, 277
- epidemiology and, 280–284
- interaction, 282–283
- mechanisms, 276–278
- metabolism, 273–274
- physiology, 273–274
- vitamin D and, 288–289

Calcium Polyp Prevention Study, 281, 283
Calcium-sensing receptor (CaSR), 274, 282
- calcium and, 277–278

Cancer family syndrome (CFS), 153
Cancer Prevention Study II, 10
Carcinogenesis in PJS, 191–192
Carcinogens
- acrylamide, 259–260
- biotransformation of, 257, 260–261
- chemical, 258–260
- HAAs, 258
- heterocyclic amines, 261
- induced models of tumorigenesis, 38
- nitrosamines, 259
- N-Nitroso compounds, 261
- PAHs, 258–259
- polycyclic hydrocarbons, 261

Carcinoma, thyroid, 127
Carotenoids, 7
Case-control studies, 10
Caspase 3, 65
CaSR. *See* Calcium-sensing receptor
β-catenin, 28–29, 59, 60, 99, 102, 126
- transgenic mice, 30–31

CBS. *See* Cystathionine β-synthase
CDC4, 57, 66
CDKN2A, 74
Cdx2 mice, 39
C. elegans, 190
Cell cycle, 277, 278
CFS. *See* Cancer family syndrome
CGH. *See* Comparative genomic hybridization
Chemical carcinogens, 258–260
Cholecystectomy, 16
Chromosomal instability pathway, 52–54, 110, 128
- *APC* and, 58–65
- genetic basis of, 54–58
- inherited syndromes with, 55
- MAP cancer development, 170

Chromosome 3p21.3, 142
Chromosome 8q24, 215

Chromosome 9q22.32, 98
Chromosome 17p, 210
CHRPE. *See* Congenital hypertrophy of retinal pigment epithelium
CIMP. *See* CpG island methylator pathways
CNVs. *See* Copy number variants
Colitis-associated colon cancer (CACC), 36
Colorectal Adenoma Prevention trial, 249
Colorectal cancer (CRC)
  accelerated evolution of, 111–114
  biotransformation of carcinogens in, 264–266
  classification of, 111
  descriptive epidemiology, 5–6
  DNA methylation in diagnosis of, 85–86
  environmental risk factors, 7–15
  epidemiology, 279–288
  genetic and molecular events in, 6
  genetic variability in NSAID targets, 245–250
  imprinting in, 77
  medical conditions, 15–16
  multi-pathway disease, 109–110
  occurrence of, 1
  pathway features, 111
  precursor lesions, 102–103
Columnar cells, 108
Comparative genomic hybridization (CGH), 53
Congenital hypertrophy of retinal pigment epithelium (CHRPE), 126, 131, 173
Copy number variants (CNVs), 263
Cowdan syndrome, 195, 196, 197
Cox. *See* Cyclo-oxygenases
Cox-1, 31, 240, 245–246
  as target, 241
Cox-2, 31, 192, 240, 246–247
  inhibitors, 16
  as target, 241
Coxib, 239
CpG dinucleotides, 71, 72
CpG island, 72, 73
  hypermethylation of, 78
  methylator phenotype, 79–82
CpG island methylator pathways (CIMP), 51–52, 74, 79–82, 102, 106, 125
  epidemiology of, 83
  histological context of, 82–83
$cPLA_2$. *See* Cytosolic phospholipase A_2
CRC. *See* Colorectal cancer
Crohn disease, 15, 36
Cronkhite-Canada syndrome, 196, 197, 198
CSF. *See* Cytostatic factor
cSHMT. *See* Cytosolic SHMT
C-terminal, 29, 38, 59
CTNNB1, 30, 59, 61–62
Cyclin E regulators, 57
Cyclin G, 210
Cyclo-oxygenases (Cox), 31, 192. *See also* Cox-1; Cox-2
  inhibition, 242
CYP1A1, 263, 265
CYP1A2, 263, 266
Cyp1A-Cre, 29
CYP1B1, 266
CYP2A7, 264
CYP2C9, 244–245, 249, 250
CYP2C11, 275
CYP2D6, 275
CYP2D25, 275
CYP2E1, 261, 262, 265, 266
CYP2R1, 275
CYP3A4, 262, 275, 278, 286
CYP24, 286
CYP27A1, 275
CYP27B1, 275, 279, 286
CYP213, 275
Cystathionine β-synthase (CBS), 224
Cytochrome P450, 260
Cytokine-deficient mice, 36–37
Cytosine-5, 71
Cytosolic phospholipase A_2 ($cPLA_2$), 31
Cytosolic SHMT (cSHMT), 224, 229
Cytostatic factor (CSF), 62

D

D2S123, 143
D5S346, 143
D17S250, 143
D1822V, 284
Dental abnormalities, 126, 173
Descriptive epidemiology, 5–6
Desmoid tumors, 125–126
DHFR. *See* Dihydrofolate reductase
Diabetes mellitus, 15
Diagnosis, DNA methylation in, 85–86
Diet, low-fat, 9
Differentially methylated region (DMR), 77
Dihydrofolate reductase (DHFR), 225
1,2-dimethylhydrazine, 38
Ding, 58
DLD1, 52, 56, 64, 65
DLG-binding domain, 59
D. melanogaster, 190
DMR. *See* Differentially methylated region
DNA hypomethylation, 75–76

DNA methylation, 102. *See also* Methylation
in diagnosis of, 85–86
neoplasia and, 73–75
polycomb proteins and, 77–79
DNA methyltransferase 1 (DNMT1), 32, 77, 229–230
DNMT1. *See* DNA methyltransferase 1
Dominant negative *N*-cadherin mice, 39–40
Drosophila melanogaster, 39
Duodenal polyps, 125

E

E466X, 170
E1317Q, 132
EB1, 60, 62, 63
EBV. *See* Epstein-Barr virus
E-cadherin, 277
E. coli, 140–141
EED, 78
EGFR pathway, 241
Embryonic stem (ES) cells, 61, 204
Endogenous oxidative damage, 173
ENG, 198, 199
ENU. *See* Ethylnitrosourea
Environmental risk factors, 7–15
alcohol, 13
animal fat, 11
anthropometry, 12–13
diet, 7
fiber, 7–10
fruits, 7–10
infection, 14–15
meat, 10
micronutrients, 7–10
occupation, 15
physical activity, 12–13
reproductive factors, 14
saturated fat, 11
tobacco, 13–14, 259
total dietary fat, 11
vegetables, 7–10
EPHB2, 184
EPHX1. *See* Epoxide hydrolase
EPIC study, 8
Epidemiology
calcium and, 280–284
CRC, 279–288
of FAP, 123–124
vitamin D and, 284–288
Epidemiology, descriptive, 5–6
Epigenetic changes, 75–83
Epimutations, heritable, 151–153
Epoxide hydrolase (EPHX1), 261, 264
Epstein-Barr virus (EBV), 72
ES cells. *See* Embryonic stem cells
Estrogen, 17
Ethylnitrosourea (ENU), 28
Exercise, 12–13
EZH2, 78, 79

F

Fabp, 39
Fabp1, 30
Familial adenomatous polyposis coli (FAP), 16, 28, 53, 64, 96, 97, 99, 183
adrenal tumors and, 127
APC gene, 128–129
attenuated, 127, 171
clinical presentation of, 124
congenital hypertrophy and, 126
dental abnormalities, 126
desmoid tumors and, 125–126
duodenal polyps and, 125
epidemiology of, 123–124
equivalence of, 100–102
extracolonic manifestations of, 131–132
gastric polyps, 124–125
genotype-phenotype association, 130–132
germline mutations leading to, 129–130
hepatobiliary tumors, 127
micro-adenomas in, 100
modifier genes, 132–133
osteomas and, 126
retinal pigment epithelium, 126
thyroid carcinoma and, 127
Familial Colorectal Cancer Type X, 143, 154, 180–181
AC and, 181
Family G, 154
FANCC. *See* Fanconi anemia type C
Fanconi anemia type C (FANCC), 150
FAP. *See* Familial adenomatous polyposis coli
Fat
animal, 11
low-fat diet, 9
saturated, 11
total dietary, 11
Fbw7, 57
Fiber, 7–10
FISH. *See* Fluorescent in situ hybridization
Flavonoids, 7
Fluorescent in situ hybridization (FISH), 52, 64
FOBT, 86
FOCM. *See* Folate-mediated one-carbon metabolism

FokI, 287
Folate, 7, 9
Folate-mediated one-carbon metabolism (FOCM), 219, 220, 225
Folylpolyglutamyl synthase (FPGS), 225
Founder mutations, 149–151
FPGS. *See* Folylpolyglutamyl synthase
Fruits, 7–10
Fusion pathways, 111–114
Fusion polyps, 111–114

G
G280A, 225
G-354T, 225
G382D, 170
G418, 76
G-579T, 230
G638A, 264
G990R, 282
G1196A, 225
G1294C, 263
GADD45, 210
Gastric polyps, 124–125
Gastrointestinal tumorigenesis, 28
GEM. *See* Genetically engineered mice
Gene discovery, 141–142
Genetically engineered mice (GEM)
  carcinogen-induced models of tumorigenesis, 38
  β-catenin transgenic, 30–31
  Cdx2, 39
  cytokine-deficient, 36–37
  DNA mismatch repair and, 34–36
  dominant negative N-cadherin mice, 39–40
  immune-deficient, 36
  models, 38–40
  mucin-deficient, 36
  P13k-deficient, 39
  $Rb^{MI/MI}$, 38
  Smad$^{-/-}$, 33–34
  TGFβ1$^{-/-}$, 32–33
  TGF-beta signaling pathway and, 32–34
  Wnt signalling pathway, 27–30
Genetic events, 6
Genetic variability
  in NSAID-metabolizing enzymes, 244–250
  in NSAID targets, 240–244, 245–250
Genomic instability, 139–140
Genotype-phenotype association, 130–131
Germline mutations
  APC gene, 126, 129–130
  BRCA1, 210
  BRCA2, 210
  epimutation, 84–86
  FAP, 129–130
  hMLH1, 151, 157
  hMSH2, 157
  p53, 212
GGH. *See* γ-glutamyl hydrolase
Glioblastomas, 158
γ-glutamyl hydrolase (GGH), 225
Glutathione S-transferases, 260
Gorlin syndrome, 195, 196, 197
Growth-factor signaling, 278–279
GSK-3B, 30, 65, 128
GSTA1, 264
GSTM1, 261, 264, 265, 266
GSTP1, 261, 264
GSTT1, 264, 265, 266
Guanine, 169

H
H3K27, 78
H19, 77
HAAs. *See* Heterocyclic aromatic amines
Hamartomatous polyps, 189
  comparisons of, 197
  dermatologic findings in, 196
  features of, 191
Haploinsufficiency, BLM, 206
HATH1, 103, 104
HCT116, 56, 57, 61–62, 63, 64
HEK, 62
HeLA, 65
Helicobacter pylori, 15, 72, 124–125
Hepatobiliary tumors, 127
Hereditary Breast and Ovarian Cancer, 150
Hereditary Hemorrhagic Telangiectasia (HHT), 199
Hereditary NonPolyposis Colon Cancer (HNPCC), 143, 144, 179, 183
  incidence of, 154–155
  Lynch syndrome, 153–154
Heritable epimutations, 151–153
Heterocyclic amines, 10, 261
Heterocyclic aromatic amines (HAAs), 258
Heterozygotes, 173–174
HHT. *See* Hereditary Hemorrhagic Telangiectasia
hMLH1, 142, 145, 146, 147, 149, 150, 159
  germline mutations in, 151, 157
  hypermethylation, 142, 152
  methylation, 152
hMSH2, 139, 141, 142, 146, 147, 149, 158, 159
  germline mutations in, 157
  missense mutation, 150
hMSH3, 142
hMSH6, 141, 146, 155, 158, 159

HNPCC. *See* Hereditary NonPolyposis Colon Cancer
Homologous recombination (HR), 204
hPMS2, 148
HPP1, 83
HPS. *See* Hyperplastic polyposis syndrome
HPV. *See* Human papilloma virus
HR. *See* Homologous recombination
HRAS, 76
HRDC domain, 203
HT29 cells, 52, 62
Human Genome Project, 220
Human papilloma virus (HPV), 72
Human Variome Project (HVP), 149
HVP. *See* Human Variome Project
15-Hydroxyprostaglandin dehydrogenase (PGDH), 243
Hypermethylation, 74
  of CpG island, 78
  hMLH1, 142, 152
  widespread and concordant, 79
Hyperplastic polyposis syndrome (HPS), 82, 183–184
  phenotypic dichotomy in, 185
  serrated pathway syndrome and, 185
Hyperplastic polyps, 5, 14, 104, 106–109
  BRAF mutations, 107–108
  goblet cell type, 107
  KRAS mutations, 107–108
  microvesicular type, 108
  mixed, 110
  variant, 109
Hypomethylation
  DNA, 75–76
  genome-wide, 75–76
  IBD. *See* Inflammatory bowel disease

I

ICG-HNPCC. *See* International Collaborative Group on Hereditary NonPolyposis Colorectal Cancer
ICR. *See* Imprinting control region
IDLs. *See* Insertion/deletion loops
IGF2, 77, 81
IHC. *See* Immunohistochemistry
Il-2, 36–37
IL33, 215–216
Immune-deficient GEM, 36
Immunohistochemistry (IHC), 85, 144
Immunomodulation, 279
Imprinting, 72
  loss of, 77
Imprinting control region (ICR), 77
Inactive X-chromosome, 71
Indoles, 7
Infection, 14–15
Inflammatory bowel disease (IBD), 15, 36
Insertion/deletion loops (IDLs), 141
International Collaborative Group on Hereditary NonPolyposis Colorectal Cancer (ICG-HNPCC), 143, 179
Iowa Women's Health Study, 13
Isothiocyanates, 7, 265

J

"Just-right" signaling, 60
Juvenile polyposis, 1, 195–200
  clinical diagnostic criteria for, 196
  dermatologic findings in, 196

K

KCNQ1, 77
53-kDa nuclear phosphoprotein, 210
Knudson's two-hit hypothesis, 60
KRAS, 6, 51, 80, 98, 103, 105, 106, 107, 110
  hyperplastic polyps with mutations, 107
  mutation, 114, 128–129
K-ras, 171
  mutations, 284
  L237M, 241, 245

L

Large genomic rearrangements, 146–148
LCA. *See* Lithocholic acid
LEF. *See* Lymphoid-enhancer factor
Lesions
  allied, 106–109
  precursor, 102–103
Leu474Phe, 224
LFL families. *See* Li-Fraumeni-Like families
LFS. *See* Li-Fraumeni Syndrome
Li-Fraumeni-Like (LFL) families, 209
Li-Fraumeni Syndrome (LFS), 209–212
  database, 212
  patients, 211
LINES, 76
Lithocholic acid (LCA), 278
Log-linear per-allele effect, 215
LOH. *See* Loss of heterozygosity
LOI. *See* Loss of imprinting
Loss of heterozygosity (LOH), 39, 52, 142, 191
  STK11, 192
Loss of imprinting (LOI), 77
LoVo, 52, 62

Low-fat diet, 9
Lymphoid-enhancer factor (LEF), 128
Lynch, Henry, 179, 180
Lynch syndrome, 1, 34–35, 84, 97, 112, 139, 141, 179
  AC and, 181
  clinical features, 157–159
  clinical variants of, 158
  diagnosis of, 140
  founder mutations, 149–151
  heritable epimutations and, 151–153
  histopathology, 155–156
  HNPCC v., 153–154
  incidence of, 154–155
  large genomic rearrangements, 146–148
  missense mutations, 148–149
  molecular screening for, 143–145
  MSI and, 142–143
  mutation profile, 146
  penetrance, 158
  tumor spectrum, 157
  MAD2, 56, 57, 66

M

MAD2LI, 64
MADH4, 198, 200
MALT. *See* Mucosa-associated lymphoid tissue
MAP. *See MUTYH*-associated polyposis
MCA. *See* Methylated CpG-island analysis
MCR. *See* Mutation cluster region
MDM-2, 210
Meat, 10, 258, 259
Medical conditions, 15–16
Medications, 16–17
Metachronous adenoma, 8–9
Methionine synthase (MTR), 223, 228–229
Methionine synthase reductase (MTRR), 223, 229
5,10-Methylenetetrahydrofolate, 221, 222–223, 226–227
Methylated CpG-island analysis (MCA), 80
Methylation. *See also* DNA methylation; Hypermethylation; Hypomethylation
  hMLH1, 152
  MGMT, 112
Methylation, *hMLH1,* 152
MGMT, 112
  methylation, 112
Micro-adenomas
  in FAP, 100
  sporadic, 100–102
Micronutrients, 7–10
Microsatellite instability (MSI), 14, 34, 51, 79, 81, 84, 102, 110, 139, 283
  discovery of, 140
  high, 110, 143, 156
  low, 143
  in Lynch syndrome, 142–143
  in MAP cancer development, 170–171
Microsatellite stable (MSS), 143
Microtubules (MT), 62
MINT, 80, 83
MINT31, 185
Mismatch repair (MMR), 34–36, 76, 85
  defective, 143, 145, 157, 158
  disabling, 151
  eukaryotic, 141–142
  mechanism, 140–141
  mutation profile, 146
  pathway, 148
  postreplicative, 139
Missense mutation, 148–149
  hMSH2, 150
Mitotic spindle arrest, 55–58
MLH. *See* MutL homologs
MLH1, 7, 34, 35, 75, 78, 80, 82, 84, 85, 112, 145, 179
MLH3, 34
MLPA. *See* Multiplex ligation-dependent probe amplification
MMR. *See* Mismatch repair
Modifier genes, 132–133
Molecular events, 6
Molecular screening, for Lynch syndrome, 143–145
Mouse models, PJS, 191
MRE11, 58
mRNA, 58
  splicing, 148
MSH. *See* MutS homologs
MSH2, 34, 35, 36, 75, 84, 85, 145, 147, 179
MSH3, 34, 141
MSH6, 34, 35, 36, 84, 145
  mutations, 181
MSI. *See* Microsatellite instability
MsPath, 157
MSS. *See* Microsatellite stable
MST1, 106
MT. *See* Microtubules
MTH1, 170
MTHFR, 219
MTR. *See* Methionine synthase
MTR2756AG, 228

MTRR. *See* Methionine synthase reductase
MTS. *See* Muir-Torre syndrome
MUC2, 36, 103
MUC5AC, 103
Mucin-deficient mice, 36
Mucosa-associated lymphoid tissue (MALT), 15
Muir-Torre syndrome (MTS), 158
Multiplex ligation-dependent probe amplification (MLPA), 129, 147–148
Multivitamin supplements, 9
Mutation 1, 150
Mutation 2, 150
Mutation cluster region (MCR), 129
Mutation profile, 146
Mutator phenotype, 139, 153
MutHLS, 141
mutL, 142
MutL homologs (MLH), 141
mutS, 142
MutS homologs (MSH), 141
MUTYH, 98, 169–170, 171
MUTYH-associated polyposis (MAP), 1, 53
  base-excision repair and, 169–170
  cancer development, 170–171
  clinical features of, 171–173
  extracolonic cancer risks in, 173
  heterozygotes, 173–174
  polyps in, 172, 174
MYC, 76
MYC oncogene, 215
MYH, 184

N
N131β-catenin, 30
N-acetylation, 263
N-acetyltransferases (NATs), 260
NADPH quinine oxidoreductase (NQO), 260
N-APC-deletion constructs, 63
NAT1, 261, 263
NAT2, 261, 263, 264–265
National Cancer Institute, 143
National Institute on Environmental Health Environmental Genome Project (NIEHS-EGP), 241, 244
NATs. *See* N-acetyltransferases
N-cadherin mice, 39–40
NDMA. *See* *N*-nitrosomethylamine
Neoplasia
  DNA methylation and, 73–75
  epidemiology of CIMP in, 83
  epigenetic changes in, 75–83
Neurofibromatosis, 196, 197
NEUROG1, 81
NHEJ. *See* Nonhomologous end-joining
NIEHS-EGP. *See* National Institute on Environmental Health Environmental Genome Project
Nitrosamines, 259
N-methyl-*N*-nitrosourea, 38
N-nitroso compounds (NOCs), 259, 261
N-nitrosomethylamine (NDMA), 259
NOCs. *See* N-nitroso compounds
Non-*APC*-associated polyposis, 133
Nonhomologous end-joining (NHEJ), 205
NOREI, 106
NQO. *See* NADPH quinine oxidoreductase
NSAIDs, 16, 31, 56, 239–240, 273, 282
  genetic variability in metabolizing enzymes, 244–245, 245–250
  genetic variability in targets, 240–250
N-terminal, 30, 60
  of *APC,* 63
Nuclear atypia, 57
Nurses' Health Study, 10
NYD-SP25 isoform 3, 215

O
O-acetylation, 261
Obesity, 12–13
Occupation, 15
OCs. *See* Oral contraceptives
ODC. *See* Ornithine decarboxylase
OGG1, 169, 170
1103delC, 170
1395delGGA, 170
129/BL6, 29
Oral contraceptives (OCs), 17
Ornithine decarboxylase (ODC), 243–244, 248
Osteomas, 126, 173
8-oxo-7,8-dihydro-2'-deoxyguanosine (8-oxodG), 169–170
8-oxodG. *See* 8-oxo-7,8-dihydro-2'-deoxyguanosine

P
P13K/Akt signaling, 31
P13k-deficient mice, 39
p14, 113
p21, 210
p53, 36, 51, 171, 226
  germline mutations, 212
  inactivation, 58
  role of, 209–212

P281L, 170
PAHs. *See* Polycyclic aromatic hydrocarbons
Parathyroid hormone (PTH), 274
PCNA, 204
PCR techniques, 86
Peristalsis, 12
Peroxisome proliferator activity receptor delta (PPARδ), 31
Peutz-Jeghers syndrome, 1, 189, 196, 197
  hamartoma, 189
  manifestations of, 189–190
  mouse model, 191
PGDH. *See* 15-Hydroxyprostaglandin dehydrogenase
PGE_2. *See* Prostaglandin E2
PGIS. *See* Prostacyclin synthase
Phase II enzymes, 262
Phenols, 7
Phenotypic dichotomy in HPS, 185
Physical activity, 12–13
Pilomatricomas, 173
PLCO. *See* Prostate, Lung, Colorectal, and Ovarian Cancer Screening Trial
PLK1, 64
PMH. *See* Postmenopausal hormones
PMS1, 34, 142
PMS2, 34, 75, 84, 141, 142, 145, 146, 147
Poly(A), 286
Polycomb proteins, 77–79
Polycyclic aromatic hydrocarbons (PAHs), 258–259
Polycyclic hydrocarbons, 261
Polymorphisms
  APC, 132
  investigating, 220–221
  metabolizing enzyme gene, 286–287
  in one-carbon metabolism, 221–226
  thymine/cytosine, 286
Polypectomy, 95
Polyposis, non-*APC*-associated, 133
Polyps, 5
  gastric, 124–125
  hamartomatous, 189, 191, 196, 197
  hyperplastic, 5, 14, 104, 106–109
  MAP, 172
  sporadic juvenile, 197
Postmenopausal hormones (PMH), 17
PPARδ. *See* Peroxisome proliferator activity receptor delta
PPARγ gene, 287
PRC2 complex, 78, 83
Precursor lesions, 102–103
Pro12 Ala, 288
Prostacyclin synthase (PGIS), 243
Prostaglandin E2 (PGE_2), 31, 241, 242
Prostaglandin synthesis, 247
Prostate, Lung, Colorectal, and Ovarian Cancer Screening Trial (PLCO), 8, 282
Provitamin D, 274
PTH. *See* Parathyroid hormone
PtK cells, 61
pVillin-$KRAS^{V12G}$, 64

Q
Q377X, 170
Q1011E, 282
Quinolone, 38

R
R8W, 241
R227W, 170
R231H, 170
R260Q, 170
RAD51, 204
Randomized controlled trials (RCTs), 239
RASSF1, 106–109
RASSF2, 106–109
RASSF5, 106–109
Rb. *See* Retinoblastoma
$Rb^{MI/MI}$, 38
RCTs. *See* Randomized controlled trials
Reactive oxygen species (ROS), 169
Recql5, 204
RecQL4, 204
RecQL5, 204
RecQ-like family, 204
Reduced folate carrier (RFC), 224–225, 229
Reproductive factors, 14
RET, 127
Retinal pigment epithelium, 126, 131, 173
Retinoblastoma (Rb), 38
Retinoid X receptor (RXR), 275
RFC. *See* Reduced folate carrier
RING finger domain-containing protein, 216
RKO, 62
RNAi, 64
ROD, 58
ROS. *See* Reactive oxygen species
Rothmund-Thompson syndrome, 204
RPA, 204
RQC domain, 203
rs6983267, 215
rs10505477, 215
RUNX3, 81
RXR. *See* Retinoid X receptor

S
Saccharomyces cerevisiae, 141
S-adenosylmethionine, 229
Saturated fat, 11
SCE. *See* Sister chromatid exchange
Schistosoma japonicum, 14–15
SEER. *See* Surveillance, Epidemiology, and End Results
Serine hydroxymethyltransferase (SHMT), 223–224
Serrated adenomas, 14, 104–105, 108
Serrated pathway syndrome (SPS), 183
  HPS and, 185
Ser(284)Thr, 229
Sessile serrated adenomas (SSA), 82, 108
SHMT. *See* Serine hydroxymethyltransferase
SIFT, 241
Silver-Russell syndrome, 72
SINES, 76
Single nucleotide polymorphisms (SNP), 152, 263, 282
SIR. *See* Standardized incidence ratio
siRNA, 64, 65
Sister chromatid exchange (SCE), 203
677IT, 221, 226
Smad$^{-/-}$, 33–34
SMAD2, 32, 33, 199
SMAD3, 33–34, 198, 199, 279
SMAD4, 32, 33–34, 171, 198, 200
  inactivation of, 128–129
SMAD7, 199
SNP. *See* Single nucleotide polymorphisms
SOCS1, 81
Southern blot analysis, 148
Sporadic juvenile polyp, 197
SPS. *See* Serrated pathway syndrome
SR-alpha, 29
SSA. *See* Sessile serrated adenomas
Standardized incidence ratio (SIR), 180
STK11, 78, 190
  functions of, 190
  LOH, 192
STK15, 57
STMN1, 64
Sulfotransferases (SULTs), 260
SULT1A1, 264, 265
Surveillance, Epidemiology, and End Results (SEER), 210
SUZ12, 78
SW480, 52, 62, 64
SW837, 52

T
T-124G, 225
T-283C, 230
Taql, 283, 286
T-cell factor (TCF), 128
T-cells, 36
TCF. *See* T-cell factor
TCF4, 277
TCII. *See* Transcobalamin II
TGFβ1$^{-/-}$, 32–33
TGF-beta, 27, 83, 198, 199, 279
  GEM and, 32–34
  signaling, 198
Therapy, DNA methylation in, 85–86
Thymidylate synthase, 221–222, 227–228
Thymine/cytosine polymorphism, 286
Thyroid carcinoma, 127
TILs. *See* Tumor-infiltrating lymphocytes
Tobacco, 13–14, 259
Top-down model, 99–100
Total dietary fat, 11
TP53, 80, 110
TPD52L3, 215
Transcobalamin II (TCII), 225
TS polymorphisms, 227
Tubular adenoma, 96
Tubulin, 62
Tubulovillous adenoma, 96
Tumorigenesis, 38
Tumor-infiltrating lymphocytes (TILs), 156
Tumors
  adrenal, 127
  desmoid, 125–126
  hepatobiliary, 127
  suppressor genes, 139–140
Turcot Syndrome, 158
2rpt/2rpt genotypes, 227
Type-C genes, 80
Tyr113His, 264

U
U2OS, 64–65
Ubiquitin-like PHD, 216
UDP-glucuronosyl transferase (UGTs), 260
UGT1A6, 244, 248–249, 265
UGT1A7, 265
UGTs. *See* UDP-glucuronosyl transferase
Ulcerative colitis, 15, 36
5′-UTR, 221
UW-FHCRC-VDR, 241

V

V232F, 170
Variable number tandem repeat (VNTR), 243
VDR. *See* Vitamin D receptor
VDREs. *See* Vitamin D response elements
Vegetables, 7–10
Villin-CreER, 29
Villous adenoma, 96, 103–104
Vitamin B12, 227
Vitamin D, 1, 11–12, 278–279
  adenomas and, 285
  calcium, 288–289
  cell cycle and, 278
  epidemiology and, 284–288
  growth-factor signaling, 278–279
  immunomodulation and, 279
  metabolism, 274–275
  metabolizing enzyme gene polymorphisms, 286–287
  observational studies, 284–287
  physiology, 274–275
Vitamin D receptor (VDR), 274
  genotypes, 287–288
Vitamin D response elements (VDREs), 282
Vitamin E, 9
VNTR. *See* Variable number tandem repeat
Vogelstein model, 79–80, 104

W

Werner syndrome, 204
Wnt signalling pathway, 27–30, 38, 59, 61, 277
  genes modifying, 31–32
Women's Health Initiative, 9, 281

X

Xenobiotic metabolizing enzymes, 260
Xenobiotic response elements (XRE), 262
Xenopus, 62, 190
XRE. *See* Xenobiotic response elements

Y

Y90X, 170
Y165C, 170

Z

ZW10, 58
ZWILCH, 58